For EU product safety concerns:
IN TECH d.o.o., Prolaz Marije Krucifikse Kozulić 3, 51000 Rijeka, Croatia,
info@intechopen.com or visit our website at intechopen.com.

Meet the editors

Aldemaro Romero is the Dean of the College of Arts and Sciences at Southern Illinois University Edwardsville. He received his bachelor's degree in Biology from the University of Barcelona, Spain, and his Ph.D. in Biology from the University of Miami, Florida. He has published more than 660 pieces of work including 15 books, peer-reviewed articles and articles in non-peer-reviewed publications. His field experience on marine mammals includes several countries in the Caribbean area. His academic interests range from environmental and evolutionary biology to history and philosophy of science and science communication. He was been awarded numerous grants and prizes for his research and science communication works.

Edward O. Keith received his bachelor's degree in Wildlife Biology and his master's degree in Physiology and Biophysics from Colorado State University in 1975 and 1978, respectively, and his doctoral degree in Biology in 1984 from the University of California at Santa Cruz. He held faculty positions at the University of Miami and Southeastern University of the Health Sciences (later Nova Southeastern University). He was an associate professor at the Oceanographic Center of that institution by the time of his death. He taught numerous courses not only on marine mammalogy but also on biochemistry, physiology, and related subjects. His research centered on pinnipeds and he was an active member of the Society for Marine Mammalogy. He published numerous peer-reviewed papers.

Contents

Preface

The highly specialized nature of marine mammals when compared with their terrestrial counterparts, the environment in which they live, and the impact humans have had on them today and throughout history, have made of the study of these creatures something unique in itself. Probably with the exception of primates, no other animal group has been perceived as so distinctive. Therefore, it is not surprising that many researchers have also taken a particular approach to their research.

This volume is aimed at providing a glimpse at such diversity of views and approaches while delivering valuable information in marine mammalogy. Given the increasing concern regarding issues of anthropogenic factors affecting these animals it is not surprising that the majority of chapters deal with environmental issues.

In the first chapter of this volume I looked at a question that has baffled some biologists and historians of science alike: how come the suggestion by Aristotle that cetaceans were closely related to their terrestrial counterparts ("viviparous quadrupeds") rather than fish was ignored for about 2,000 years. Interestingly enough it was a non-evolutionist naturalist, Carl Linnaeus, who would create the taxon Mammalia and include cetaceans among them. I advanced the idea that a combination of environmental classification and scholasticism led to their misclassification for centuries.

The late Ed Keith (see his obituary at the end of this book) presents a matrix model of fasting metabolism in the northern elephant seal (*Mirounga angustirostris*). He provides evidence that pups maintain a paradoxical fasting hyperglycemia while fasting for 6-8 weeks after nursing for about 30 days. He discusses this apparent contradiction as possibly related to differences in time scale between the enzymatic reactions occurring among these animals versus the time scale of the actual fasting period.

Marsili et al. introduce the original term of "Test Tube Cetaceans" in the title of their chapter describing cetacean fibroblast cell cultures obtained from the skin biopsies of stranded cetaceans, as the "test tube cetaceans" to evaluate susceptibility to genotoxicity of different environmental contaminants. They suggest how to evaluate the presence of DNA damage by Comet assay in these cetaceans after treatment with different genotoxic compounds.

The chapter by Alava and Gobas on biomagnification and trophic transport of persistent organic pollutants (POPs) in the food chain of the Galapagos sea lion (*Zalophus wollebaeki*) provides us with what is probably the first biomagnification assessment of POPs in a tropical marine ecosystem of the southeastern Pacific. Despite the fact that they studied a population of marine mammals in a protected area (the Galapagos Marine Reserve) and far away from the usual sources of industrial pollutants, the authors found that endangered species at the top of the food web are not immune to the health risks associated with the long range environmental transport of POPs. Therefore, this problem, which has been extensively documented for other areas of the world, now appears commonplace wherever you look for it.

Wise et al. provide another study on the topic of marine mammal toxicology. They use skin biopsy applications in free ranging marine mammals and found it as a very useful tool for studying marine mammal toxicology and conservation. They developed a cell line to measure baseline DNA damage levels while serving as a species-specific model for evaluating the impacts of marine pollutants on DNA. Thus, using a skin biopsy they were able to assess both exposure and impact of exposure.

Mouton and Botha take an ecological approach to a topic that has become more and more the focus of attention among some marine mammalogists: cutaneous lesions in cetaceans due to human impacts on the environment. They review previous reports to evaluate the microbes that seem to be the causative agents, as well as contributing factors such as anthropogenic activities. They found that anthropogenic activities play a role in allowing contact and consequent adhesion of opportunistic microbes from the natural environment, as well as from sewage entering the marine environment. They also point out at toxic pollutants intruding on the physiology of these mammals by compromising their immune systems, rendering them susceptible to a host of health threats. They conclude that skin lesions among cetaceans may be indicative of an ecosystem under severe pressure and a result of human activities.

Arbiza et al. report evidence of influenza virus and *Mycobacterium pinnipedii* infections among individuals of two pinnipeds (*Arctocephalus australis* and *Otaria flavescens*) on the coasts of Uruguay. They confirmed that fur seals could act as reservoirs of human influenza strains that circulated in the past, and also suggest that influenza A and B viruses may be transmitted from humans to seals. This is most likely the result of interactions during capture and research activities, as well as in rehabilitation centers and sometimes with divers that swim near the seal islands. Furthermore, they suggest that keepers and veterinarians at zoos, aquaria and rehabilitation centers are at increased risk of infection because of their extensive contact with the animals.

Ohishi et al. looked at morbillivirus, a causative agent of mass die-offs of marine mammals. Given that a notable biological feature of morbillivirus is its high level of host specificity, they researched SLAM (the principal cellular receptor for morbilliviruses allowing entry and propagation) and found that 32 amino acid residues on the interface of SLAM V domain, which are potentially involved in the interaction with viruses. These amino acid residues are thought to be important for host–virus specificity. They hypothesize that recent climate change may increase the opportunities for new contacts among wild mammals and for the transmission of viruses and propose a new approach to assess the viral sensitivities of wild mammals by analyzing the host receptors.

In her chapter Brito described early sealing and whaling in the Northeast Atlantic from Portuguese activities. Using relatively unknown Portuguese records of encounters and hunting of monk seals in the Atlantic and medieval and early modern whaling in the Iberian Peninsula, she was able to identify and understand environmental changes integrated in a time of resources exploitation and highly predatory perceptions towards the marine environment.

In my chapter on Yankee whaling in the Caribbean basin I analyze available information at providing a historical context for understanding the economic and political issues that influenced the development of this activity and their cultural and ecological impact. Yet, that influence was not uniform and the heterogeneity in its impact was due to a number of historical and cultural circumstances. At the end I conclude that the development of whaling in the Wider Caribbean area was the result of how multiple factors interplayed.

This book ends with a sad note. It is the obituary of the co-editor of this volume. Edward O. Keith was an excellent scholar and teacher and an even better human being whose life was cut short just when the book was almost ready to be sent for publication. His colleagues and friends will sorely miss him.

Aldemaro Romero
College of Arts and Sciences, Southern Illinois University Edwardsville,
Peck Hall, Edwardsville, IL,
USA

Section 1

History of Marine Mammalogy

Chapter 1

When Whales Became Mammals: The Scientific Journey of Cetaceans From Fish to Mammals in the History of Science

Aldemaro Romero

Additional information is available at the end of the chapter

http://dx.doi.org/10.5772/50811

1. Introduction

Cetacea (whales and dolphins) is a natural group that has for centuries generated a great deal of misunderstanding and controversy regarding its proper place in natural classification. As late as 1945 Simpson wrote that "Because of their perfected adaptation to a completely aquatic life, with all its attendant conditions of respiration, circulation, dentition, locomotion, etc., the cetaceans are on the whole the most peculiar and aberrant of mammals."

Although both molecular and paleontological data have provided a much better understanding of the placement of this group among mammals, there is no question that despite being studied and dissected by dozens of naturalists since Aristotle, these animals were always misclassified. This group provides an interesting case study for intellectual inertia in the history of science. In other words, why did so many scientists misplace this group in the natural classification despite the fact that they themselves were gathering critical information that showed the close relationship these animals had to what we know today as mammals?

The aim of this chapter is to explore this question. To that end I will (1) survey the naturalists who studied cetaceans providing clues of their true nature, (2) describe the intellectual environment in which their conclusions were made, and (3) discuss the factors behind this intellectual inertia.

For the purpose of this chapter I have only taken into consideration works that had some scientific basis and/or that in some ways influenced the process of placing cetaceans as mammals. Authors are enumerated based on the date of the major publication they produced on cetaceans. For synonyms in names of marine mammals through time see Artedi (1738) and Linnaeus (1758).

2. Ancient times

2.1. Aristotle

Aristotle[1] was the son of Nicomachus, the personal physician of King Amyntas of Macedon and Phaestis, a wealthy woman[2]. Nicomachus may have been involved in dissections (Ellwood 1938, p. 36), a key tool in Aristotle's biological studies, particularly on marine mammals. Aristotle lost both his parents when he was about 10 and from then on he was raised of his uncle and/or guardian Proxenus, also a physician (Moseley 2010, p. 6). Early Greek physicians known as asclepiads usually taught their children reading, writing, and anatomy (Moseley 2010, p. 10).

In 367 BCE Aristotle moved to Athens to study at Plato's Academy, and later travelled throughout Asia Minor and studied living organisms while at the island of Lesbos (344-342 BCE) where he collected a lot of information about marine mammals. He later created his own philosophical school, the Lyceum, in Athens where most of his written work was produced between 335 and 323 BCE.

Aristotle is the first natural historian from whom we have any extensive work. One of his surviving opuses is *Historia Animalium* (inquiry about animals)[3]. There he classified animals as follows (beginning from the top): "blooded" animals (referring to those with red blood, vertebrates) with humans at the top, viviparous quadrupeds (what we would call terrestrial mammals), oviparous quadrupeds (legged reptiles and amphibians), birds, cetaceans, fishes, and then "bloodless" animals (invertebrates). He named each one of these groups a "genus."

Humans
Viviparous quadrupeds (terrestrial mammals)
Oviparous quadrupeds (reptiles and amphibians)
Birds
Cetaceans
Fish
Malacia (squids and octopuses)
Malacostraca (crustaceans)
Ostracoderma (bivalve mollusks)
Entoma (insects, spiders, etc.)
Zoophyta (jellyfishes, sponges, etc.)
Higher plants
Lower plants

Based on the "kinds" of animals and the varieties he described we can distinguish somewhere between 550 and 600 species. Most of them he had observed directly and even

[1] *b*. Stagira, Chalcidice, Macedonia, today's Greece, 384 BCE; *d*. Chalcis, Euboea, Ancient Greece, today's Greece 322 BCE

[2] Biographical information on Aristotle is largely based on Barnes (1995).

[3] We used the text available at http://classics.mit.edu/Aristotle/history_anim.html

dissected but others were based on tales and he warned about the accuracy of those descriptions. For example, although he mentioned information in numerous occasions provided to him by fishers, many times (but not always) he debunks some of the fallacies he heard based on his own observations, particularly when it came to reproduction.

Of what we would consider today as mammals (including cetaceans) he described about 80 and about 130 species of fishes, which, again, underlines the extensive work, he did on marine creatures, mostly while living at Lesbos. Under the genus "Cetacea" he included at least three species: (1) "dolphins" probably a combination of striped dolphin (*Stenella coeruleoalba*, the most frequent species in the Mediterranean), the common dolphin (*Delphinus delphis*), and the bottlenose dolphin (*Tursiops truncatus*); (2) the harbor porpoise (*Phocoena phocoena*) which he described as "similar to dolphins but smaller and found in the Black Sea" ("Euxine") (*HA* 566b9)[4]; and, (3) the fin whales (*Balaenoptera physalus*) another common species in the Mediterranean at that time.

The motives behind Aristotle classification system, particularly animals, were not biological in nature but rather philosophical. For him these creatures were evidence for rational order in the universe. This approach meant that species were rigid elements of the world and, thus, he never contemplated mutability or anything close to evolution, despite the fact that earlier Greek philosophers such as Anaximander envisioned the mutability of species. Furthermore, Aristotle's motive for conducting this categorization was done in such a way that we can then identify the causes that explain why animals are organized the way they are. His investigation into those causes is carried out in other surviving biological works (e.g., *Parts of Animals*). When describing species he adhered to his teleological doctrine of purposiveness in nature.

Aristotle was able to distinguish between homology and analogy, recognizing cetaceans as a natural group with many similarities with other mammals ("viviparous quadrupeds"). He considered cetaceans as "blooded" animals, adding, "viviparous such as man, and the horse, and all those animals that have hair; and of the aquatic animals, the whale kind as the dolphin and cartilaginous fishes" (*HA* 489a34-489b3). He also wrote: "all creatures that have a blow-hole respire and inspire, for they are provided with lungs. The dolphin has been seen asleep with its nose above water as he snores (*HA* 566b14). All animals have breasts that are internally or externally viviparous, as for instance all animals that have hair, as man and the horse; and the cetaceans, as the dolphin, porpoise and the whale -for these animals have breasts and are supplied with milk" (*HA* 521b21-25). Among the species he described were dolphins, orcas, and baleen whales, noting that "the [whale] has no teeth but does have hair that resemble hog bristle" (*HA* 519b9-15). Thus, he was the first to separate whales and dolphins from fish.

However, Aristotle placed whales and dolphins below reptiles and amphibians, because their lack of legs, despite his physiological and behavioral observations that they were related more closely to "viviparous quadrupeds" than to fish.

---

[4] These citations for *Historia Animalium* follow the Bekker' pagination.

Aristotle followed his teacher Plato in classifying animals by progressively dividing them based on shared characters. This is an embryonic form of today's classification more fully developed by Linnaeus. The reason he ordered the different "genera" the way he did was because he considered "vital heat" (characterized by method of reproduction, respiration, state at birth, etc.) as an index of superiority placing humans at the very top. Men were superior to women because they had more "vital heat." On this he followed Hippocrates's ideas, since the Greek physician thought there was an association between temperature and soul.

Yet he was not fully satisfied by this approach given that a number of "genera" had characters that were shared across groups, particularly when compared with their habitats. For example, both fishes and cetaceans had fins, but they differ markedly on other characters such as reproduction (oviparous vs. viviparous) or organs (gills vs. lungs, respectively).

Many of Aristotle's observations about cetaceans remain accurate. In terms of internal anatomy he mentioned that they have internal reproductive organs (*HA* 500a33-500b6), that dolphins, porpoises, and whales copulate and are viviparous, giving birth to between one and two offsprings having two breasts located near the genital openings that produce milk (*HA* 504b21), that dolphins reach full size at the age of 10 and their period of gestation is 10 months, show parental care, some may live up to 30 years and this is known because fishers can individually identify them by marks on their bodies (*HA* 566b24), and that dolphins have bones (*HA* 516b11).

Regarding behavior and sensory organs he said that dolphins have a sense of smell but he could not find the organ (*HA* 533b1), that dolphins can hear despite the lack of ears (*HA* 533b10-14), produce sounds when outside the water (*HA* 536a1), that dolphins and whales sleep with their blowhole above the surface of the water (*HA* 537a34), are carnivorous (*HA* 591b9-15), and swim fast (*HA* 591b29).

He held that cetaceans are not fishes because they have hair, lungs (*HA* 489a34), lack gills, suckle their young by means of mammae, they are viviparous (*HA* 489b4), and that their bones are analogous to the mammals, not fishes. Still they he calls them "fishes" (*HA* 566b2-5).

These basic Aristotelian biological descriptions persisted for good and for bad until Charles Darwin's evolutionary work. On one hand his descriptions were so accurate that Darwin admired Aristotle, to the point that he said privately that the intellectual heroes of his own time "were mere schoolboys compared to old Aristotle."[5] Yet the fact that Aristotle saw the natural world as fixed in time with no room for evolution and that he kept calling cetaceans "fishes," would delay intellectual progress for many centuries when it came to the classification of these animals.

---

[5] Darwin Correspondence Database, http://www.darwinproject.ac.uk/entry-11875 accessed on 25 Feb 2012.

Aristotle's influence on naturalists' classification of life would extend until Darwin's times when evolutionary views replaced the fixity of species as elements in nature.

2.2. Pliny the elder

Pliny the Elder[6] was the son of an equestrian (the lower of the two aristocratic classes in Rome) and was educated in Rome. After serving in the military he became a lawyer and then a government bureaucrat. In these positions he travelled not only throughout what is Italy today but also what it would later became Germany, France and Spain as well as North Africa (Reynolds 1986).

He wrote a 37-volume *Naturalis Historia*[7] (ca. 77-79) in which according to himself he had compiled "20,000 important facts, extracted from about 2000 volumes by 100 authors" and was written for "the common people, the mass of peasants and artisans, and only then for those who devote themselves to their studies at leisure" (Preface 6). This is the earliest known encyclopedia of any kind, which has been interpreted as a Roman invention in order to compile information about the empire (Naas 2002, Murphy 2004). It was a rather disorganized book, whose prose has been criticized by many (Locher 1984). Pliny seemed to be more interested in what appeared to be curiousities than what were facts. This is a big collection of facts and fictions, based mostly said on things said by others.

He devoted 9 of the 37 volumes to animals and ordered them according to where they live. Volume IX (*Historia Aquatilium*) of *Naturalis Historia* is devoted to aquatic creatures, whether living in oceans, rivers or lakes, whether vertebrate or invertebrate, real or mythical. Based on their size he categorized as "monster" anything big, whether it is a whale, a sawfish or a tuna (IX 2,3).

He grouped together all known species of cetaceans (*cete*) but constantly mixed their descriptions with those of other marine mammals such as seals as well as with cartilaginous fishes, such as some sharks (*pristis*). Pliny mentioned the three species cited by Aristotle: dolphins (*delphinus*, probably a combination of striped dolphin [*Stenella coeruleoalba*] and the common dolphin [*Delphinus delphis*], IX 12-34), porpoises (*porcus marinus*, the harbor porpoise [*Phocoena phocoena*], IX 45) and whales (*ballaena*, possibly a combination of large toothless whales [mysticetes] IX 12-13). Then he added a few more: the *thursio* or *tirsio* (probably the bottlenose dolphin, *Tursiops truncatus* IX 34), the *physeter* (probably the sperm whale [*Physeter macrocephalus*] IX 8) found in the "Gallic Ocean" (probably the Bay of Biscay, IX 3, 4), the *orca* (probably the killer whale [*Orcinus orca*] IX 12-14), and the river dolphin from India (possibly *Platanista gangetica*, IX 46). He also mentioned some mythical creatures such as *Homo marinus* (Sea-Man, IX 10) and the *Scolopendra marina* (IX, 145) a mythical

---

[6] *b.* as Gaius Plinius Secundus 23/24 CE in what is now Como, Italy; *d.* 25 August 79 CE near Pompeii, Italy.

[7] I used the version available at:
http://www.perseus.tufts.edu/hopper/text?doc=Plin.+Nat.+toc&redirect=true

organism whose legend may be based on polychaetes, marine annelids characterized by the presence of many legs (Leitner 1972, p. 218).

Pliny recognized that neither whales nor dolphins have gills, that they suckle from the teats of their mothers, and that they are viviparous. In addition to these true facts copied from Aristotle, he mentioned exaggerations such as whales of four jugera (ca. 288 m) in length that because of their large size "are quite unable to move" (IX 2,3). In addition to some of the biological facts mentioned by Aristotle, Pliny adorns his narrative with all kind of casual tales about interactions between cetaceans and humans.

By lumping together all kinds of aquatic organisms it is hard to distinguish what he called "fish" and what he did not (see for example IX 44-45). His classification took a step back from Aristotle because he did not try for a comprehensive classification of animals. He failed to compare organisms based on shared or divergent characters. Many times he ordered creatures based on size, from the largest to the smallest. Yet, his work had great influence for 1700 years, which was unfortunate because he was an uncritical compiler of other people's writings (even if they were contradictory). Pliny also created a number of unfounded impressions about the reality of nature. His only positive contribution was that he established the norm of always citing the sources of his information (in actuality 437 authors, whose works, in some cases, are no longer available).

3. Medieval times

During the middle ages, little progress was made in the sciences. Students were urged to believe what they read and not to question conventional wisdom. Logic determined truth, not observation. Free thought was non-existent and minds were filled with mythological explanations for the unknown. Marine mammals were depicted as monsters and little new information was generated.

4. The renaissance

The Renaissance was a time of awakening and the religious ideology began to be questioned. The translations of the works of Aristotle and Pliny into Latin and the introduction of the printing press helped to spread the little knowledge accumulated until that time about natural history in the western world. For example, by 1500 about 12 editions of Aristotle's *Historia Animalium* and 39 of Pliny's *Historia Naturalis* had seen the light, which is evidence of the popularity of these works. During this age of discovery the finding of species that were never mentioned neither by Aristotle nor the Bible, opened up scientific curiosity about new creatures around the world. Thus, people once again began to seek new knowledge. However, in these times, naturalists were more compilers of information than investigators despite the fact that they were performing more dissections that in turn uncovered new taxonomic possibilities. Still, scientists relied on environmental aspects to classify animals. Collecting was a primary activity during this era (Alves 2010, p. 54).

4.1. Pierre Belon

Belon[8] was the first author studying marine mammals in this historical period. Little is known about his family and early years. He traveled extensively throughout Europe and the Middle East, including the Arabian Peninsula and Egypt. Among the places he visited were Rome where he met two other ichthyologists, Rondelet and Salviani (see below). He studied medicine at the University of Paris and botany at the University of Wittenberg, Germany. He served as a doctor and apothecary for French kings, as well as a diplomat, traveler, and as a secret agent (he was murdered under strange circumstances) (Wong 1970).

His *L'Histoire Naturelle des Estranges Poissons Marins* (1551) was the first printed scholarly work about marine animals. This book was expanded and published in French in 1555 as *La Nature et diversité des poissons* including 110 species with illustrations for 103 of them.

Belon not only reproduced information from Aristotle and Pliny but also added his own observations including comparative anatomy and embryology. For him "fish" was anything living in the water. He divided "fishes" in two large groups: the first was "fish with blood" (as Aristotle had done) that included not only actual fishes but also cetaceans, pinnipeds, marine monsters and mythical creatures such as the "monk fish," as well as other aquatic vertebrates such as crocodiles, turtles, and the hippopotamus. He called a second group "fishes without blood" and consisted of aquatic invertebrates (see also Delaunay 1926).

He ordered what we know as cetaceans today in a vaguely descending order based on size: *Le balene* (mysticete whales, although in the illustration he depicted a cetacean with teeth), *Le chauderon* (sperm whale? although he mentions the sawfish), *Le daulphin* (common dolphins on which he devoted 38 pages of this 55-page book), *Le marsouin* (porpoise), and *L'Oudre* (bottlenose dolphin) (for a rationale on the identification of these species see Glardon 2011, p. 393-398). He dissected common dolphins (*D. delphis*) and porpoises (*P. phocaena*) acquired at the fish market in Paris brought in by Normandy fishers, and probably a bottlenose dolphin (*T. truncatus*) as well.

He described these marine mammals as having a placenta, mammae, and hair on the upper lip of their fetus. Belon wrote that apart from the presence of hind limbs, they conform to the human body plan with features such as the liver, the sternum, milk glands, lungs, heart, the skeleton in general, the brain, genitalia. He also dealt with issues of breathing and reproduction (although from the description it is clear that he never saw one of these animals giving birth, since he depicted the newborn surrounded with a membrane). He drew the embryo of a porpoise and the skull of a dolphin (Fig. 1). Despite all this he did not make the connection between cetaceans and "viviparous quadrupeds" and based his entire classification on environmental foundations, as he made clear in the introduction of his work.

4.2. Edward Wotton

Wotton[9] was the son of a theologian who did general studies at Oxford and studied medicine and Greek at Padua (1524-6). He was a practicing physician who published *De*

---

[8] *b.* 1517, Soultière, near Cerans, France; *d.* April 1564, Paris, France.

[9] *b.* 1492, Oxford, England; *d.* 5 October 1555, London, England.

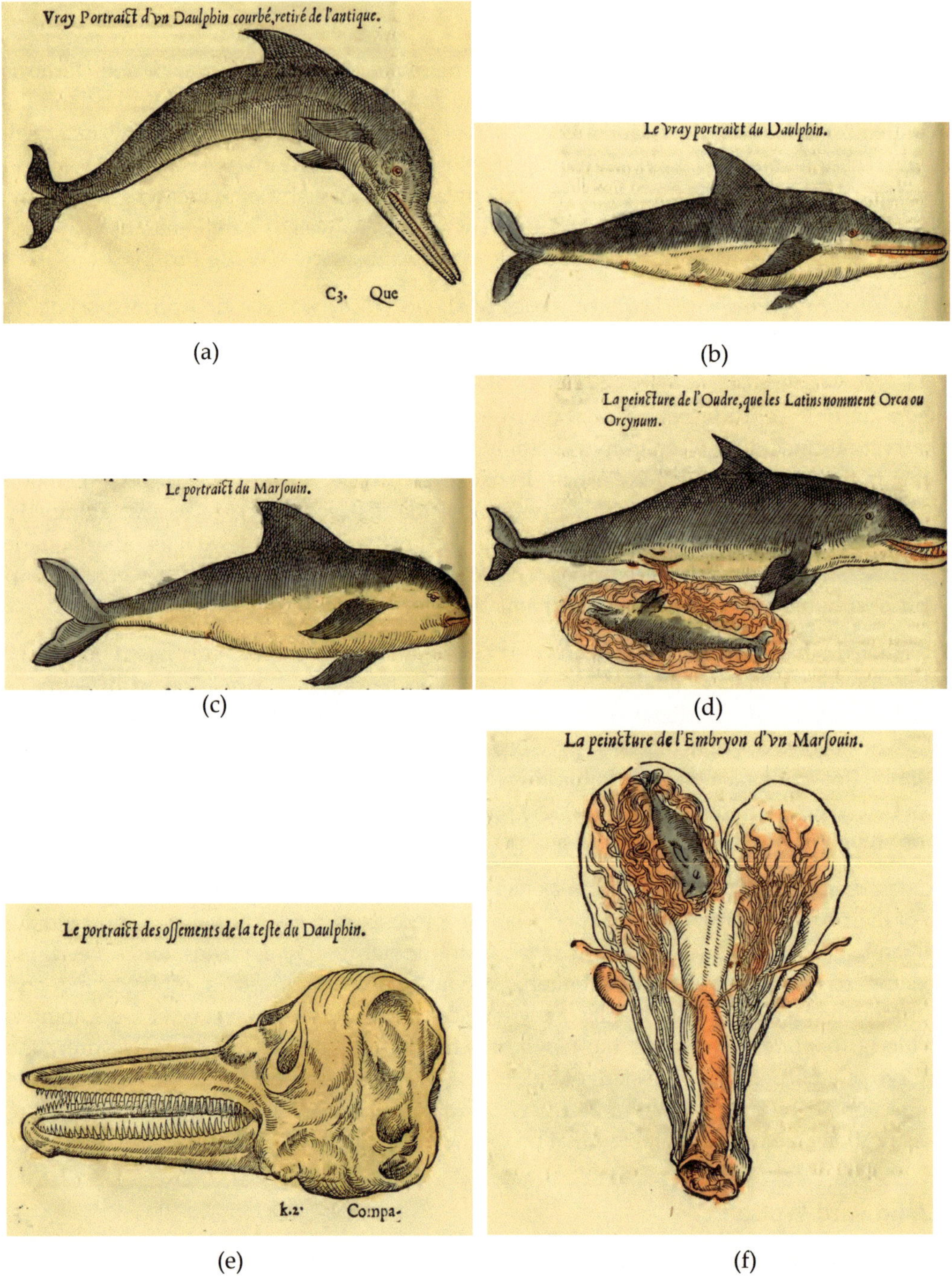

Figure 1. Illustrations of marine mammals by Belon (1551): (a) and (b) are representations of the common dolphin (*Delphinus delphis*); (c) a porpoise (*Phocaena phocaena*); (d) a bottlenose dolphin (*Tursiops truncatus*, although he uses the name of "Orca") presumably giving birth; (e) the skull of a dolphin; (f) a porpoise fetus in a placenta, showing that he had actually dissected these animals.

Differentiis Animalium Libri Decem (1552), probably the first published book on natural history of the Renaissance. This was a 10-part ("books") treatise that followed the classification structure by Aristotle while adding some comments from Pliny. In Book 8 (pp. 171-173) he placed *Cete* together with fishes because of the medium they inhabit. Except for entomology he did not conduct any original observations on animals nor include any illustrations. His contemporaries noted his lack of originality (Nutton 1985).

The list of cetacean species included *Delphino* (dolphins), *Phocaena* (porpoises), *Balaena* (mysticete whales), *Orca* (either the bottlenose dolphin or the killer whale) and *Physeter* (the sperm whale).

4.3. Guillaume Rondelet

Rondelet[10] was the son of a drug and spice merchant. He studied medicine at the University of Montpellier, one of the best medical schools in Europe at that time. While in Paris he studied anatomy under Johannes Guinther, who also taught Vesalius. Rondelet would later become Professor of medicine and Chancellor at Montpellier (Keller 1975). He probably acquired his interest in ichthyology at a young age while living in Montpellier (about 12 km from the coast) because his family owned a farm that was a stopping place for carts of fish from the Mediterranean (Oppenheimer 1936). During his trips as personal physician to Francois Cardinal Tournon (who was also the patron of Belon) to the Atlantic coasts of France, he became acquainted with the whaling industry. Rondelet met several contemporary ichthyologists while in Rome (1549-1550) such as Belon, Hippolyto Salviani, and Ulyssis Aldronvandi (Gudger 1934). Guillaume Pellicer, Bishop of Montpellier, who was also interested in fishes but never published on ichthyology, may have influenced Rondelet (Oppenheimer 1936, Dulieu, 1966).

He enjoyed dissecting and did so frequently for both teaching and research purposes. He published *Libri de Piscibus Marinis in quibus verae Piscium effigies expressae sunt* (1554) with a second part titled *Universae Aquatilium Historiae pars altera* (1555) about both marine and freshwater animals. Both were later translated into French as *L'histoire entière des poissons* (1558, 599), a monograph for teaching purposes.

After writing about food, habitat, morphology, and physiology, he described 145 freshwater and 190 marine species that included at least seven species of cetaceans: *delphino* (common dolphin), *phocaena* (porpoise), *tursione* (bottlenose dolphin, although the illustration more resembles a porpoise), *balaena vulgo* and *balaena vera* (two different species of mysticetes whose true identities are difficult to ascertain), *orca* (killer whale), and *physetere* (sperm whale) (Fig. 2). He also included among cetaceans the *priste* (sawfish) and mythical animals such as Pliny's *scolopendra cetacea*, the *monstruo leonino* (a lion covered with scales and with a human face), the *pisce monachi habitu* (a fish that looks like a monk), and the *pisce Episcopi habitu* (a fish that looks like a bishop) of which he was skeptic. All together his book contained more species than previous published works. Each species description included

[10] *b*. 27 September 1507, Montpellier, France; *d*. 30 July 1566, Réalmont, Tarn, France

the animal's name in different languages, their morphology (external and internal), feeding habits, and use as food for humans. Species were differentiated similarly to Aristotle as blooded and non-blooded. Although Aristotle inspired the entire book, including teleological considerations in his discussions, Rondelet added some original ideas, especially concerning anatomy and descriptions of the small cetaceans he dissected. Rondelet made correlations between form, function, and environment.

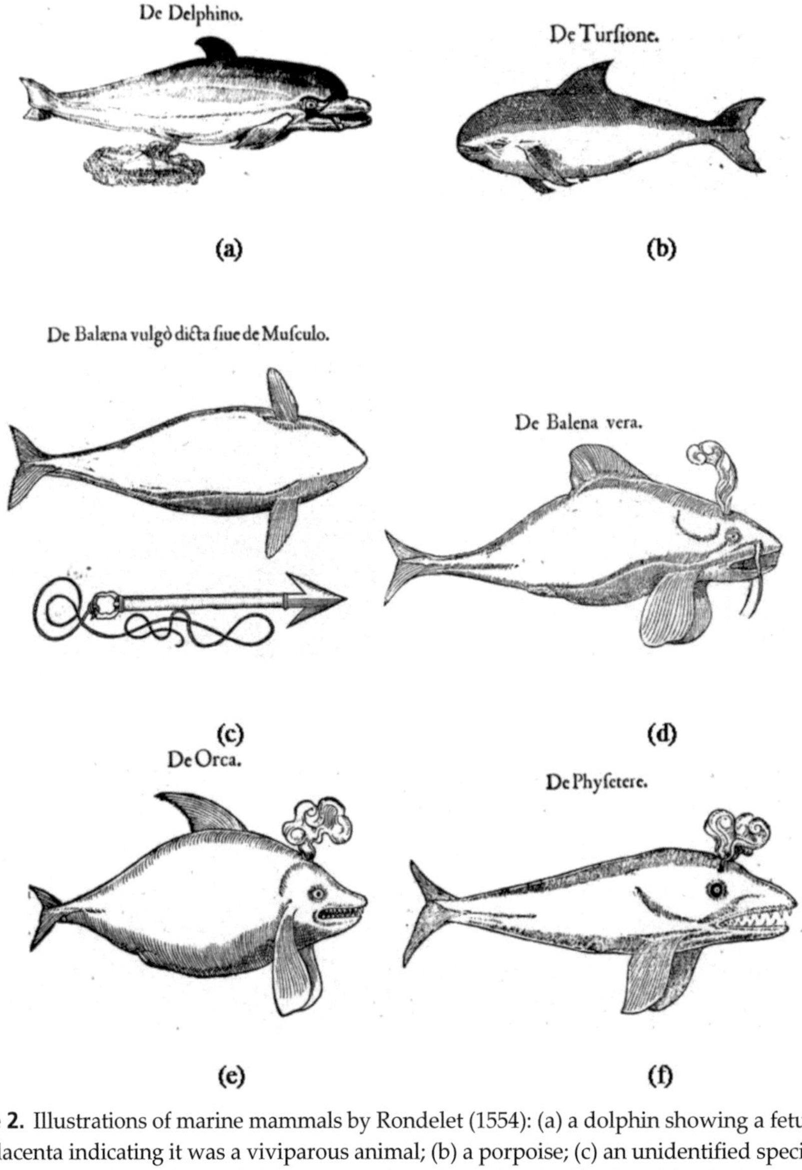

Figure 2. Illustrations of marine mammals by Rondelet (1554): (a) a dolphin showing a fetus surrounded by a placenta indicating it was a viviparous animal; (b) a porpoise; (c) an unidentified species of mysticete, probably a right whale because may have been observed by Rondelet during a whaling operation in the Atlantic; (d) an unidentified species of mysticete that he never saw as evidenced by the depiction of barbels above the mouth; (e) orca (*Orcinus orca*); (f) a sperm whale (*Physeter macrocephalus*).

Despite noting differences, he grouped marine mammals with fish based on habitat. For example, he noted that fishes with scales lack lungs and have a three-chamber heart while what we know today as marine mammals have hearts with four chambers. He compared the anatomy of a dolphin to that of the pig and humans. Based on this and his descriptions of other internal organs, he considered marine mammals to be a type of aquatic quadruped. Yet, he did not propose a system of classification. He did not advance the notion of valid classification, but because of the quality of his descriptions his work remained as the main reference for about 100 years.

4.4. Conrad Gessner

Gessner[11] probably developed an interest in zoology after seeing the carcasses of furred animals at his father's workshop where several furriers worked. He also lived with a great-uncle, an herbalist, who furthered his interest in natural history (Bay 1916, Gmelig-Nijboer 1977, p. 17, Wellisch 1984, p. 1). He was an avid traveler who studied theology and medicine in Bourges, Paris, Montpellier, and Basel (Fischer 1966) and had great facility for classical languages. During his travels Gessner met with Belon and Rondelet. He is considered as the "father of bibliography" because of his work on compiling information about books (Bay 1916). Gessner himself had a very large private library of more than 400 volumes (which was a very large private collection for his time) of which 19% of the volumes were on natural history and 13 of them were on zoology (Leu et al. 2008, pp. viii, 1, 13, 21). He published *Historiae Animalium* (1551-1558), an encyclopedic (4 volumes, 4,500 pages treatise) but uncritical compilation of information and bibliography in which he intended to itemize all of God's creations. In addition to classic authors such as Aristotle and Pliny, Gessner obtained information from whomever he could correspond. He classified cetaceans among 'aquatic animals,' i.e., including fishes. The fourth volume (*Piscium & Aquatilium*) of 1297 pages was published in 1558 and was about the aquatic animals. A fifth volume on reptiles and arthropods was not published until 1587, posthumously. *Historiae* was added to the list of prohibited books because Gessner was Protestant. Yet, the 14 editions in different languages of this book reveal its popularity.

Gessner followed Aristotle's classification of animals when it came to their grouping by volume (Vol. 1: viviparous quadrupeds; Vol. 2: oviparous quadrupeds; Vol. 3: birds; Vol. 4: aquatic animals; Vol. 5: serpents). He ordered them alphabetically, like a "Dictionarium," in each volume, which did not provide a rational classification based on relationships of any kind; on the other hand this alphabetical order facilitated its use as an encyclopedic source. Gessner's intention was to collect any piece of information ever written about each animal by any author in history, he cited nearly 250 authors including Rondelet (*Libri de Piscibus Marinis*, 1554), Belon (*De Aquatilibus*,1553), and Salviani (*Aquatilium Animalium*, 1554). The latter only mentioned marine mammals *in passim*.

---

[11] *b.* 16 March 1516, Zürich, Switzerland; *d.* 13 December 1565, Zürich.

(a) (b)

(c)

Figure 3. Some of the "Cetis" described by Gessner (1558): (a) and (b) two examples of marine monsters; (c) a whale attacking a ship and another being flensed during whaling operations. Both show mysteces with teeth, which indicates that Gessner never saw these animals. This exemplifies that Gessner was an uncritical compiler of information.

Information included names of the animals in various languages (some times more than a dozen) comprising epithets and etymology (even inventing common names in other languages when those names were not available), physical features, geographic distribution, the animal's way of living including diseases and their cures, behavior, utility towards man (e.g., for food or medical purposes), and tales. His work was full of illustrations: some were very accurate showing that he had first-hand knowledge of the animal in questions while other were bizarre or just invented, especially when dealing with mythical creatures.

Gessner included a 16-page-folio discussion about the dolphin very much along the lines of Aristotle and Pliny. As an uncritical compiler he included contradictory or totally false information such as mythical species and even "monsters." In volume 4 he relied heavily on Belon and Rondelet. For example, *Monachus marinus* (sea monk, IV, p. 519) description was copied from Rondelet who, in turn, had received the description from Marguerite, Queen of Navarre, who heard it from Emperor Charles V's ambassador, who had claimed to see the monster himself (Kusukawa 2010). He did not add much to what was already known. Among marine mammals he mentioned are the *Balaena* (mystecete whales, IV, p. 128) depicted more as sea monster than as an actual whale, *Cetis diversis* (IV, p. 207), an amalgam of marine monsters based on Olaus Magnus's descriptions of sea monsters from seas from northern Europe, *Hominis marinis* (IV, p. 438), a collection of humanoid sea monsters such as the sea-monk and the sea-bishop. To certain extent he was skeptical of accuracy of some of these descriptions by other authors.

Many of the figures were made by others and copied directly from other books including those of "cetaceous" animals as was the case of a whale which was copied from Olaus Magnus' map of the Northern Lands (IV, p. 176) (Fig. 3).

4.5. Ulysses Aldrovandi

The last author who published anything of significance about marine mammals during the Renaissance was Aldrovandi[12]. He was born to a noble and wealthy family, which allowed him to initially dedicate his life to his own pursuits. He was educated in Bologna, Padua, and Rome, receiving degrees in law and medicine although he never practiced those professions. He was appointed as the first professor of natural history in the University of Bologna. Although he was a pious Catholic, because of what he read he was charged with heresy. After producing himself in Rome, he was acquitted. While in Rome he met Rondelet and accompanied him to the fish markets where he became interested in ichthyology (which included the study of marine mammals) collecting specimens for his own museum. He traveled extensively throughout Italy and made a collection of about 11,000 animal specimens for pedagogical purposes; most of them can be found today at the Bologna Museum to which he bequeathed not only his specimens but also his library and unpublished manuscripts as well (Alves 2010, pp. 56-82). He also conducted dissections (Impey and McGregor 1985). He was a true encyclopedist following the tradition of the University of Bologna at that time (Tugnoli Pattaro 1994). He wrote extensively but the quality of his animal descriptions and illustrations were poor from the scientific viewpoint

[12] *b.* 11 September 1522, Bologna, Italy; *d.* 4 May 1605, Bologna.

(Fig. 4). Aldrovandi was an uncritical compiler who included legends of mythical animals in his writings similar to the medieval bestiaries and in the tradition of Pliny.

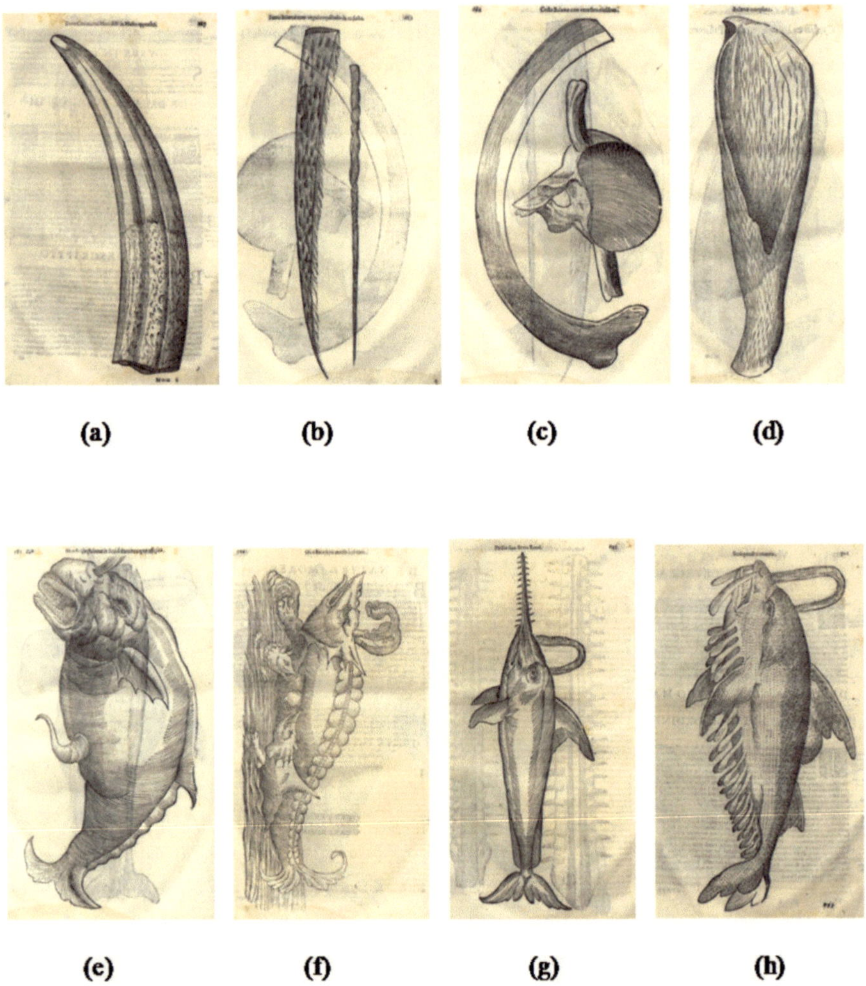

Figure 4. Depiction of some marine animals by Aldrovandi (1613): Some show that he actually saw some of those skeletal pieces such as (a) a tooth possible from a sperm whale, (b) a baleen and the prominent tooth of a narwhal (*Monodon monoceros*), (c) a rib and a vertebra, possibly of a large whale, and (d) a scapula. In other cases he illustrated whales with human-like emotions (e); whales with feet (f); sawfishes with cetacean characteristics (g); and Pliny's "Scolopendra cetacea" (h), which perpetuated the notion that such animal existed. Overall he was a very uncritical compiler when it came to marine mammals.

He published *De piscibus libri V, et De cetis lib. vnus* (1613) where he defined "Pisces" as animals covered with scales and "aquatilis" as "anything else that lives in the water" while recognizing that cetaceans are air-breathing creatures. The species that he mentioned were the ones cited by his predecessors: *Balaena, Physeter, Orca, Delphino, Phocaena,* and *Tursione,* while including the *Manate Indorum, Phoca, Pristi* (the sawfish), and the mythical *Scolopendra*

Cetacea. From the illustrations (Fig. 4) it is clear he never saw any of these animals with the exception of some of their skeletal parts. As an uncritical compiler of information he did not add anything new to the knowledge of these creatures and, yet, was cited by later authors.

5. Modern science

In this period, observation and experimentation moved to the forefront of science. Classification was based on similarities and differences in characters. During this time English physicians travelled to Padua, Bologna and Paris to be trained in human dissection since the status of medicine in England was still poor. People involved in these kind of activities had a background in either medicine (or "physic" as it was called then) and/or theology (Kruger 2004). During this time the center of gravity of science moved from the Mediterranean world to northern Europe, mostly England.

5.1. Johann Jonston

The first researcher of the biology of marine mammals in this period was Johann Jonston[13]. Although born in Poland, Jonston's father was Scottish and his mother German. He was educated in St Andrews, Frankfurt, Cambridge, and Leiden, receiving a medical degree from the last two institutions. He traveled extensively throughout Europe teaching, and despite offers for academic positions, he decided to make a living as an independent scholar (Miller 2008). He published *Historiae naturalis de Piscibum Partem* in 1657. Jonston was another encyclopedist who when it came to natural history was more a compiler than anything else, relying heavily on Gessner and Aldrovandi while adding some new information from New World creatures from George Marcgrave. Thus, he did not offer any significant critical view to his sources although his descriptions were briefer than those of his predecessors. He gave no hint of biological classification for marine mammals and also added further mistakes and legends (even 'monsters'). He slightly modified Aldrovandi's classification of fishes by adding 'pelagic' fishes. Yet his books were widely read and translated.

He dealt with cetaceans on pages 213-224 of his *Historiae* and included the same species as Aldrovandi: *Balaena, Physetere, Orca, Delphino, Phocaena* and the mythical *scolopendra cetacea*, the sawfish, pinnipeds, and the manatee among the cetaceans.

5.2. Walter Charleton

Charleton[14] was the son of a church rector of modest means. He was educated at Oxford as a physician at that time when medical education in England emphasized scholastic approaches to knowledge and British colleges had inadequate anatomical staff and teaching facilities. The practical elements of practicing medicine were not acquired until after assisting a more experienced practicing medical doctor.

---

[13] *b*. 15 September 1603, Szamotuly, Poland; *d*. 8 June 1675, Legnica, Poland.

[14] *b*. 2 February 1619, Shepton Mallet, Somerset, England; *d*. 24 April 1707, London, England.

Charleton was a follower of epicurean atomism (materialism) (Kargon 1964) and an eclectic (Lewis 2001), whose interest in natural history was more or less theological because, as he said, men were obligated into "naming & looking into the nature of all Creatures" (Boot 2005, p. 119). In other words, just as Ray and Willoughby did later, natural science was the search a divine pattern in nature, part of the research agenda of the Royal Society – to which Charleton belonged (Rolleston 1940, Sharpe 1973). His publications showed him more as a compiler than as an innovator. His major contribution to science was the discovery that tadpoles turn into frogs (Booth 2005, p. 1).

He published two books dealing with animal classification: *Onomasticon zoicon* (1668) and *Exercitationes de Differentiis & Nominibus Animalium* (1677) works that listed the names of all known animals (including some fossils) in the western world in several languages with a somewhat taxonomy discussion, including remarks about these animals habits and habitats that contained anatomical descriptions of two animals that he had dissected. As Belon did over a century before, he divided "fishes" as either "with blood" (vertebrates) and "without blood" (invertebrates). He grouped under "Cetaceos" not only actual cetaceans but also the sawfish, seals, walruses, manatees, hippopotamus and the mythical "scolopendra cetacea." The actual cetaceans described were *Balaena vulgaris* (probably the right whale), *Physeter, & Physalus* (probably the fin whale but also other species), *Cetus dentatus* (the sperm whale), *Pustes* (indeterminate species, maybe the beluga), *Orca* (the killer whale), *Monoceros* (the narwhal), *Delphinus* (probably a composite of delphinidae), and *Phocaena* (the porpoise).

5.3. Edward Tyson

Tyson[15] was born into an affluent merchant family. He performed numerous dissections as a college student, obtained his medical degree at Oxford University and was a lecturer of Anatomy at the Barber-Surgeons Hall in London. Tyson was the first of the comparative anatomists in the modern sense. He did extensive dissections and was the first to use a microscope as part of his anatomical studies. His description of the highly convoluted cetacean brain as well as his recognition of the many homologies with "viviparous quadrupeds", rather than the fishes that they externally resembled, constituted a major landmark contribution to the history of biology (Kruger 2003).

In *Phocaena, or, The anatomy of a porpess dissected at Gresham Colledge, with a preliminary discourse concerning anatomy and natural history of animals* (1680), he noted that "What we have here is a signal Example of the same between Land-Quadrupeds and Fishes; for if we view a *Porpess* on the outside, there is nothing more than a fish; for if we view a *Porpess* on the inside, there is nothing less. (...) It is viviparous, does give suck, and hath all its Organs so contrieved according to the standard of them in Land-Quadrupeds; that one would almost think of it to be such, but it lives in the Sea, and hath but two fore-fins." Adding later "The structure of the viscera and inward parts have so great an Analogy [*sic*] and resemblance to those of Quadrupeds, that we find them here almost the same. The greatest

[15] *b*. 20 January 1651 Clevedon, near Bristol, Somerset, England; *d*. 1 August 1708, London, England.

difference from them seems to be in the external shape, and wanting feet. But here too we observed that when the skin and flesh was taken off, the forefins did very well represent an Arm, there being the *Scapula*, an of *Humeri*, the *Ulna*, and *Radius*, and bone of the *Carpus*, the *Metacarp*, and 5 *digiti* curiously joynted. The Tayle too does very well supply the defect of feet both in swimming as also leaping in the water, as if both hinder feet were colligated into one, though it consisted not of articulated bones but rather Tendons and Cartilages."

Tyson's description of the internal anatomy of the porpoise is remarkable, particularly when it comes to its nervous system (Kruger 2003). In many ways he thought that the "porpess" was the transitional link between terrestrial mammals and fish.

In his monograph Tyson surveyed contributions from previous authors. He corresponded with John Ray (see below). Ray had also dissected a porpoise (an exercise on which he reported in a published form in 1671), nine years before Tyson but was far more superficial and added very little to what other authors such as Rondelet had done. Tyson met Ray around 1683 and the latter invited Tyson to contribute to Willughby's *De Historia Piscium* (Montagu 1943, p. 103).

Tyson was critical of encyclopedic approaches and relying on classical authors when it came to natural history. He set new standards in terms of direct observation and comparative anatomy. He also established an understanding of homology not seen since Aristotle. He proved to be a very competent observer of internal anatomy and he saw comparative anatomy as a means to explain the Great Chain of Being (or *scala naturae* or ladder of nature) as proposed by Plato and Aristotle.

5.4. Samuel Collins

A contemporary of Tyson was Samuel Collins[16]. The son of the rector of Rotherfield, Sussex, who got his education at Cambridge, Collins travelled to several universities in France, Italy and the Netherlands finally getting his medical degree at the University of Padua, later becoming physician of Charles II. He taught anatomy at the Royal College of Physicians[17]. Collins published *A Systeme of Anatomy* (London 1685), which was the earliest attempts to illustrate the brains of a broad variety of mammals, birds, teleosts, and elasmobranchs in a remarkable two-volume folio edition of 1,263 pages. It included 73 full-page illustrations of very high quality. There he described a female porpoise. However, it seems that he had used Tyson's previous descriptions and unfortunately says nothing about the brain of this cetacean. Had he had examined the brain of the porpoise he would have noted the great similarities of this organ with those among the "viviparous quadrupeds." Collins did not discuss the similarities between the other internal organs of the porpoise and those called mammals today either. He acknowledged Tyson' previous contributions in this matter.

---

[16] *b*. 1618, Rotherfield, Sussex, England; *d*. 11 April 1710, Westminster, Middlesex, England.

[17] Biographical information obtained from: http://munksroll.rcplondon.ac.uk/Biography/Details/950 and accessed on 2 April 2012.

In addition to Tyson, Collins's anatomy draws largely upon the works of Thomas Willis. In the opening Epistle-Dedicatory to James II he claimed that various chapters "are illustrated by the Dissection of other Animals (which I have performed with Care and Diligence, speaking the wonderous Works of the Glorious Maker) rendering the Parts of Man's Body more clear and more intelligible." In volume two of his huge work he described numerous folio copper plates containing the most extensive comparative anatomy of the brain then existing, an expansive account of the functional significance of his findings, as well as practical clinical commentary.

5.5. John Ray

Ray[18] was the first naturalist who truly represented this new era of careful observation. His father was a blacksmith and his mother was an herbal healer. He studied at the University of Cambridge, pursuing comparative anatomy although initially his main interest was botany. He taught Greek, mathematics and humanities at Cambridge but abandoned his teaching position after refusing to comply with the Act of Uniformity of 1662. He was a very religious person who undertook the study of nature to understand God's creation (Raven 1950). Fairly early he developed a plan with his student and patron, Francis Willughby[19] to produce a joint general natural history. To that end Ray and Willughby went on an extended tour of England and Europe (1662-1666), including the medical school at Montpellier. Although they did not always travel together both collected specimens, got involved in dissections and acquired books and illustrations (Kusukawa 2000), an endeavor bankrolled by Willughby. When Willughby died, Ray took over his parts of the general natural history. Willughby left him an annuity of £60 and Ray stayed on as tutor to Willughby's children until 1675, when Willughby's mother, also his patron, died, and the widow immediately terminated the relationship. Ray inherited a small farm that also contributed to the family's maintenance while he earned money from his productive publishing. Therefore Ray had the financial freedom to pursue his intellectual interests.

Ray's first published work on cetaceans was *Dissection of a Porpess* (1671). He does a much better job in describing the internal anatomy of this animal when compared with Rondelet but does not get into the detail that Tyson achieved later. During the narrative of his findings he keeps noticing that a porpoise has a lot in common with the "quadrupeds". Yet he persisted calling them "fishes."

Ray published *Historia piscium* (1686), under Willughby's name 14 years after his patron death, though Ray himself contributed the vast majority of the content. He carried out the first serious attempt to achieve a systematic arrangement, the success of which can be attributed by the fact that it served as a basis for the systematics work of the following

---

[18] *b.* 29 November 1627, Black Notley, near Brainton, Essex, England; *d.* 17 January 1705, Black Notley, England.

[19] *b.* 22 November 1635, Middleton, Warwickshire, England; *d.* 3 July 1672, Middleton.

century. His approach was based on direct observation, collaboration with other researchers, and critical reading of previous authors.

Historia Piscium is divided into two parts that were printed separatedly: the first is the narrative and the second, titled *Ichthyographia*, were the illustrations. Many libraries today have both bound together. As sources Ray used authors mentioned earlier in this chapter: Rondelet, Salviani, Gessner, Aldrovandi and Belon, among others. Yet, far from merely compiling information from them, Ray insisted in very comprehensive descriptions of species and discarded all monsters and mythical creatures mentioned by his predecessors. Ray not only removed narratives of marine invertebrates but also other aquatic animals such as the crocodile and the hippopotamus. He divided his subject matter into three groups: cetaceans, cartilaginous fishes, and bony fishes. He recognized that when it comes to reproduction and internal anatomy cetaceans are identical to the "viviparous quadrupeds." Still, he kept cetaceans within the "piscium" despite the fact that he was well aware that they were biologically distinct from fishes.

In his narrative of species Ray moved away from in the practical aspects related to these animals. Aspects such as usage for medical purposes were very common among previous authors because of their medical background. Yet, Ray was very keen at compiling names on the belief that a universal language could be construct based on the knowledge of nature. As Kusukawa (2000) has argued convincingly, Ray believed that there was a need for "a construction of a universal language based on a table that properly expressed the natural order and relations between things." Hence a precise description and classification was the route to achieve that goal. The final product counted not only on the intellectual support of the Royal Society's members who provided constructive criticism and moral support but also their financial support. The cost of publishing *Historia Piscium* was not only very high, mostly because of the expense of the illustrations (187 plates), but also the 500 copies printed sold poorly. As a consequence the Society could not print Isaac Newton's *Principia*.

Ray's third publication related to marine mammals was *Synopsis Methodica Animalium Quadrupedum et Serpentini Generis* (1693). By then he was totally convinced that cetaceans were not fishes: "For except as to the place on which they live, the external form of the body, the hairless skin and progressive swimming motion, they have almost nothing in common with fishes, but remaining characters agree with the viviparous quadrupeds." He placed today's terrestrial mammals (including the manatee) among the 'hairy animals' very close to the *Cetaceum genus* (cetaceans).

In *Synopsis* Ray included a section called *Pisces Cetacei seu Belluae marinae* where he expressed that these animals breath and give birth like the "oviparous quadrupeds." He grouped them into two categories according to the presence of teeth much as we do today separating odontocetes from mysticetes. Ras was the first in doing so. The species he cited were *Balaena vulgaris* (Rondelet), *Balaena* (Fin-Fish), *Physeter* or *Balaena physeteris*, *Orca* (Rondelet & Belon), *Cete* (Sperm whale), Pot Walfish, *Albus piscis cetaceus* (white fish), *Monoceros cetaceo* (*Narhual islandis*), *Delphino antiquorum* (dolphin, from Rondelet), *Phocaeno* (Rondelet & Belon), dissecting a specimen of the latter in 1669.

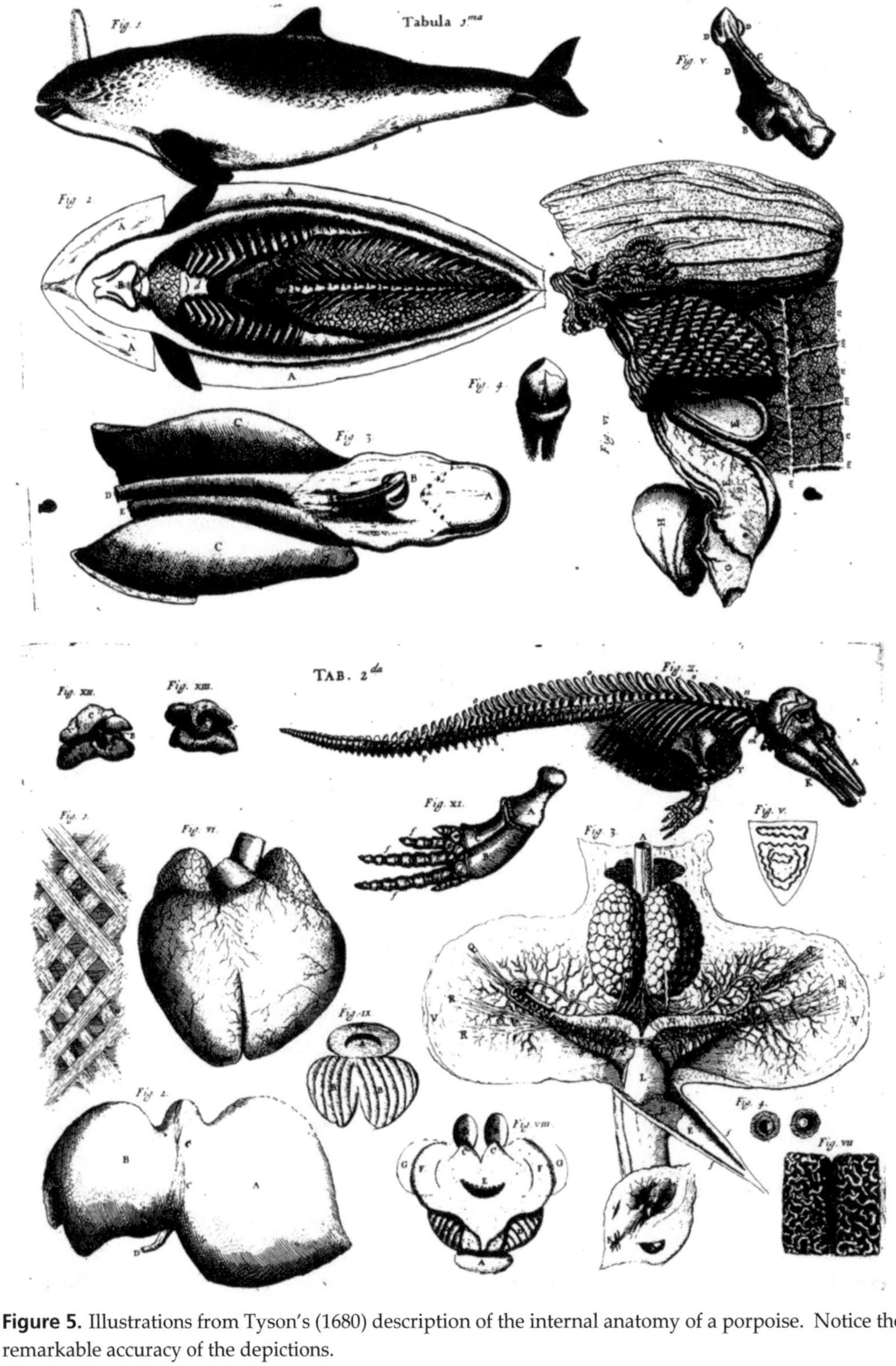

Figure 5. Illustrations from Tyson's (1680) description of the internal anatomy of a porpoise. Notice the remarkable accuracy of the depictions.

Ray developed a division of animals characterized by having blood, breathing by lungs, two ventricles in the heart, and being viviparous. Ray subdivided this group into aquatic (cetaceans) and terrestrial or quadruped including sirenians (manatees and dugongs). He rejected tales of fabulous animals while perfecting Aristotle's classification by diving vertebrates into those having hearts with two ventricles (mammals and birds) from those with a single ventricle (reptiles, amphibians and fish). He also advanced the understanding of other groupings. He established the significance of the generic principle, defined species, and was a leading contributor to the gigantic task of classification.

Ray came close to recognizing mammals as a separate group based on "warm-blood," vivipary, and hair. He conceded the relationship of cetaceans with viviparous quadrupeds; described genera and species; established ordinal classification of mammals; systematic phrases and names; used of descriptive phrases as well as monomial names (a taxonomic name consisting of a single word); a dichotomous ("A is B or not B") classification of mammals. Yet, he lacked the vision or intellectual courage to reunite marine mammals with their terrestrial relatives and still placed the former with the fish "in accordance with common usage." Still he was possibly the best naturalist of the seventeenth century.

5.6. Peter Artedi

Artedi[20] was the son of a parish priest who developed an interest in fishes from an early age. He studied medicine at the University of Uppsala, devoting most of his time at studying natural history. At 29 years of age he went to London for a year to study natural history collections and described the sighting of a whale in November 1734, probably downstream of the London Bridge. He then moved to Leiden, The Netherlands, to complete his medical studies and there he met Linnaeus, whom he knew from their native Sweden, forging a lifelong personal and professional relationship. Linnaeus introduced him to an Amsterdam chemist, Albert Seba, and Artedi started working on Seba's fish collection. Artedi died at the age of 30 by drowning in an Amsterdam canal. After his death, Linnaeus recovered his manuscripts and published *Ichthyologia* (1738) without amending Artedi's original work. Despite the fact that this was an unfinished work, it was a fundamental publication that marked the origin of ichthyology as we know it today. After a long (96 pages) introduction describing previous authorities on ichthyology the second part deals with the taxonomic terminology he used, particularly regarding the concept of genus and distinguishing between species and varieties. His system set the basis for the modern systematic classification of living organisms later established by Linnaeus. In part three he went into the classification of species including detailed description of them, some of which he had dissected himself. For this Artedi is considered the father of ichthyology (Wheeler 1962, 1987, Broberg 1987).

---

[20] *b*. 10 March 1705, Anundsjö, Västernorrland, Sweden; *d*. 27-28 September 1735, Amsterdam, The Netherlands.

Artedi separated actual fishes from cetaceans (which he called "plagiuri") based on the plane of the caudal fin. He described 7 genera and 14 species including the manatee and the "siren" as follows:

Order: Plagiuri

Physeter
- *Balaena major* (Ray, p. 15)
- *Balaena macrocephala* (Ray, p. 16)

Delphinus
- *Delphinus* (*Phocaena*) (Art. Syn. 104)
- *Delphinus* (*Delphin*) (Art. Syn. 105)
- *Delphinus* (*Orca*) (Art. Syn. 106)

Balaena
- *Balaena vulgaris* (Ray p. 6, 16)
- *Balaena edentula* Fin-Fish (Ray p. 6, 10)
- *Balaena tripinnis* (Ray 16)
- *Balenae* (*Balaena tripinnis*) (Ray 17)

Monodon
- *Monoceros pisces* (Will. 42, Ray 11, Charleton 168)

Catodon
- *Balaena minor* (Ray p. 15)
- *Balaena major* (Ray p. 17, Will. P. 41)

Trichechus
- *Manatus* (Rondelet p. 490, Gessner p. 213, Charleton 169, Aldrovandi 7 28, Jonston 223)

Siren
- *Homo marinus*

Artedi established the basic classification of fishes that lasted for about 200 years and separated cetaceans into a totally different order than fishes; he apparently knew that they were different, but still tradition was difficult to break and thus he included them into his ichthyological treatise. He also established the basic branching of animal groups into Class, Maniples (Families), Genera, and Species, a system that was to be closely followed by Linnaeus (Wheeler 1987, Broberg 1987). His work set the foundations for what Linnaeus would culminate as the definitely recognition of cetaceans as distinct group within mammals.

5.7. Carolus Linnaeus

Linnaeus (or Linné)[21] had as a father a country person who loved plants. Linnaeus followed a medical career but was actually more interested in botany than in anything else. Linnaeus met Artedi in 1729 and their interests were complementary: Artedi, a zoologist interested mostly on fishes, and Linnaeus, interested in botany. He would later edit Artedi's book in ichthyology that was published in 1738. What Linnaeus learned from Artedi set the basis for a better classification not only of plants but also animals in general.

Even some of Linnaeus students were developing a better understanding of cetaceans as being really close to "viviparous quadrupeds." That was the case of Pehr Löfling[22], one of Linnaeus' students who came very close to making major contributions to the true nature of dolphins and manatees based on his observations of these animals in South America. In his description of Amazon freshwater dolphin or boto, Löfling was clear about when writing that whales and dolphins were different from fishes: "Pisces per pulmonibus spirantibus." However, his early death and the fact that his manuscripts were never published prevented him from gaining recognition in the scientific community (Romero *et al.* 1997).

With all of this background, the botanist Linnaeus was ready to revolutionize biological classification and in the 10th Edition (1758) of his fundamental work *Systema Naturae,* he introduced the term Mammalia, and included *Cete* among them. For Linnaeus, mammals were united by having hair, being viviparous, and producing milk. He coined the term *cetacea* and separated them from fishes and grouped them with the rest of the viviparous quadrupeds based on the following characteristics: two-chamber heart, breathing by lungs, hollow ears, internal fertilization, and production of milk.

Thus, Linnaeus revolutionized the science of systematics by developing a fully natural system of classification, using consistently the binomial nomenclature, and designing species with Latinized names (genus and species). He developed a hierarchy (class, order, genus, species) as proposed by Artedi, with species defined as similar individuals bound together by reproduction, which also set the basis of the biological species concept. The use of telegraphic speech-like (very short sentences) diagnosis for species descriptions and the standardization of synonymies (same species with different names) in order to reach a taxonomic consensus made his classification even more useful since from now on one could find clarity on what a particular species was tracing its description to other authors. He also doubled the number of species described by Ray. Thus, despite the fact that he was not a zoologist *per se* nor was involved in dissection of animals, he was far from a compiler in that he applied critical thinking to the way he ordered nature.

[21] *b.* 23 May 1707, Stenbrohult, Småland, Sweden; *d.* 10 January 1778, Uppsala, Sweden.

[22] *b.* 1729, Valbo, Sweden; *d.* 22 February 1756, Guayana, Venezuela.

This progress is even more remarkable when considering that Linnaeus was far from an evolutionist. For him species were fixed except for small variations due to climatic/local conditions. Yet, Linnaeus was, without question, the founder of systematics and the one who laid the foundations for the naturalists to become specialists and, therefore, opened the door for the first group of marine mammal specialists, now that these creatures were not longer considered "fishes." It was not until Linnaeus that the science of taxonomy made the strides that have lead us to where we are today in our understanding of the natural world. Linnaeus understood biological principles and placed animals in groups based on homologies rather than using environment to drive classification, and this was what allowed him to recognized cetaceans as a distinct group within mammals.

6. Conclusion

Persuing at the information provided above there are a number of discernable patterns. One is the preponderance of pre-Linnean researchers interested in marine mammals who had a medical background of some sort. That is not surprising because medicine was the closest thing to science as a career existed until the eighteenth century. Also, being interested in medicine created more opportunities to dissect animals and, therefore, understanding of their internal anatomy that was particularly crucial in establishing the homology between cetaceans and the "viviparous quadrupeds." Yet, this positive influence was marred by the proliferation of encyclopedists who, for the most part, were uncritical compilers of other authors' information. However, the major impediment to any attempts to develop a natural classification for cetaceans was the insistence on classifying them by virtue of the environment in which they live, something that even diverted the thoughts of keen observers such as Ray and Artedi, despite of abundant evidence to the contrary having been collected since Aristotle.

Finally, we should not overlook the role played by intellectual inertia in the development of science. As Horder (1998) clearly demonstrated, scientists need to know the history of their field to avoid errors of the past, something that has also been argued for specific fields of biology (see Romero 2009, Chapter 1).

Author details

Aldemaro Romero
College of Arts and Sciences, Southern Illinois University Edwardsville, Peck Hall, Edwardsville, IL, USA

Acknowledgement

I thank Dr. Matthew Cashen for his advice interpreting Aristotle's writings and to Dr. Carl Springer for similar help with Pliny's work.

7. References

Aldrovandi, U. 1613. U. Aldrovandi … de Piscibus libri V et de Cetis lib. Uncus. Bononiae, 732 pp.

Alves, W.L. 2010. Ulisse Aldrovandi's Opera Omnia: collecting natural wonders. Thesis: Departmental Honors in Art History, Wheaton College, Norton, MA. 179 pp.

Artedi, P. 1738. *Ichthyologia sive opera omnia piscibus scilicet: Bibliotheca ichthyologica. Philosophia ichthyologica. Genera piscium. Synonymia specierum. Descriptiones specierum. Omnia in hoc genere perfectiora, quam antea ulla. Posthuma vindicavit, recognovit, coaptavit & edidit Carolus Linnaeus, Med. Doct. & Ac. Imper. N.C.* Wishoff, Leiden.

Barnes, J. 1995. Life and work, pp. 1-26, *In*: J. Barnes (Ed.). *The Cambridge Companion to Aristotle*. Cambridge: Cambridge University Press.

Bay, J.C. 1916. Gesner, the father of bibliography. *Papers of the Bibliographical Society of America* 10:53-88.

Belon, P. 1551. L'histoire naturelle des estranges poissons marins : avec la vraie peincture & description du daulphin, & de plusieurs autres de son espece. De l'imprimerie de Regnaud Chaudiere. Paris.

Belon, P. 1551. La nature et la diversité des poisons. Avec leurs pourtraicts representez au plus près naturel. Ch. Estienne: Paris.

Booth, E. 2005. 'A subtle and mysterious machine'. The medical world of Walter Charleton (1619-1707). Dordrecht: Springer.

Broberg, G. 1987. Petrus Artedi in his Swedish context. Proc. V Congr. Europ. Ichthyol., Stockholm 1985, pp. 11-15.

Charleton, W. 1668. *Onomasticon zoicon plerorumque Animalium Differentias et Nomina Propria pluribus Linguis exponens. Cui accedunt Mantissa Anatomica et quaedam de Variis Fossilium Generibus*. Jacobum Allestry, London. 363 pp.

Charleton, W. 1677. *Exercitationes de Differentiis & Nominibus Animalium*. Oxford. 343 pp.

Collins, S. 1685. *A Systeme of Anatomy* treating of the body of man, beasts, birds, fish, insects and plants. Illustrated with many schemes, consisting of variety of elegant figures, drawn from the life, and engraven in seventy four Folio copper plates and after every part of man's body hath been anatomically described, its diseases, cases and cures are concisely exhibited. London, 2 vol.

Delaunay, P. 1926. Pierre Belon naturaliste. Le Mans: Imprimerie Mannoyer.

Dulieu, L. 1966. Guillaume Rondelet. *Clio Medica*. 1:89-111.

Ellwood, C.A. 1938. *A history of social philosophy*. New York: Prentice Hall.

Gudger, E.W. 1934. The five great naturalists of the sixteenth century: Belon, Rondelet, Salviani, Gesner, and Aldrovandi: a chapter in the history of ichthyology. Isis 22:21-40.

Fischer, H. 1966. Conrad Gessner (1516-1565) as bibliographer and encyclopedist. The Library 21:269-281.

Gessner, C. 1558. Historiae Animalium. Liber IIII qui est de Piscium & Aquatilium natura… Tiguri

Glardon, P. 2011. L'Histoire naturelle au XVIe Siècle. Genève: Libraire Droz S.A.

Gmelig-Nijboer, C.A. 1977. Conrad Gessner's "Historia Animalium" an inventory of Renaissance Zoology. Utrecht: Communicationes Biohistoricae Ultrajectinae.

Horder, T. J. 1998. Why do scientists need to be historians? *Quarterly Review of Biology* 73:175-187.

Impey, O. & A. MacGregor. 1985. *The origins of museums. The cabinet of curiosities in sixteenth- and seventeenth-Century Europe*. Oxford: Clarendon Press.

Jonston, J. 1657. *Historiae naturalis de piscibus et cetis libri V, cum aeneis figuris, Johannes Jonstonus,... concinnavit.* J. J. Schipperi, Amsterdam. 306 pp.

Kargon, R. 1964. Walter Charleton, Robert Boyle, and the acceptance of Epicurean Atomism in England. *Isis* 55:184-191.

Keller, A.G. 1975. Rondelet, Guillaume, pp. 527-528, *In*: C.C. Gillispie (Ed.). Dictionary of Scientific Biography, Vol. 9. New York: Scribner's Sons.

Kruger, L. 2003. Edward Tyson's 1680 account of the 'porpess' brain and its place in the history of comparative neurology. *Journal of the History of the Neurosciences* 12:339-349.

Kruger, L. 2004. An early illustrated comparative anatomy of the brain: Samuel Collins' A Systeme of Anatomy (1685) and the emergence of comparative neurology in 17th century England. *Journal of the History of the Neurosciences* 13:195-217.

Kusukawa, S. 2000. The *Historia Piscium* (1686). Notes Rec. Royal Society London 54:179-197.

Kusukawa, S. 2010. The sources of Gessner's pictures for the *Historia animalium*. *Annals of Science* 67:303-328.

Leu, U.B.; R. Keller & S. Weidmann. 2008. *Conrad Gessner's private library*. Leiden: Brill.

Leitner, H. 1927. *Zoologische terminologie beim älteren*. Hildesheim: Verlag.

Lewis, E. 2001. Walter Charleton and early Modern Eclecticism. *Journal of the History of Ideas* 62:651-664.

Linnæus, C. 1758. Systema naturæ per regna tria naturæ, secundum classes, ordines, genera, species, cum characteribus, differentiis, synonymis, locis. Tomus I. Editio decima, reformata. – pp. [1–4], 1–824. Holmiæ. (Salvius).

Locher, A. 1984. The structure of Pliny the Elder's Natural History, pp. 20-29, *In*: *Science in the Early Roman Empire: Pliny the Elder, his sources and influence* (R. French and F. Greenaway, Eds.). London: Croom Helm.

Miller, G.L. 2008. Beasts of the New Jerusalem: John Jonston's natural history and the launching of millenarian pedagogy in the seventeenth century. History of Science 46:203-243.

Montagu, A. 1943. *Edward Tyson, M. D., F. R. S., 1650-1708, and the rise of human and comparative anatomy in England; a study in the history of science.* Philadelphia: The American Philosophical Society.

Moseley, A. 2010. *Aristotle*. London: Continuum International Publishing Group.

Murphy, T. 2004. *Pliny the Elder's Natural History: The Empire in the Encyclopedia.* Oxford: Oxford University Press.

Naas, V. 2002. *Le projet encyclopédique de Pline L'Ancien*. Rome: Ecole Française de Rome.

Nutton, V. 1985. Illustrations from the Welcome Institute Library. Medical History 29:93-97.

Oppenheimer, J.M. 1936. Guillaume Rondelet. *Bulletin of the Institute of the History of Medicine.* 4: 817-834.

Raven, C.E. 1950. John Ray naturalist. His life and works. Cambridge: University Press.

Ray, J. 1671. An account of the dissection of a porpess, promised Numb. 74, made, and communicated in a Letter of Sept. 12, 1671, by the learned M. John Ray., having therein observ'd some things omitted by Rondeletius. Philosophical Transactions (Royal Society) 6:2274-2279.

Ray, J. 1686. Franciscus Willughbei…De historia piscium libri quatuor. Oxonii, 343+30 pp.

Ray, J. 1713. Synopsis methodica avium & piscium; opus posthumum…Londini, 166 pp.

Reynolds, J. 1986. The Elder Pliny and his times, pp. 1-10, *In*: *Science in the Early Roman Empire: Pliny the Elder, his sources and influence* (R. French and F. Greenaway, Eds.). London: Croom Helm.

Rolleston, H. 1940. Walter Charleton, D.M., F.R.C.P., F.R.S. Bulletin of the History of Medicine 8:403-406.

Romero, A. 2009. Cave Biology: Life in Darkness. Cambridge: Cambridge University Press.

Romero, A., A.I. Agudo and S.J. Blondell. 1997. The scientific discovery of the Amazon river dolphin *Inia geoffrensis*. *Marine Mammal Science* 13(3):419-426.

Rondelet, G. 1554-1555. Libri de piscibus marinis…Universae aquatilium historiae pars altera…Lugduni, 583 pp.

Rondelet, G. 1558. La Premiere Partie de l'Histoire entiere des Poissons … Lyon, France. 599 pp.

Sharpe, L. 1973. Walter Charleton's Early Life, 1620-1659, and Relationship to Natural Philosophy in Mid-Seventeenth Century England. Annals of Science 30:311-340.

Simpson, G. G., 1945. The principles of classification and a classification of mammals. *Bulletin of the American Museum of Natural History* 85: 163 – 216.

Tugnoli Pattaro, S. 1994. Scienziati pionieri all'Universita di Bologna: il caso Aldrovandi. *Forum Italicum* 8:1130124.

Tyson, E. 1680. *Phocaena, or, The anatomy of a porpess dissected at Gresham Colledge, with a preliminary discourse concerning anatomy and natural history of animals*. Benj. Tooke, London. 48 pp.

Wellisch, H.H. 1984. *Conrad Gessner. A Bio-Bibliography*. Zug: IDC.

Wheeler, A.C. 1962. The life and work of Peter Artedi, (pp. vii-xxi) In: Peter Artedi Ichthyologia. New York: Wheldon & Wesley.

Wheeler, A.C. 1987. Peter Artedi, founder of modern ichthyology. Porc. Congr. Europ. Ichthyology., Stockholm 1985. , pp. 3-10.

Wong, M. 1970. Belon, Pierre. Pp. 595-596, In: C.C. Gillispie (Ed.). Dictionary of Scientific Biography, Vol. 1.

Wotton, E. 1552. De differentiis animalium libri decem ... cum amplissimus indicibus, in quibus primùm authorum nomina, unde quaequae desumpta sunt, singulis capitibus sunt notata & designata: deinde omnium animalium nomenclaturae, itémque singulae eorum partes recensentur, tam Graecè, quàm Latinè. Paris.

Section 2

Physiology

Chapter 2

A Matrix Model of Fasting Metabolism in Northern Elephant Seal Pups

Edward O. Keith

Additional information is available at the end of the chapter

http://dx.doi.org/10.5772/51863

1. Introduction

Northern elephant seals (*Mirounga angustirostris*) undergo regular periods of aphagia during their annual life cycle, as do many other phocids (Le Boeuf and Laws 1994). After nursing for about 30 days, the weaned pup fasts for 6-8 weeks, maintaining a fasting hyperglycemia, hyperlipidemia, hypoketonemia, and hypoinsulinemia (Champagne, *et al.* 2005). In most mammals, fasting is accompanied by hypoglycemia, and thus the fasting hyperglycemia in these animals is paradoxical. Previous studies of glucose metabolism in these animals (Keith and Ortiz 1989) indicate that the hyperglycemia results from both low rates of glucose utilization, due to very low insulin levels (Kirby, *et. al.* 1987), as well as high rates of glucose carbon recycling through both lactate and glycerol. Other studies indicate that fatty acids are the major energy substrate during this time (Castellini, *et. al.* 1987), and that these animals conserve nitrogen by having very low urea turnover and excretion rates (Houser and Costa 2001). Figure 1 shows a 10 compartment conceptual flow diagram of metabolite flux in fasting northern elephant seal pups as simulated in this study.

Mathematical models of biochemical systems are a prerequisite for a true understanding of the complexity of metabolic and physiologic systems. A model can be defined in both a physical and mathematical sense as a set of equations that describe the behavior of a dynamic system, and the response of the system to a given stimulus (Jeffers 1982). Many types of models exist. The models that most closely approximate reality are often the most complex, and it is often difficult to derive unbiased or valid estimates of model parameters. Matrix models offer a way to sacrifice some of the "reality" to gain the advantages of mathematical deduction and prediction (Jeffers 1978). Matrix models are ideally suited to simulate the results of isotope tracer experiments and linear compartment analysis models (Shipley and Clark 1972).

2. Materials and methods

The amount of carbon residing in each pool in Figure 1, and the fluxes between the pools, were determined using single injection radiotracer methods as described (Pernia, *et. al.* 1980, Keith and Ortiz 1989, and Castellini, *et. al.* 1987). If the animal is in steady state, the change in pool size (Q) for each compartment will be:

$$Q_{(t)} = Q_{(0)}e^{-kt}$$

where k is the fractional turnover rate, t is time, and e is the base of the natural logarithms. These parameters may be estimated by injecting a known amount of tracer (q) labeled in some way (such as ^{14}C or ^{3}H) but which is metabolically indistinguishable from the metabolite of interest (tracee).

Blood samples are taken over time, and the specific activity of the tracer determined and plotted on semi-log paper. If the assumption of first order kinetics holds, the plot will be linear, and the slope of the line is k. The assumption of instantaneous mixing allows extrapolation of the line back to time = 0, and the estimation of the specific activity at time = 0 (SA_0). The size of the pool (Q) can then be estimated (Katz, *et. al.* 1974):

$$Q = \frac{q}{SA_0}$$

The magnitude of the entry rate (R_0) can then be estimated:

$$R_0 = k * Q$$

If the specific activity curve is plotted on regular paper and integrated, the Stewart-Hamilton equation provides a stochastic estimate of the irreversible loss rate (L) (Shipley and Clark 1972):

$$L = \frac{q}{\int_0^{inf} SA(t)dt}$$

Recycling rate (R) is the part of the entry rate (R_0) which leaves the pool of interest and returns to it during the experiment. It is the difference between the entry rate (R_0) and the irreversible loss rate (L) (Nolan and Leng 1974):

$$R = R_0 - L$$

Once the sizes of the pools and the flux rates between them are known, differential equations can be written to describe the rate of change of each compartment (Shipley and Clark 1972):

$$dQ_i / dt = (\sum_{j=1}^{10} Q_j * k_{ij}) + (Q_i * k_{ii}) \text{ for } i = 1 \rightarrow 10$$

Where: Q_j = Size of pool j (source of flux to pool i).

- k_{ij} = Rate constant for flow to i from j.
- Q_i = Size of pool i (pool of interest).
- k_{ii} = Sum of all rate constants for flows from compartment i:

$$\text{i.e.}_{ji} \text{ for } j = 1 \rightarrow 10$$

If the coefficients of the differential equations (k_{ij}) are entered into a source/destination matrix ($\underline{A}$), the matrix may be used to predict the magnitude of the flux between each compartment:

$$\underline{F} = \underline{A} * \underline{Q}_0$$

Where $\underline{F}$ is the vector of the sum of the fluxes to and from all pools, $\underline{Q}_0$ is the vector of initial pool sizes, and $\underline{A}$ is the coefficient matrix derived above. The future state of the system then becomes:

$$\underline{Q}_{t+1} = \underline{F} * \underline{Q}_0$$

Iteration of the above two operations allows computation, in difference equation format, of the future state of the system at any time desired. However, use of the matrix exponential permits analytical determination of the state of the system at any future time in one operation:

$$\underline{Q}_{t+1} = e^{\underline{A}t} * \underline{Q}_0$$

Initially the system dynamics were simulated for an 8-week period, approximately as long as the duration of the fasting period of the northern elephant seal pups. In most cases, the simulation revealed monotonic (linear) declines in the sizes of the pools, with the obvious exception of the sink pool. However, in the case of the carbohydrate pools, i.e. glucose, glycerol and lactate, the pool sizes increased rapidly during the first part of the simulation, and then declined monotonically. This suggested that the estimates for the initial condition sizes of the pools were too low. In order to correct for this, a second, shorter (24 hr) simulation was conducted, after correcting the initial condition sizes of these pools by extrapolating the linear part of the pool size decline later in the simulation back to time zero.

A matrix may be described in terms of its characteristic equation, which will have the same order as the number of rows (= columns) in a square matrix (Jeffers 1978, Swartzman 1987). The roots of this characteristic equation are the eigenvalues (λ), which can be used to assess the stability of the system described by the matrix (Heinrich et al. 1977, Edelstein-Keshet 1988). Simply put, if all of the eigenvalues, or their real parts, are negative, the system is stable. If one or more eigenvalues are positive, the system is unstable. Zero value eigenvalues indicate a closed system (Edelstein-Keshet 1988, Halfon 1976, and Swartzman 1987). Other matrix parameters relevant to stability analysis are the trace (τ) and determinant (Δ) of the matrix. The trace is the sum of the eigenvalues, and the determinant in the product of the eigenvalues (Heinrich et al. 1977). A dimensionless τ-Δ parameter plane may be envisioned which relates the magnitude of these two parameters to model stability, or the type of instability (Edelstein-Keshet 1988). All matrix calculations described herein were conducted using MATLAB®.

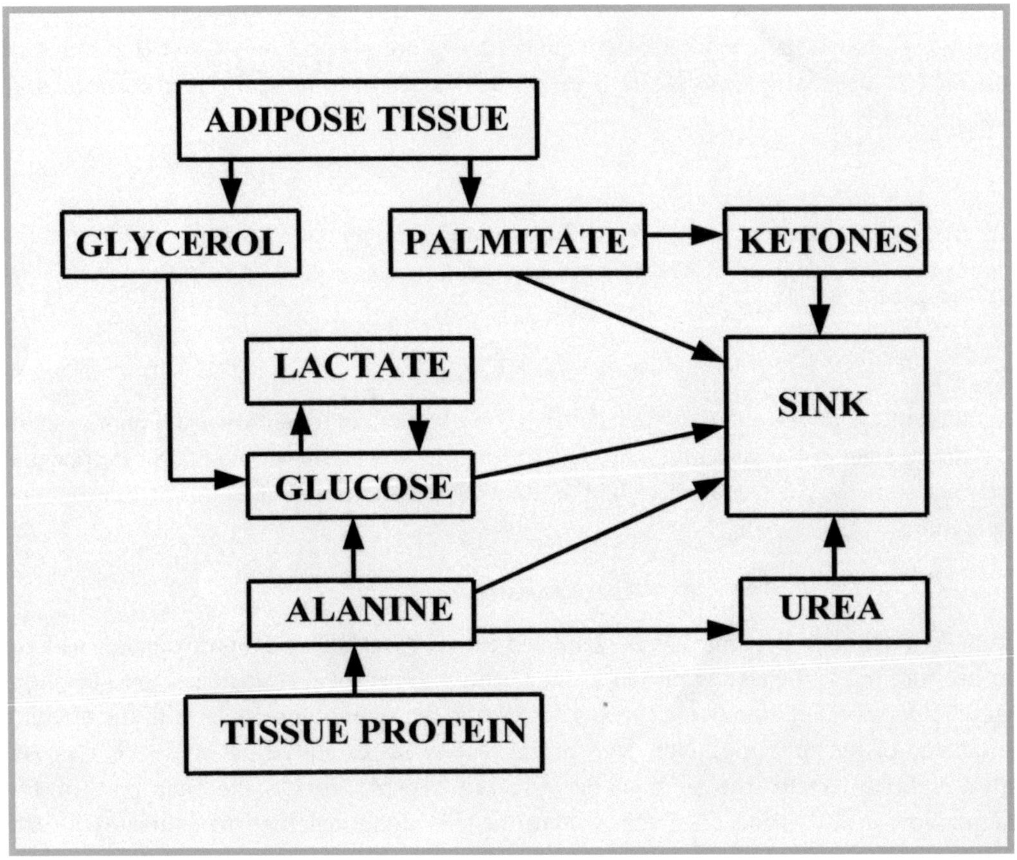

Figure 1. A 10-compartment flow diagram of metabolite flux in fasting northern elephant seals as simulated in this study. The boxes (or pools) represent the moles of carbon present in the animal as each metabolite, and the arrows represent the interchange of carbon between metabolites. There are no inputs to the model because the animal is fasting.

3. Results

Table 1 shows the initial conditions of the vector $\underline{Q}_0$ which contains the initial conditions of the system, and the first vector of fluxes calculated by multiplying $\underline{Q}_0$ by the source/destination matrix $\underline{A}$, which is contained in Table 2. The characteristic equation of this matrix is: $0 = 1.0x^{10} - 0.186x^9 + (9.34x10^{-3})x^8 + (9.50x10^{-5})x^7 + (3.78x10^{-7})x^6 + (6.45x10^{-10})x^5 + (4.18x10^{-13})x^4 + (7.95x10^{-17})x^3 + (7.88x10^{-22})x^2 + (2.15x10^{-28})x + 0$.

	Q_i	FLUX
1. Adipose	1.2500e+002	-1.2750e-003
2. Glycerol	2.3100e-0 01	1.3800e-005
3. Palmitate	6.0000e-003	1.5000e-006
4. Lactate	7.5000e-003	3.9000e-005
5. Glucose	4.0000e-002	4.6722e-004
6. Alanine	4.4000e-003	-2.4304e-004
7. Urea	4.1700e-001	-1.2070e-004
8. Protein	2.0000e+002	-5.6160e-005
9. Ketones	3.5000e-003	8.1250e-006
10. Sink	1.0000e-000	9.3508e-004

Table 1. Initial condition values in moles of carbon for the 10 compartment model shown in Figure 1. The flux is the sum of the flows into and out of each pool during the first time step.

	SOURCE POOL									
DESTINATION POOL	**ADIPOSE**	**GLYCEROL**	**PALMITATE**	**LACTATE**	**GLUCOSE**	**ALANINE**	**UREA**	**PROTEIN**	**KETONES**	**SINK**
ADIPOSE	-1.02e-5	0	0	0	0	0	0	0	0	0
GLYCEROL	5.10e-6	-2.70e-3	0	0	0	0	0	0	0	0
PALMITATE	5.10e-6	0	-1.06e-1	0	0	0	0	0	0	0
LACTATE	0	0	0	-1.20e-3	1.20e-3	0	0	0	0	0
GLUCOSE	0	2.70e-3	0	1.20e-3	-4.50e-3	3.30e-3	0	0	0	0
ALANINE	0	0	0	0	0	-6.80e-2	0	2.808e-7	0	0
UREA	0	0	0	0	0	1.00e-3	-3.00e-4	0	0	0
PROTEIN	0	0	0	0	0	0	0	-2.808e-7,	0	0
KETONES	0	0	3.25e-3	0	0	0	0	0	-3.25e-3	0
SINK	0	0	1.03e-1	0	3.30e-3	3.40e-2	3.00e-4	0	3.25e-3	-1

Table 2. Source/destination matrix used to simulate the kinetics of metabolite flux in fasting northern elephant seal pups. Non-diagonal elements represent the rate constants for flow from the column pool to the row pool, in terms of time^{-1}. The diagonal elements are the sum all rate constants for flows from the pool, i.e. the sum of the column.

The roots of this equation are the eigenvalues (λ) of the matrix: 0.00; -8.10×10^{-4}; -4.89×10^{-3} ; -2.70×10^{-3}; -3.25×10^{-3}; -1.06×10^{-1}; -1.02×10^{-5}; -3.00×10^{-4}; -6.80×10^{-2}; -2.81×10^{-7}. Notice that there is one zero value, indicating that this is a closed system. The remainder of the eigenvalues are all negative, indicating a stable system. The trace (τ), or sum of the eigenvalues, of matrix $\underline{A}$ is -0.186 and the determinant (Δ), or product of the eigenvalues of matrix $\underline{A}$ is zero. Figure 2 shows the dimensionless τ-Δ parameter plane. The point for matrix A would lie on the y-axis, just below the x-y intercept, indicating that the system approaches a saddle point of stability. Notice that τ^2 is 0.0346, which is greater than 4Δ, which is zero, indicating again that the matrix lies in the portion of the phase-plane corresponding to a saddle point condition (Heinrich et al. 1977).

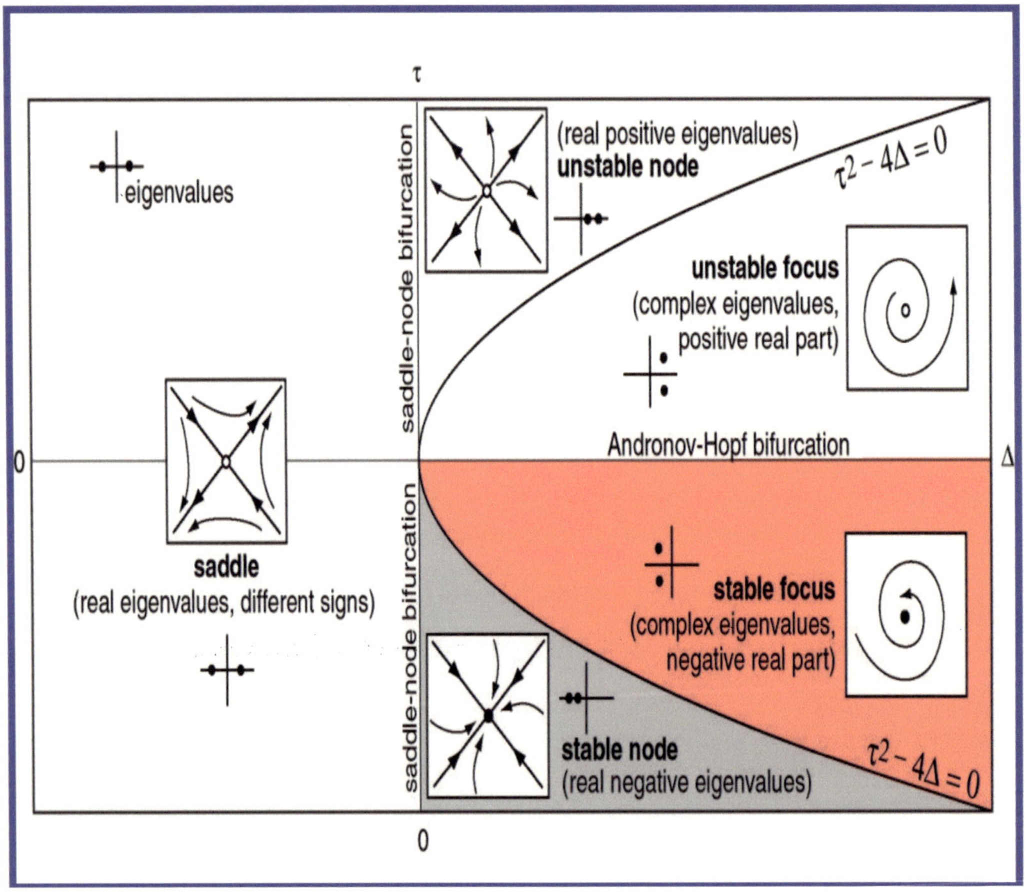

Figure 2. Dimensionless Cartesian phase-plane representing the range of trace (τ) and determinant (Δ) values possible, and the regions of qualitative behavior of linear systems. The point corresponding to the system simulated here lies on the y-axis just to the left of the x-y intercept.

Figure 3 shows the changes in the size of the adipose tissue and sink pools over an 8-week simulation. These pools are plotted together because they were of similar size. As expected, the adipose tissue pool declined at a constant rate, while the sink pool accumulated carbon asymptotically.

Figure 4 shows the changes in the size of the various lipid and lipid-derived pools over an 8-week simulation. The ketone pool rose slightly early on, and then declined slowly, as did the palmitate pool. At this resolution, it appeared that the glycerol pool declined at a constant rate throughout the simulation but when compared to the glucose and lactate simulations (Figure 5) it became clear that this pool was poorly initialized.

Figure 5 shows the changes in carbohydrate and carbohydrate-derived pools over an 8-week simulation. Apparently the glucose and lactate pools were initialized incorrectly, as reflected in their rapid increase in the first week of the simulation. Once they became stable, they both declined at the same rate, which was equivalent to the rate of decline of the glycerol pool. This similarity in rates of decline suggests that these pools are closely linked through recycling pathways, and may in fact represent a subsystem of the model.

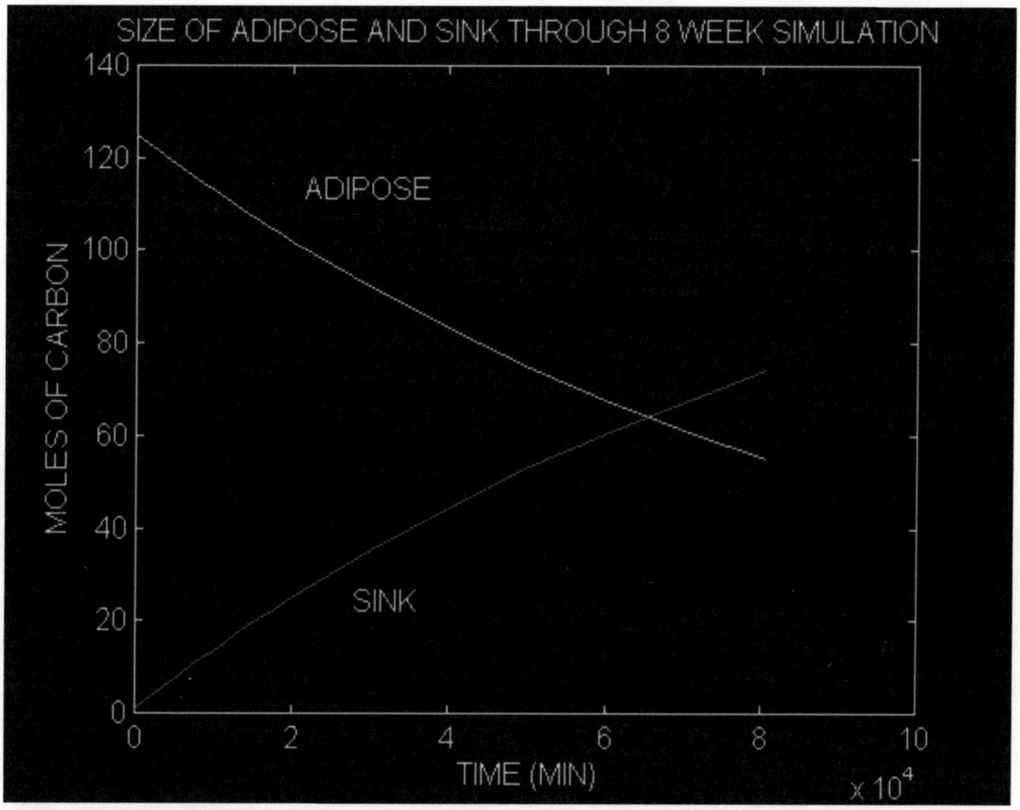

Figure 3. Linear plot of the changes in the size of the adipose tissue and sink pools over an 8-week simulation. These pools are plotted together because they were of similar size. As expected, the adipose tissue pool declined at a constant rate, while the sink pool accumulated carbon asymptotically.

Figure 6 shows the changes in nitrogen containing pools over an 8-week simulation. The tissue protein pool declined at a very slow rate, suggesting almost no protein catabolism. This is supported by the decline and continued low level of the urea and alanine pools. The continued low urea level indicates again that protein catabolism is occurring at a very low rate.

Figure 7 shows the time course of the contents of the palmitate, glycerol, and ketone pools over a 24-hour simulation. Figure 8 shows the time course of the contents of the alanine, urea, and tissue protein pools over an 8-week simulation. Notice that the tissue protein pool declined almost imperceptibly, while the urea and alanine pools declined initially and then became stable.

Figure 9 shows the time course of the contents of the glucose, glycerol, and lactate pools over a 24-hour simulation. The glucose and lactate pools were apparently poorly parameterized, and grew rapidly to a zenith, and then declined steadily. The glycerol pool declined continuously.

Figure 10 shows the time course of the contents of the glucose, glycerol, lactate, alanine and ketones pools after re-estimation of the initial conditions by extrapolating the linear parts of Figures 7, 8 and 9 to time zero.

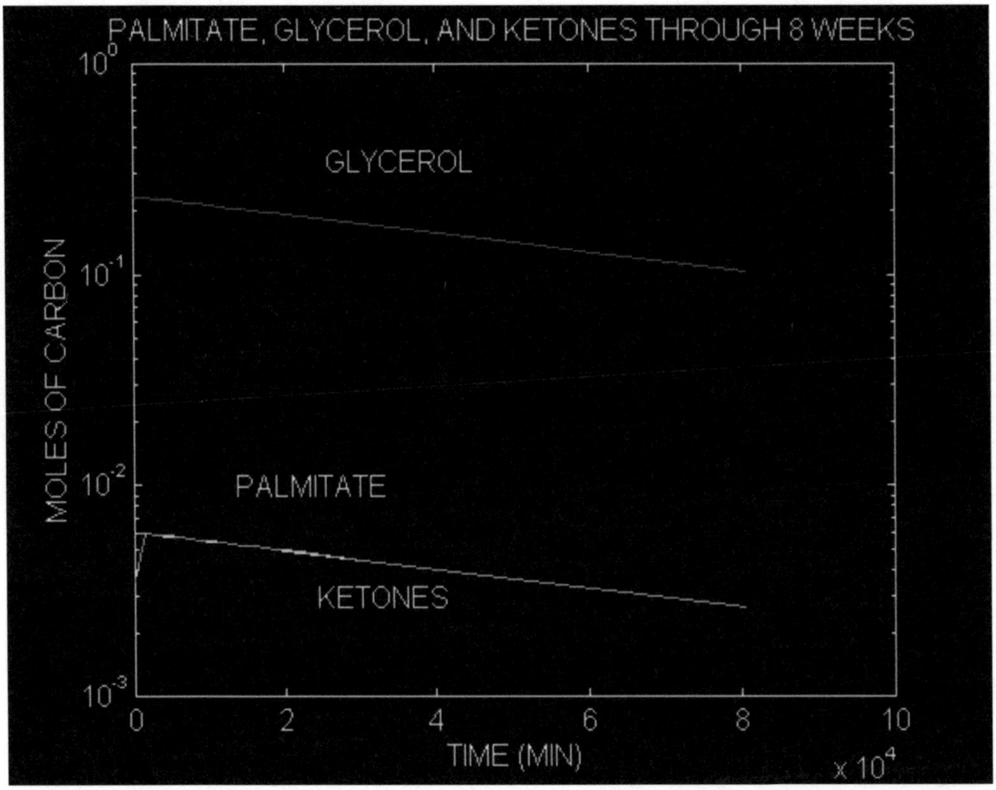

Figure 4. Log-normal plot of the changes in the size of the various lipid and lipid-derived pools over an 8-week simulation. The ketone pool rose slightly early on, and then declined slowly, as did the palmitate pool. The glycerol pool appeared to decline at a constant rate throughout the simulation.

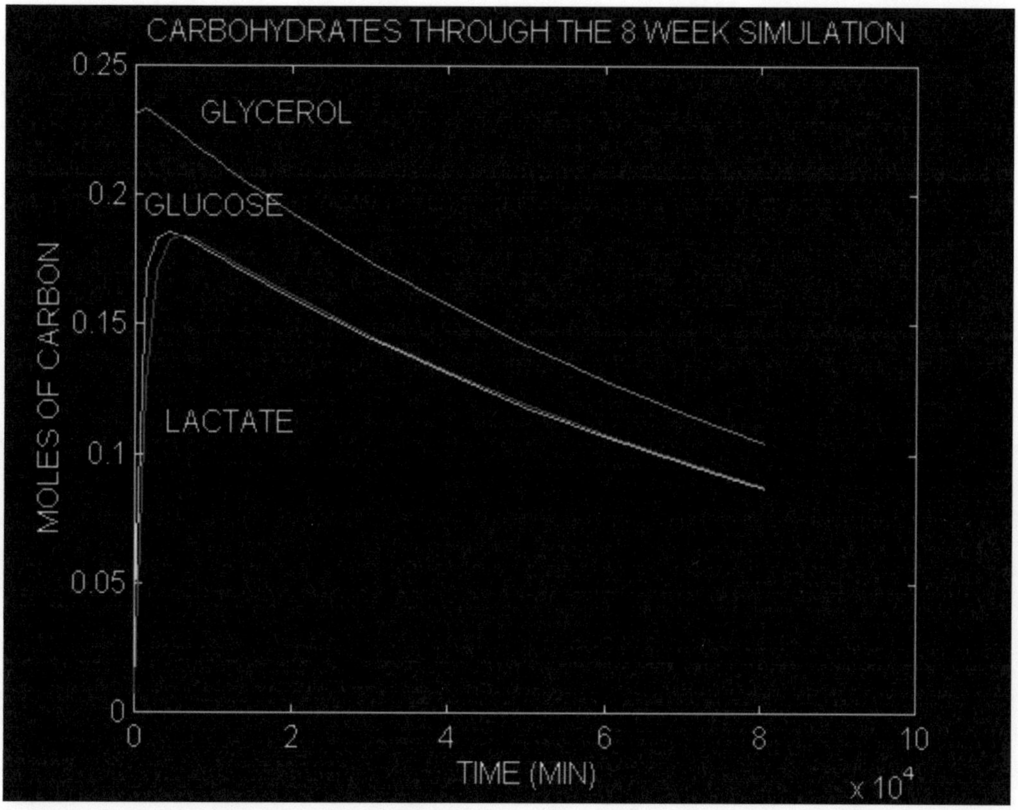

Figure 5. Linear plot of the changes in carbohydrate and carbohydrate-derived pool over an 8-week simulation. Apparently the glucose and lactate pools were initialized incorrectly, as reflected in their rapid increase in the first week of the simulation. Once they became stable, they both declined at the same rate, which was equivalent to the rate of decline of the glycerol pool. This similarity in rates of decline suggests that these pools are closely linked through recycling pathways, and may in fact represent a subsystem of the model.

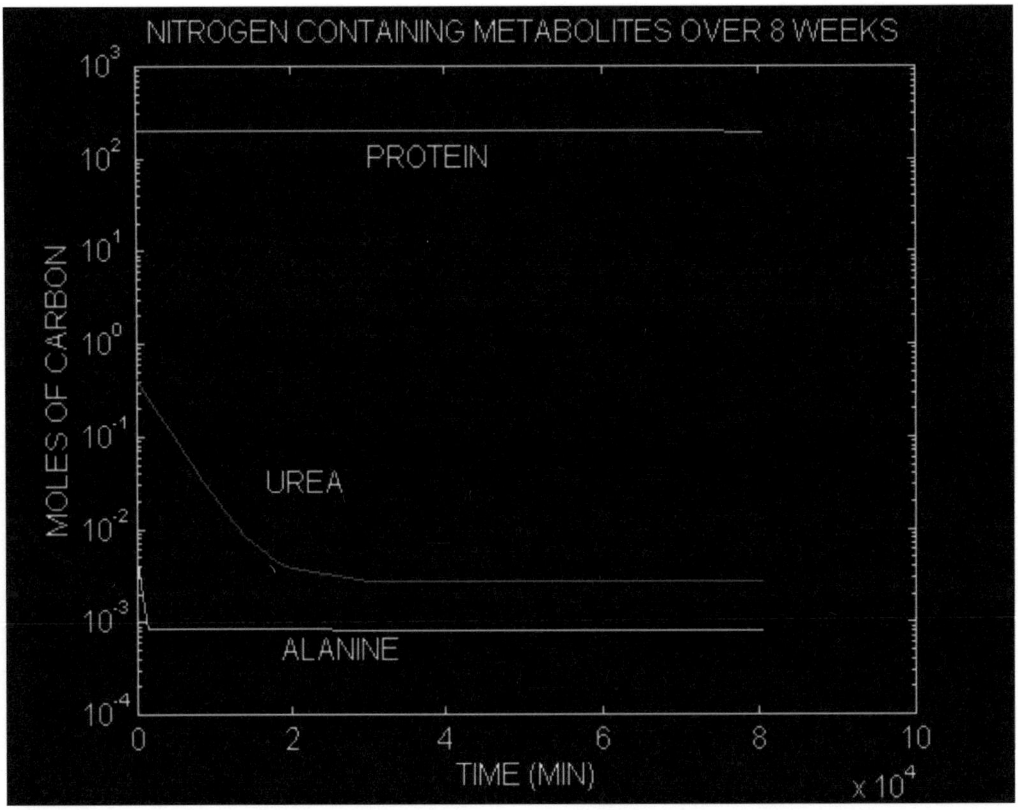

Figure 6. Changes in nitrogen containing pools over an 8-week simulation. The tissue protein pool declined at a very slow rate, suggesting almost no protein catabolism. This is supported by the decline and continued low level of the urea and alanine pools. The continued low urea level indicates again that protein catabolism is occurring at a very low rate.

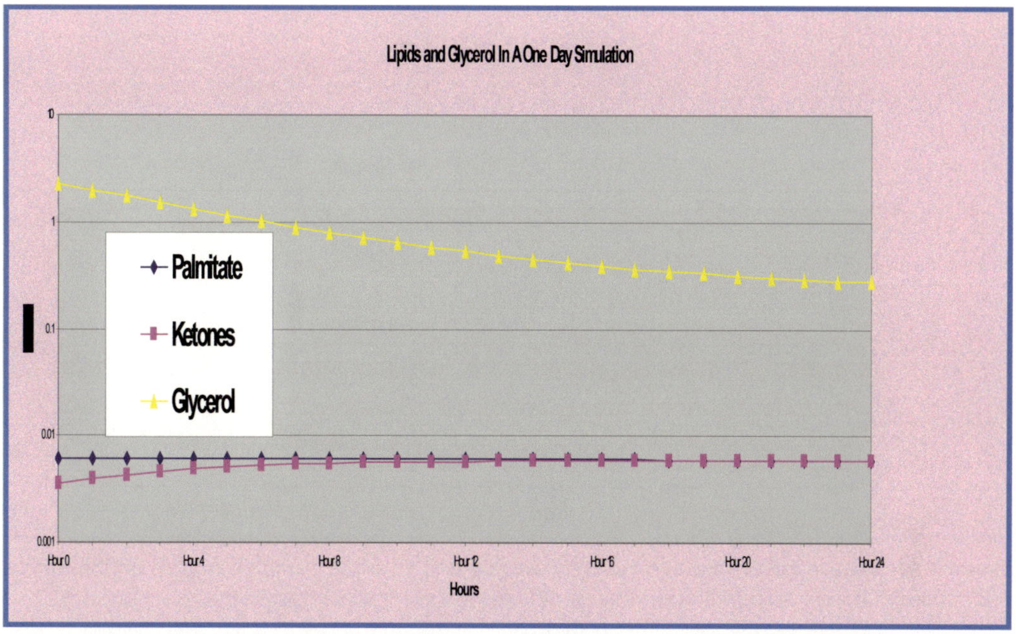

Figure 7. Linear plot of the changes in the size of the various lipid and lipid-derived pools over a 24-hour simulation. The ketone pool rose slightly early on, and then declined slowly, as did the palmitate pool. The glycerol pool declined at a constant rate throughout the simulation.

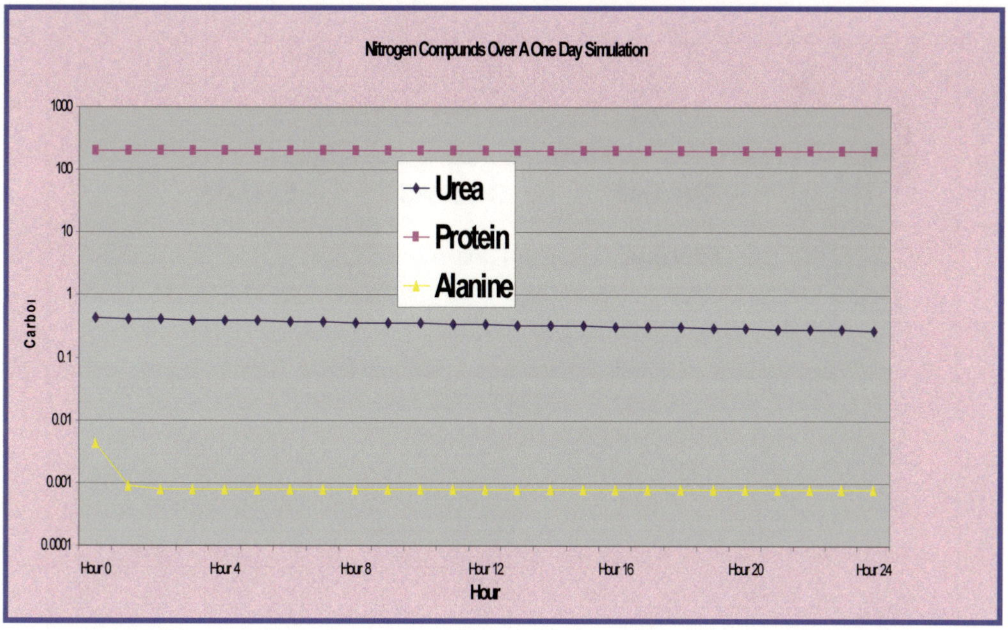

Figure 8. Changes in nitrogen containing pools over time. The tissue protein pool declined at a very slow rate, suggesting almost no protein catabolism. This is supported by the decline and continued low level of the urea and alanine pools. The continued low urea level indicates again that protein catabolism is occurring at a very low rate.

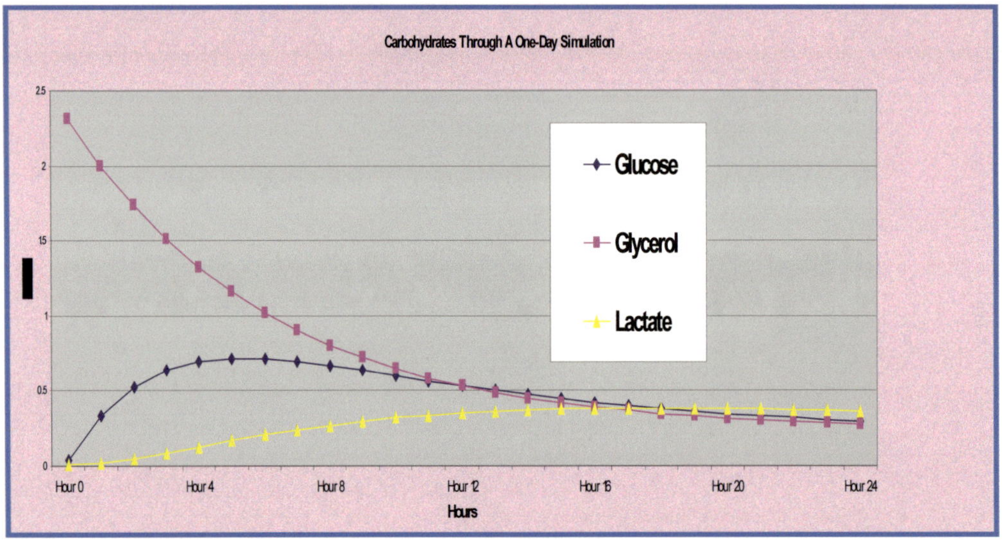

Figure 9. Changes in carbohydrate and carbohydrate-derived pool over 24 hours. Apparently the glucose and lactate pools were initialized incorrectly, as reflected in their rapid increase in the first hours of the simulation. Once they became stable, they both declined at the same rate, which was equivalent to the rate of decline of the glycerol pool. This similarity in rates of decline suggests that these pools are closely linked through recycling pathways, and may in fact represent a subsystem of the model.

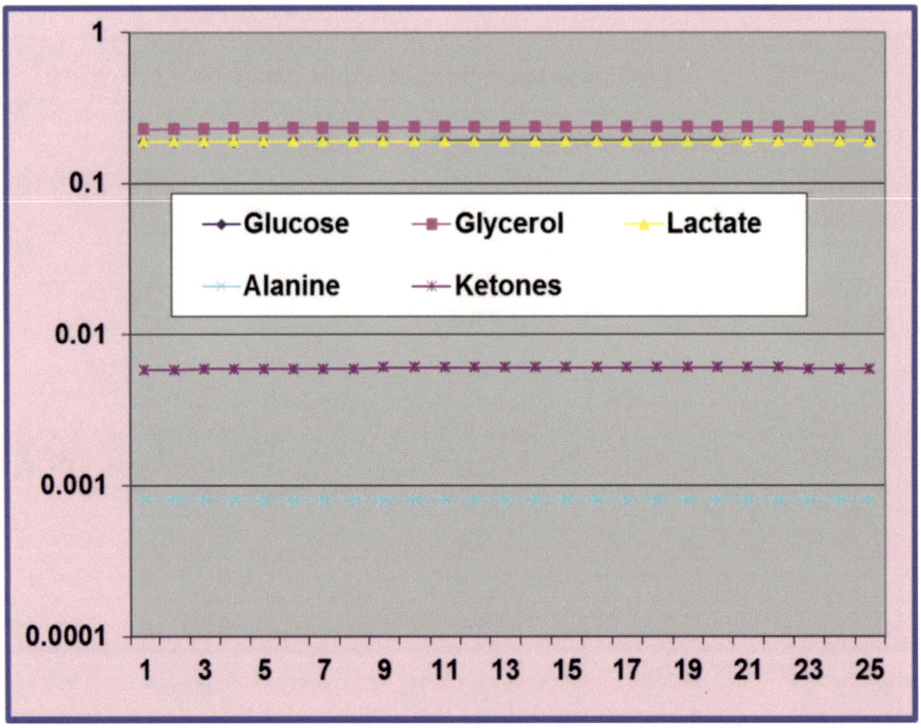

Figure 10. Changes in the glucose, glycerol, lactate, alanine and ketones pools over 24 hours after re-estimation of the initial conditions by extrapolating the linear parts of Figures 6, 7, and 8 to time zero.

4. Discussion

As initially formulated, the value of the rate constant for efflux from the sink pool ($k_{10,10}$) was set at zero, because there should be no efflux from the sink. Eigenanalysis of this matrix yielded 9 negative eigenvalues and one zero eigenvalue, indicating a stable, closed system (Keith 1999). An eigenvalue of zero is to be expected from a matrix with a zero on the main diagonal (Edelstein-Keshet 1988, Swartzman 1987). For this reason, the zero eigenvalue can be considered an artifact of matrix construction, and not truly representative of the system being simulated. Therefore, a second analysis was conducted in which the value of the rate constant for efflux from the sink pool was set to one, which allowed all of the contents of the sink pool to exit at every time step. Eigenanalysis of this matrix yielded 10 negative eigenvalues, indicative of a now open and still stable system. The trace (τ) of this matrix was -0.186, and the determinant (Δ) was 2.174×10^{-32}. In this case τ^2 was 0.0346 and 4Δ was 8.696×10^{-32}. Taken together, these values indicate that the system without a sink lies near a stable node condition in the phase-plane. However, this is difficult to reconcile with the biological reality that a fasting elephant seal with no food or water inputs is not at equilibrium, and cannot survive forever (Ortiz *et. al.* 1978).

The genesis of this apparent contradiction may lie in differences in time scale or time constants. Differential equations with widely different time scales are "stiff" (Heinrich et al. 1977) and will have eigenvalues of different orders of magnitude. This is apparent here where the eigenvalues range from -1.06×10^{-1} to -2.81×10^{-7}. The reciprocals of the eigenvalues are the relaxation times (Heinrich et al. 1977) and these likewise vary over six orders of magnitude, indicating that there are fast-reacting variables and slow-reacting variables in the simulation. Such hierarchical time structure may obscure predictions of model stability because the eigenvalues only characterize the system in the close time-neighborhood of the steady state where linear approximation is appropriate (Heinrich et al. 1977). Thus, predictions of model stability based on the signs of the eigenvalues may contradict a prediction of model instability based on relaxation times and slow-moving versus fast-moving variables in the system (Heinrich et al. 1977) over the duration of the actual fast of the animal.

Elevated palmitate levels are consistent with field data indicating that the major energy substrate during fasting (Castellini, *et. al.* 1987, Keith 1984). The decline in ketone levels through the simulation is consistent with field data indicating low levels of ketone bodies in the plasma of fasting northern elephant seal pups (Costa and Ortiz 1982). Declines in alanine, tissue protein, and urea levels are also consistent with field data (Pernia, *et. al.* 1980), and are validated by data which show that the lean body mass of the animal doesn't change during significantly during the fasting period (Ortiz, *et. al.* 1978). Lack of significant protein catabolism, with concomitant low urea levels, indicates that the animals do not maintain their elevated blood glucose levels at the expense of gluconeogenesis from amino acids (Keith 1984). Close similarities in the rates of decline of the glucose, glycerol, and lactate pools in the later parts of the simulation may be indicative of the extensive glucose carbon recycling which occurs in these animals during fasting. There is extensive interchange of carbon between these three pools, as indicated by high levels of Cori cycle and glucose-glycerol cycle activity (Keith 1984, Keith and Ortiz 1989). It is postulated that this high degree of recycling may be one reason for the ability of these animals to avoid ketoacidosis, a major deleterious consequence of fasting in many other mammals, and thus allow them to undergo a prolonged fast during this vulnerable period in their life history.

Author details

Edward O. Keith
Farquhar College of Arts and Sciences, Nova Southeastern University, Fort Lauderdale, FL, USA

5. References

Castellini, M.A., D.P. Costa, and A.C. Huntley. 1987. Fatty acid metabolism in fasting northern elephant seal pups. Journal of Comparative Physiology B 157:445-449.

Champagne, C.D., D.S. Houser, and D.E. Crocker. 2005. Glucose production and substrate cycle activity in a fasting adapted animal, the northern elephant seal. Journal of Experimental Biology. 208:859-868.

Edelstein-Keshet, L. 1988. Mathematical models in biology. Random House, Inc., New York, NY. 325 pp.

Halfon, E. 1976. Relative stability of ecosystem linear models. Ecological Modeling 2:279-286.

Heinrich, R., S.M. Rapoport, and T.A. Rapoport. 1977. Metabolic regulation and mathematical models. Progress in Biophysics and Molecular Biology 32:1-82.

Houser, D.S. and D.P. Costa. 2001. Protein catabolism in suckling and fasting northern elephant seal pups (*Mirounga angustirostris*). Journal of Comparative Physiology B 171:635-642.

Jeffers, J.N.R. 1978. An introduction to systems analysis: with ecological applications. University Park Press, Baltimore, MD 87 pp.

Jeffers, J.N.R. 1982. Modeling. Chapman and Hall, London. 80 pp.

Katz, J., H. Rostami, and A. Dunn. 1974. Evaluation of glucose turnover, body mass, and recycling with reversible and irreversible tracers. Biochemical Journal 142:161-170.

Keith, E.O. 1984. Glucose metabolism in fasting northern elephant seal pups. Ph.D. Dissertation, University of California, Santa Cruz, CA. 101 pp.

Keith, E.O., and C.L. Ortiz. 1989. Glucose kinetics in neonatal elephant seals during postweaning aphagia. Marine Mammal Science 5:99-115.

Keith, E.O. (1999). A matrix model of metabolite flux in northern elephant seal pups undergoing natural fasting after weaning. *Comparative Biochemistry and Physiology.* 124A:S100, 1999.

Le Boeuf, B.J, and R.M. Laws. 1994. Elephant seals. Population ecology, behavior, and behavior. University of California Press, Berkeley, CA 450 pp.

Nolan, J.V., and R.A. Leng. 1974. Symposium on 'tracer techniques in nutrition'. Isotope techniques for studying the dynamics of nitrogen metabolism in sheep. Proceedings of the Nutrition Society 33:1-8.

Ortiz, C.L., D.P. Costa, and B.J. Le Boeuf. 1978. Water and energy flux in elephant seals fasting under natural conditions. Physiological Zoology 51:166-178.

Shipley, R.A., and R.E. Clark. 1972. Tracer methods for *in vivo* kinetics. Academic Press, New York, NY 239 pp.

Swartzman, G.L. 1987. Ecological simulation primer. Macmillan Publishing Company, New York, NY. 257 pp.

Section 3

Environmental/Ecological Issues

Chapter 3

"Test Tube Cetaceans": From the Evaluation of Susceptibility to the Study of Genotoxic Effects of Different Environmental Contaminants Using Cetacean Fibroblast Cell Cultures

Letizia Marsili, Silvia Maltese, Daniele Coppola, Ilaria Caliani, Laura Carletti, Matteo Giannetti, Tommaso Campani, Matteo Baini, Cristina Panti, Silvia Casini and M. Cristina Fossi

Additional information is available at the end of the chapter

http://dx.doi.org/10.5772/54429

1. Introduction

Cetacean diversity, like all biodiversity worldwide, is seriously threatened; its loss seems to be occurring at a very rapid and increasing rate [1]. In March 2010, the European Commission set a key objective for 2020: halt the loss of biodiversity and the degradation of ecosystem services in the EU [2]. To the objective of improving the effectiveness of conservation strategies it becomes important to know the health status of endangered species and then to develop methods of investigation that are not destructive and the least invasive possible. In the last few years a non destructive sampling method, the skin biopsy, was developed in cetaceans to obtain viable tissue samples from free-ranging animals [3]. With the skin biopsy it is possible to assess the effects of multiple pressures related to bioaccumulation of anthropogenic contaminants, infectious diseases, climate change, food depletion from over-fishing, bycatch, noise, shipping and collision that stress cetacean species. The evaluation of their "health status" is possible using a suite of sensitive tools, such as non-destructive biomarkers, that will enable us to detect the presence and the effects of contaminants, the reproduction alteration, the genotoxicity, the immunosuppression, the feeding ecology and the general stress [4]. Actually, it is very difficult to discern the effects of one threat from those of another when multiple threats are acting simultaneously; for example the incidence of pathology in cetaceans is closely related to the level of pollution in their environments and thus bacterial and viral infections and contaminants should be considered from a holistic point of view [5]. Regarding the effects of anthropogenic contaminants, it is known that some contaminants, such as organochlorines (OCs), polybrominated diphenyl ethers (PBDEs), bisphenol A (BPA) and phthalates are endocrine disrupting chemicals (EDCs) and immunosuppressors [6, 7, 8, 9, 10]. Others, such as polycyclic aromatic hydrocarbons (PAHs), derived from both natural (e.g., oil spills, forest fires, natural petroleum seeps) and anthropogenic (e.g., combustion of fossil fuels, use of oil for cooking and heating, coal burning) sources, are carcinogenic, teratogenic and mutagenic compounds [11] and some studies have shown that PAHs with four or more rings can induce dioxin-like activity and weak estrogenic responses [12]. Moreover PAHs have attracted scientific interest due to their genotoxicity [13]. But how is it possible to discriminate the effects of a specific toxic in a mixture of many pollutants and assess the susceptibility of a particular cetacean species to just one class of contaminants? The aim of the present study is to use cetacean fibroblast cell cultures, obtained from skin biopsy of free-ranging animals and from skin tissue of stranded animals dead within 12 h [14, 15], as an "in vitro" method, called ***"Test Tube Cetaceans"***, to investigate the effects of environmental contaminants. In particular we use *Test Tube Cetaceans* to explore the susceptibility to genotoxic effects of different environmental contaminants in these marine mammals. Cell cultures were obtained from several species of cetaceans: fin whale (*Balaenoptera physalus*) and Bryde's whale (*Balaenoptera edeni*) for mysticetes, sperm whale (*Physeter macrocephalus*), killer whale (*Orcinus orca*), Risso's dolphin (*Grampus griseus*), bottlenose dolphin (*Tursiops truncatus*), striped dolphin (*Stenella coeruleoalba*), long-beaked common dolphin (*Delphinus capensis*) and common dolphin (*Delphinus delphis*) for odontocetes. Here we present the results for three different biomarkers of anthropogenic stress in cetacean cell cultures that will enable us to assess: i) exposure to contaminants, ii) immunosuppression and iii) genotoxicity.

i. **Interspecies differences in the mixed function oxidase (MFO) induction as biomarker of exposure to different environmental contaminants:** the evaluation in fibroblast cell cultures with immunofluorescence technique of the presence and the induction of two components (CYP1A1 and CYP2B) of the cytochrome P450 monooxygenase system (MFO), among the most relevant in drug and xenobiotic metabolism, was used to evaluate interspecies sensitivities to various classes of environmental contaminants. In particular CYP1A1 is induced by planar compounds such as planar OCs (coplanar polychlorinated biphenyls (PCBs)) and PAHs [16] and CYP2B by globular compounds such as PBDEs, PCBs and OC insecticides such as dichlorodiphenyltrichloroethane (DDT) and its metabolites [17, 18].
ii. **Qualitative and quantitative major histocompatibility complex (MHC) class I chain related protein A (MICA) expression as toxicological stress marker of the immune system:** the evaluation of the qualitative and quantitative MICA protein expression in fibroblast cell cultures with the immunofluorescence technique was used as toxicological stress marker of the immune system of different species of cetaceans [19]. The genes encoding for MICA and MICB are found within the major histocompatibility complex. Although MIC products have been found in various cells/tissues, the current consensus is that MIC genes are mainly expressed in gastrointestinal epithelium,

endothelial cells and fibroblasts. MIC molecules are considered to be stress-induced antigens that are recognized by cytotoxic T cells and natural killer (NK) cells, which play an important role in the surveillance of transformed infected and damaged cells [20]. Because the cetacean skin is an important tissue of the immune system and contributes to biological structure by acting not only as a protective physical barrier, but also as a target for immune components that mount the initial defense against invading pathogens, noxious stimuli, and resident neoplastic cells, the evaluation of MICA protein expression in cetaceans can be used to evaluate the status of the immune system of different species of cetaceans.

iii. **Detection of DNA damage by Comet assay as genotoxicity biomarker:** the presence of compounds such as PAHs, OCs and heavy metals in the marine environment can damage the DNA of living cells. The loss of DNA integrity can determine genotoxic effects, such as DNA base modifications, strand breaks, depurination and cross-linkages [21]. The comet assay or single cell gel electrophoresis (SCGE) is a sensitive method used as an indicator of genotoxicity and an effective biomarker for detecting DNA damage in living cells of aquatic animals [22]. Compared to other genotoxicity tests, such as chromosomal aberrations, sister chromatid exchanges, alkaline elution, and micronucleus assay, the advantages of the Comet assay include its demonstrated sensitivity for detecting low levels of DNA damage (one break per 10^{10} Da of DNA), requirement for small number of cells (~10,0000) per sample, flexibility to use proliferating cells, low cost, ease of application, and the short time needed to complete a study [23].

2. Sampling methods

2.1. Free- ranging cetaceans

Samples of skin (epidermis, dermis and blubber) were obtained from free-ranging specimens of long-beaked common dolphin (*Delphinus capensis;* MDC12) and common dolphin (*Delphinus delphis;* DDL1) using an aluminium pole armed with biopsy tips (0.7 cm ø, 3.0 cm length), while skin biopsies from free-ranging specimens of Bryde's whale (*Balaenoptera edeni;* MBE3), killer whale (*Orcinus orca;* MOO12), sperm whale (*Physeter macrocephalus;* PMAS1), bottlenose dolphin (*Tursiops truncatus;* TTA1) and Risso's dolphin (*Grampus griseus;* GGL1) were obtained using a Barnett Wildcat II crossbow with a 150-pound test bow, using a biopsy dart with modified stainless steel collecting tip (0.9 cm ø, 4.0 cm length). Biopsy samples were taken in the dorsal area near the dorsal fin, with CITES authorization (CITES Nat. IT025IS, Int. CITES IT 007) in the Sea of Cortez (MDC12, MBE3 and MOO12) and Mediterranean Sea (DDL1, PMAS1, TTA1 and GGL1). A small fragment of the biopsy was immediately stored in cell medium for the cell cultures.

2.2. Stranded cetaceans

Skin tissue of stranded cetaceans (dead within 2-12 h) were obtained from specimens found dead along the Italian coasts in the period 2005–2009 (CITES Nat. IT025IS, Int. CITES IT 007). Samples were taken under the dorsal fin of stranded specimens of fin whale (RT2 and

RT25), sperm whale (PM6), bottlenose dolphin (TurNic) and striped dolphin (RT1 and RT23), and immediately placed in cell medium.

2.3. Sex identification

Sex determination in cetaceans was carried out by genetic investigations according to Berubè & Palsboll [24].

3. Fibroblast cell cultures

The development of a non-invasive sampling method for obtaining viable tissue samples for cell cultures from skin biopsies of free-ranging and stranded cetaceans was described by Marsili *et al.* [14]. Successful cell cultures were obtained from all the animals. After the biopsy, skin samples were stored in sterile medium MEM Eagle Earle's salts w/L-glutamine and sodium bicarbonate (Mascia Brunelli, Milan, Italy) + 10% gamma irradiated fetal calf serum (Mascia Brunelli) + 1% MEM not essential aminoacids (NEAA) solution 100x (Mascia Brunelli) + 1% Penicillin/Streptomycin 100x (Mascia Brunelli) + 0.1% Amphotericin B 100x (Mascia Brunelli) at room temperature and was processed within 24 h of collection. In the laboratory, each sample was washed with Earle's balanced salt solution (EBSS; Mascia Brunelli) containing antibiotic (Penicillin/Streptomycin 100x [Mascia Brunelli]) and antimycotic (Amphotericin B 100x [Mascia Brunelli]) solutions. All specimens were handled using sterile techniques. Initially, the collected tissue was cut into small pieces with curved surgical scissors, placed in 30-mm Petri dishes and incubated with Trypsin-EDTA solution 1x (Mascia Brunelli) for 15 min at 37°C. The biopsy fragments were washed again and then placed in Falcon 25 flasks, moistened with medium. After 24 h at 37°C in an incubator with 5% CO_2, the cultures were covered with 1 ml of medium. Half of the culture medium was replaced every 48 h with fresh medium.

4. Indirect immunofluorescence tecnique

Third generation fibroblast cell cultures were exposed to the different mixtures of contaminants reported in the Table 1.

We used immunofluorescence in fibroblast cultures for a qualitative and semi-quantitative analysis of target proteins CYP1A1, CYP2B and MICA. After a first reaction with the primary polyclonal antibodies (goat anti-rabbit cytochrome P450 1A1 and goat anti-rabbit cytochrome P450 2B; Oxford Biochemical Research (Oxford MI, USA); rabbit polyclonal anti-MICA; Abcam), the cells were treated with the respective secondary antibodies (Alexa Fluor 594 goat anti-rabbit IgG (H+L) for CYP1A1 and CYP2B; Alexa Fluor 568 rabbit anti-goat IgG (H+L) for MICA; Invitrogen), labelled with red-fluorescent Alexa Fluor dye. Immunofluorescence was quantified with a specially designed Olympus Soft Imaging Systems macro, *DetectIntZ,* which works with the image acquisition, processing and analysis system, *analySIS^B* (Olympus) [15]. The image analysis procedure has the objective of quantifying, with an adimensional index generated for this purpose, the amount of Alexa

Name	Mixture	Dose 1	Dose 2	Dose 3	Dose 4	Dose 5
OC mixture	(Arochlor 1260 + pp'DDT + p'DDE) solubilized in DMSO (0.05%)	0.01 μg/ml	0.1 μg/ml	1 μg/ml	5 μg/ml	25 μg/ml
PAHs	Benzo(a)pyrene (1mM) + beta-naphthoflavone (20mM) solubilized in acetone (0.1%)	0.5μM BaP + 10μM BnF	2.5μM BaP + 50μM BnF	12.5μM BaP + 250μM BnF	/	/
Flame retardants	BDE-MXE solubilized in nonane (0.01 μg/ml)	0.01 μg/ml	0.05 μg/ml	0.1 μg/ml	/	/
BPA	BPA solubilized in ethanol (0.1%)	0.1 μg/ml	1 μg/ml	10 μg/ml	100 μg/ml	/

Table 1. The different mixtures of contaminants and doses to which cell cultures were exposed.

Fluor localized in the membrane of cytoplasmatic area of sample cells. The sample cells are imaged using DAPI and this image is presented to the operator for threshold selection of cytoplasmatic and nuclei Region of Interests (ROIs) across the field. The procedure then utilizes these ROIs to measure fluorescence intensity of Alexa Fluor sample cell and summarizes the results in a worksheet. The system generates index values which are unitless until compared with other units, such as number of cells to obtain mean fluorescence per cell or the area in which it is calculated to obtain mean fluorescence per mm^2. Images are all obtained with a magnification of 20X, a calibration of 0.65 μm/pixel and a resolution of 1360 x 1024 x 8 pixel. Exposure times were maintained fixed while reading the CYP1A1, CYP2B and MICA for each treatment. A series of images of each slide was acquired so that a minimum of 250 cells/slide could be counted. The total fluorescence revealed by the program is divided by number of cells to obtain arbitrary unity of fluorescence (AUF) per cell. Several slides for CYP1A1, CYP2B and MICA were made for each culture: one was a blank (cells treated only with primary and secondary antibodies), one was a secondary blank (cells treated only with secondary antibody), one was a chemical blank (cells treated with contaminant carrier), two were for each treatment dose of contaminants. The blank enabled the natural presence of the target proteins in cultured fibroblasts to be checked. The secondary blank enabled validation of the dose of secondary antibody without cross reaction as the primary antibody was absent.

5. Genotoxicity biomarker: comet assay

Fibroblast cell cultures (third generation) of striped dolphin (RT23) were subjected to this experimental protocol for 4 h. A cell line was exposed to a mixture of benzo(a)pyrene

(1mM) and beta-naphthoflavone (20mM), solubilized in acetone (0.1%), at three doses (Table 1), plus an acetone (0.1%) control. Fibroblast cell cultures were processed for the comet assay after Caliani *et al.* [13], with some modifications. The cells were centrifuged at 1000 g for 10 min, then were embedded in agarose (0.5% low-melting agarose) and layered on conventional slides, predipped in 1% normal melting agarose. The slides were immersed into a freshly made lysis solution (2.5 M NaCl, 10 mM Tris, 0.1 M EDTA, 1% Triton X-100, and 10% DMSO, pH 10) for at least 1 h at 4°C in the dark. The slides were then placed on a horizontal electrophoresis tray previously filled with freshly prepared cold alkaline buffer and left for 20 min to allow DNA unwinding. Electrophoresis was performed at 25 V and 300 mA for 20 min. The DNA migration was evaluated at three different pH conditions (pH 13, pH 12.1, pH 8). Slides were then neutralized in Tris (0.4 M, pH 7.5) for 3x5 min and stained with Sybr safe 1:10.000 in TE (10 mM Tris-HCl pH 7.5 and 500 mM EDTA pH 7.5) buffer. A total of 50 cells per slide were examined under epifluorescence at 40X magnification. The amount of DNA damage was evaluated as the percentage of DNA migrating out of the nucleus using an image analyser (Komet 6.0 Software, Kinetic Imaging Ltd.), connected to a fluorescent microscope (Olympus BX41).

6. Interspecies differences in the Mixed Function Oxidase (MFO) as biomarker of exposure of different environmental contaminants

Fibroblast cell cultures of fin whale (RT2), Bryde's whale (MBE3), sperm whale (PM6), killer whale (MOO12), Risso's dolphin (GGL1), bottlenose dolphin (TurNic), striped dolphin (RT1), long-beaked common dolphin (MDC12) and common dolphin (DDL1) were treated for 48 h with different environmental contaminants and the quantification of the induction of endogenous proteins such as CYP1A1 and CYP2B was used as target of toxicological susceptibility. The presence and the induction of CYP1A1 and CYP2B were evaluated with the indirect immunofluorescence and quantified with the Olympus macro, *DetectIntZ*. CYP1A1 is induced by planar compounds and CYP2B by globular compounds. The treatments were performed with OC mixture; flame retardants; PAHs; and BPA (Table 1). In the total mixture of Arochlor 1260 [25] only the 1.3033% shows a CYP1A1 inductive capacity while the remaining congeners are CYP2B inducers [26, 27, 28, 29, 30]. pp'DDE and pp'DDT are known as CYP2B inducers [31, 32] but an experiment on fibroblast cell culture of sperm whale (PM6) treated only with pp'DDT and pp'DDE has shown a capacity of these compounds to induce also the CYP1A1 (Figure 1). Examining at the bromine substitution patterns in the basic structure of the PBDE molecule, and with the support of the other studies on this topic [33] we can say that in the BDE-MXE mixture, the 18.72% is CYP1A1 inducer and the rest of congeners are CYP2B inducers. Benzo(a)pyrene and beta-naphthoflavone are important planar compounds and CYP1A1 inducers [31]. Finally BPA that may be a human-specific inducer of the CYP3A4 gene [34], but many studies have shown that BPA inhibits several P450-dependent monooxygenases activities (CYP1A2, CYP2A2, CYP2B2, CYP2C11, CYP2D1, CYP2E1 and CYP3A2) [35].

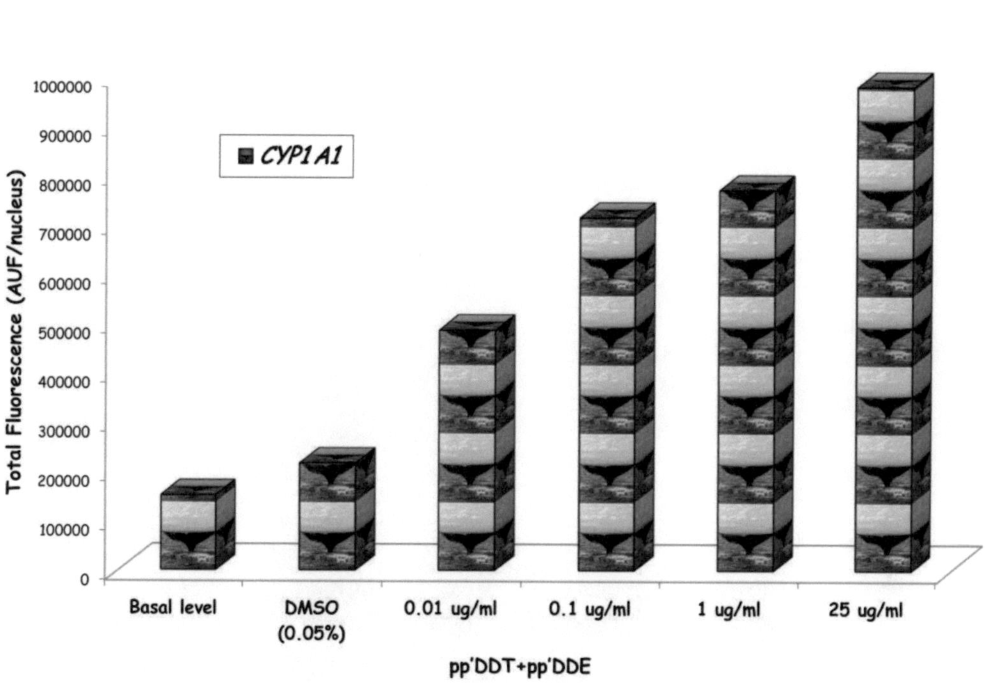

Figure 1. Basal levels of immunofluorescence (AUF/nucleus) of CYP1A1 in fibroblast cells of sperm whale treated with pp'DDT and pp'DDE.

6.1. Basal levels of CYP1A1 and CYP2B in different species

The first result of these experiments in nine cetacean species was the detection of the presence of CYP1A1 and CYP2B in fibroblast cells of all species, revealed by immunofluorescence (Figure 2A-B); a higher basal expression of both proteins was found in Risso's dolphin (GGL1) and bottlenose dolphin (TurNic), while fin whale (RT2) and sperm whale (PM6) showed the lowest levels of these proteins. The Risso's dolphin (GGL1) and the sperm whale (PM6) have a very similar diet, consisting mostly of squid. Nevertheless, they have a very different basal expression of the two cytochromes. As for the other species, very high levels of CYP1A1 were present in the Bryde's whale (MBE3). This mysticete sampled in the Sea of Cortez showed CYP1A1 levels more than 20 times greater than the other mysticete studied, the Mediterranean fin whale (RT2). Regarding the levels of contaminants detected in the blubber of different species, the bottlenose dolphin (TurNic), stranded along the coasts of the Mediterranean Sea, had very high levels of organochlorine contaminants in its blubber (DDTs = 77.4 μg/g lipid weight (l.w.); PCBs = 262.6 μg/g l.w.) that are potent inducers of CYP2B. In fact, especially in this specimen this cytochrome appears to be markedly higher than the levels shown by other species (Figure 2B). But the sperm whale (PM6) was also a stranded specimen found on the Italian coasts (Mediterranean Sea) having high values of these xenobiotics in the blubber. It seems therefore that this basal activity is more species-specific than related to the geographical location, diet, toxicological status, etc. in which the animals were found.

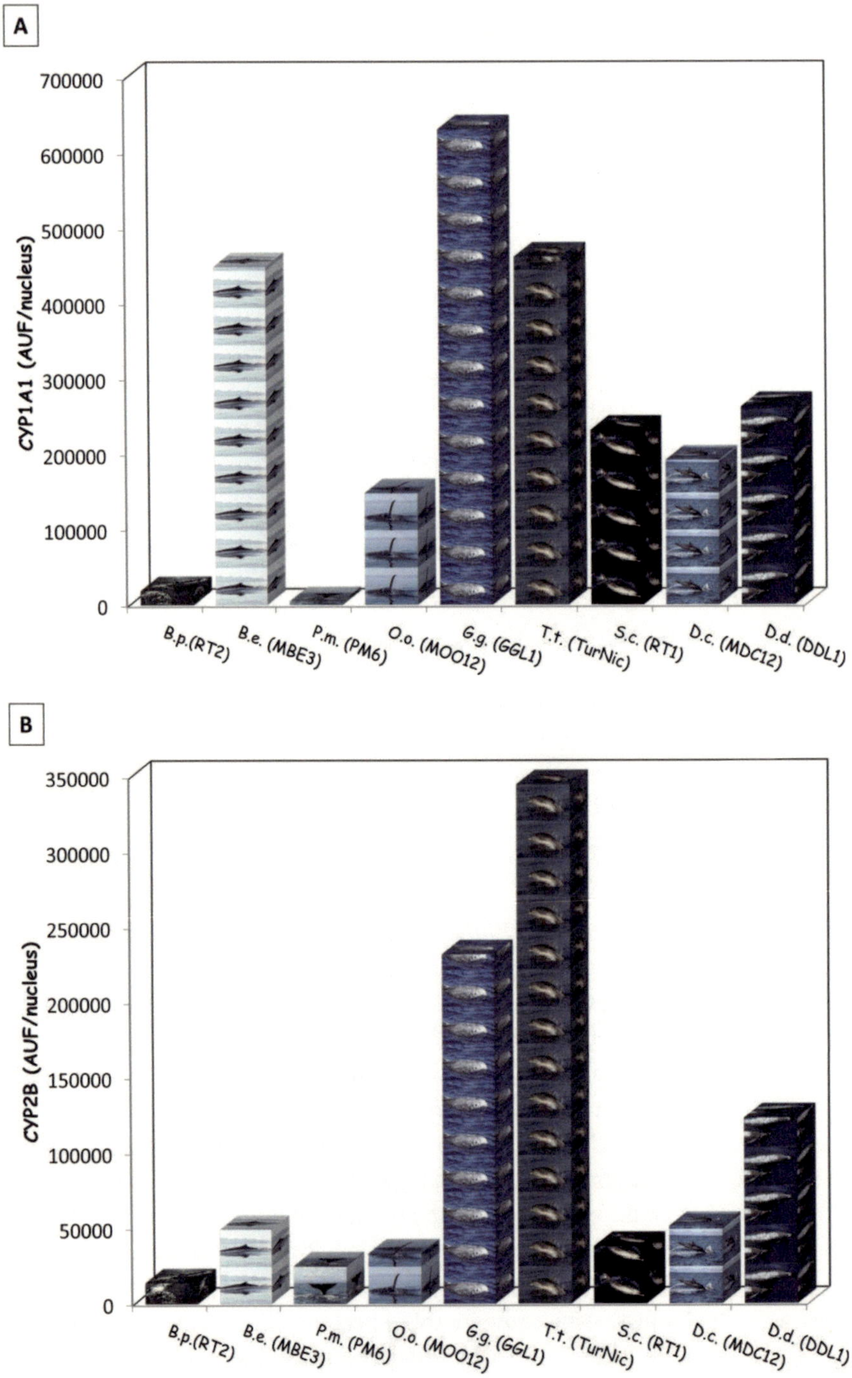

Figure 2. (A-B): Basal levels of immunofluorescence (AUF/nucleus) of CYP1A1 (A) and CYP2B (B) in fibroblast cells of fin whale (B.p.), Bryde's whale (B.e.), sperm whale (P.m.), killer whale (O.o.), Risso's dolphin (G.g.), bottlenose dolphin (T.t.), striped dolphin (S.c.), long-beaked common dolphin (D.c.) and common dolphin (D.d.).

6.2. CYP1A1 and CYP2B in different species after treatment with OC mixture

Results of the mean levels of immunofluorescence of CYP1A1 (A) and CYP2B (B), revealed in cultured fibroblasts of different species treated with OC mixture and expressed as index numbers, are reported in Table 2A-B.

A CYP1A1	*DMSO 0.05%*	*0.01 µg/ml*	*0.1 µg/ml*	*1 µg/ml*	*5 µg/ml*	*25 µg/ml*
RT2 (fin whale)	100	125.1	170.3	148.0	104.3	143.6
MBE3 (Bryde's whale)	100	31.6	106.2	55.8	/	/
PM6 (sperm whale)	100	/	/	15.7	31.9	77.4
MOO12 (killer whale)	100	427.2	288.6	207.8	696.6	49.0
GGL1 (Risso's dolphin)	100	/	/	144.5	224.1	104.7
TurNic (bottlenose dolphin)	100	63.0	96.5	81.4	/	/
RT1 (striped dolphin)	100	56.4	31.0	49.3	94.1	219.6
MDC12 (long-beaked common dolphin)	100	325.9	312.1	836.9	/	/
DDL1 (common dolphin)	100	/	/	111.4	117.3	82.9
B CYP2B	*DMSO 0.05%*	*0.01 µg/ml*	*0.1 µg/ml*	*1 µg/ml*	*5 µg/ml*	*25 µg/ml*
RT2 (fin whale)	100	112.3	110.1	110.4	116.3	146.5
MBE3 (Bryde's whale)	N.C.	N.C.	N.C.	N.C.	N.C.	N.C.
PM6 (sperm whale)	100	/	/	134.6	292.1	212.4
MOO12 (killer whale)	100	156.4	300.5	456.0	153.3	104.3
GGL1 (Risso's dolphin)	100	/	/	70.8	45.2	128.1
TurNic (bottlenose dolphin)	100	88.6	136.2	85.7	/	/
RT1 (striped dolphin)	100	399.8	756.2	305.7	141.8	368.0
MDC12 (long-beaked common dolphin)	100	205.2	127.6	189.4	/	/
DDL1 (common dolphin)	100	/	/	112.1	102.0	84.7

Table 2. (A-B): Mean values of immunofluorescence of CYP1A1 (A) and CYP2B (B) revealed in cultured fibroblasts of different species treated with OC mixture. The immunofluorescence is expressed as index numbers respect to solvent control. Different colour of box is related to different increase of these proteins. N.C. = no cells.

The results confirm the capability of this methodology to detect CYP1A1 (Table 2A) and CYP2B (Table 2B) induction with OC mixture in many species of this study; particularly we had induction of CYP1A1 and CYP2B, with respect to chemical blank (DMSO), at all doses in fin whale (RT2) (Figure 3A; D) and long-beaked common dolphin (MDC12); an induction of CYP1A1 was detected at all doses in Risso's dolphin (GGL1) and of CYP2B at all doses in sperm whale (PM6), killer whale (MOO12) and striped dolphin (RT1) (Figure 3F). No induction of CYP1A1 was detected in sperm whale (PM6) and bottlenose dolphin (TurNic), while CYP2B showed OC induction at least at one treatment dose in all species. Different induction responses were given by the different specimens: there was a dose/response induction for CYP1A1 only for long-beaked common dolphin (MDC12) and for CYP2B only for fin whale (RT2) (Figure 3D), while a bell-shaped response was present for CYP1A1 in Risso's dolphin (GGL1) and common dolphin (DDL1), and for CYP2B in sperm whale (PM6), killer whale (MOO12) (Figure 3E), striped dolphin (RT1) (Figure 3F) and common dolphin (DDL1). Discontinuous induction response was showed for CYP1A1 and CYP2B by the other specimens such as CYP1A1 in killer whale (MOO12) (Figure 3B) and striped dolphin (RT1) (Figure 3C). It is interesting to point out that all species, following treatment with the OC mixture, showed a greater response of CYP2B, compared to CYP1A1, confirming that these xenobiotics mostly with globular structure have a major ability to induce this cytochrome.

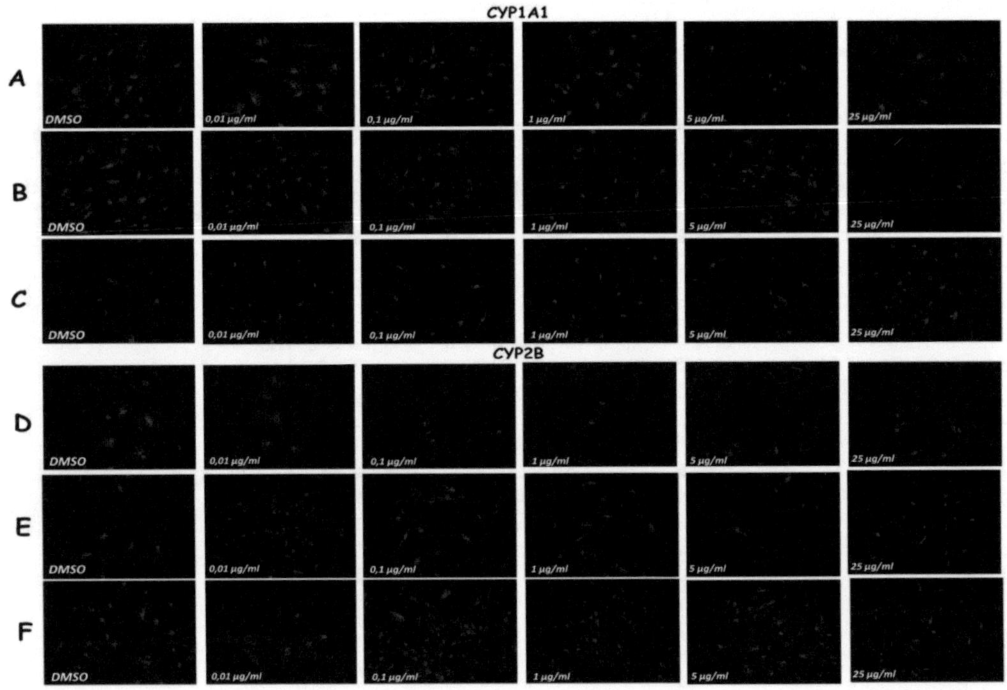

Figure 3. A-F: Immunofluorescence (AUF/nucleus) of CYP1A1 (A-C) and CYP2B (D-F) in fibroblast cells of fin whale (RT2) (A, D), killer whale (MOO12) (B, E) and striped dolphin (RT1) (C, F) treated with OC mixture. DAPI and Alexa Fluor 594 (Intensity 200ms) images of DMSO and the five OC mixture treatments.

6.3. CYP1A1 and CYP2B in different species after treatment with flame retardants

Results of the mean levels of immunofluorescence of CYP1A1 (A) and CYP2B (B), revealed in cultured fibroblasts of different species treated with flame retardants and expressed as index numbers, are reported in Table 3A-B. Marked differences in CYP1A1 (Table 3A) and CYP2B (Table 3B) induction by flame retardants were detected in different species, with higher sensitivity of responses in striped dolphin (RT1) for CYP1A1 (Figure 4A) and killer whale (MOO12) for CYP2B. To be highlighted that we have an inductive response of both cytochromes in the same animals, precisely in sperm whale (PM6), killer whale (MOO12), striped dolphin (RT1) (Figure 4A; C), long-beaked common dolphin (MDC12) (Figure 4B; D) and common dolphin (DDL1). Bottlenose dolphin (TurNic) showed only the CYP1A1 induction.

A CYP1A1	*Nonane 0.01 µg/ml*	*0.01 µg/ml*	*0.05 µg/ml*	*0.1 µg/ml*
RT2 (fin whale)	100	52.1	91.2	51.3
MBE3 (Bryde's whale)	N.C.	N.C.	N.C.	N.C.
PM6 (sperm whale)	100	113.2	105.2	52.0
MOO12 (killer whale)	100	132.2	36.6	131.1
GGL1 (Risso's dolphin)	100	73.3	98.6	82.6
TurNic (bottlenose dolphin)	100	71.1	102.5	131.9
RT1 (striped dolphin)	100	128.4	232.6	273.1
MDC12 (long-beaked common dolphin)	100	205.2	127.6	189.4
DDL1 (common dolphin)	100	135.4	133.9	154.6
B CYP2B	*Nonane 0.01 µg/ml*	*0.01 µg/ml*	*0.05 µg/ml*	*0.1 µg/ml*
RT2 (fin whale)	100	60.2	68.4	54.1
MBE3 (Bryde's whale)	N.C.	N.C.	N.C.	N.C.
PM6 (sperm whale)	100	111.9	70.5	63.1
MOO12 (killer whale)	100	314.7	110.7	52.7
GGL1 (Risso's dolphin)	100	70.2	75.9	48.6
TurNic (bottlenose dolphin)	100	71.5	69.2	94.1
RT1 (striped dolphin)	100	177.4	109.9	128.3
MDC12 (long-beaked common dolphin)	100	149.5	139.3	178.1
DDL1 (common dolphin)	100	118.4	85.8	324.5

Table 3. (A-B): Mean values of immunofluorescence of CYP1A1 (A) and CYP2B (B) revealed in cultured fibroblasts of different species treated with flame retardants. The immunofluorescence is expressed as index numbers respect to solvent control. Different colour of box is related to different increase of these proteins. N.C. = no cells.

A dose dependent induction of CYP1A1 was detected in striped dolphin (RT1) and common dolphin (DDL1) and of CYP2B only in long-beaked common dolphin (MDC12). A bell-shaped response was present for CYP1A1 and CYP2B in sperm whale (PM6), and for CYP2B in killer whale (MOO12) and striped dolphin (RT1), while discontinuous induction responses were showed for CYP1A1 and CYP2B by the other species. Also these contaminants are mainly with globular structure as OC mixture, but no differences are present in the induction of the two cytochromes.

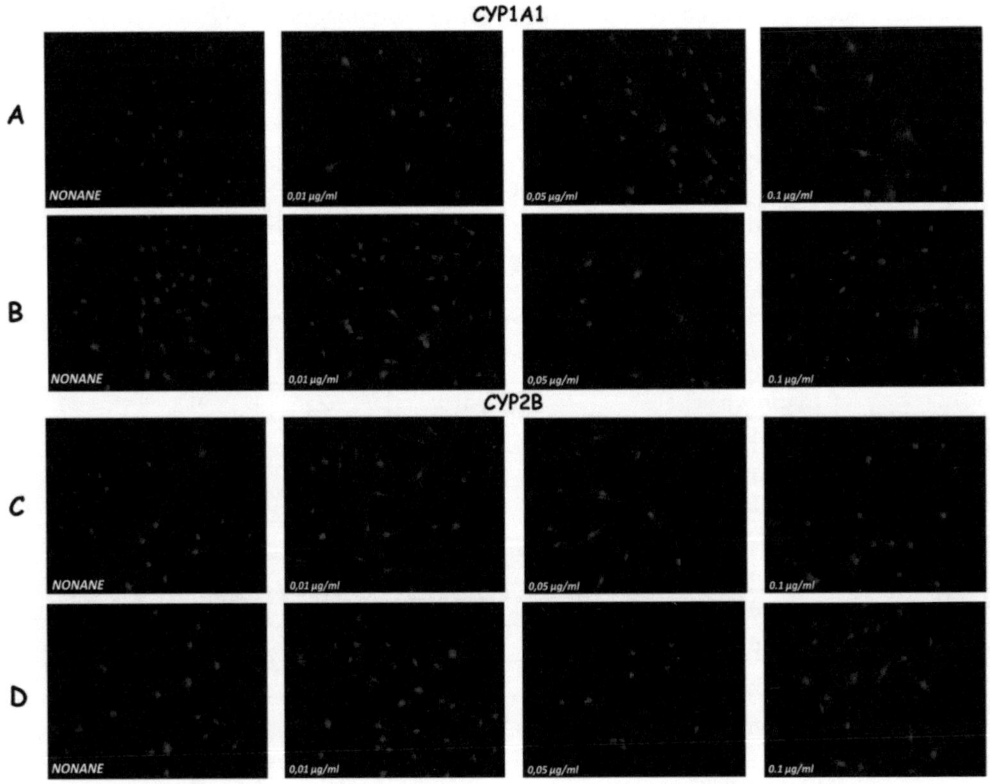

Figure 4. A-D: Immunofluorescence (AUF/nucleus) of CYP1A1 (A-B) and CYP2B (C-D) in fibroblast cells of striped dolphin (RT1) (A, C) and long-beaked common dolphin (MDC12) (B, D) treated with flame retardants. DAPI and Alexa Fluor 594 (Intensity 200ms) images of Nonane and the three flame retardant treatments.

6.4. CYP1A1 and CYP2B in different species after treatment with PAHs

In Table 4A-B we reported the results of the mean levels of immunofluorescence of CYP1A1 (A) and CYP2B (B), revealed in cultured fibroblasts of different species treated with PAHs, expressed as index numbers.

Only the fibroblasts of some species were treated with PAHs due to the fact that there was not a sufficient amount of cells from all of them to perform the various contaminant treatments. These planar contaminants are known to induce CYP1A1. In fact, an increase of CYP1A1 was detected, at least at one dose, in all specimens cultured fibroblasts exposed to PAHs (Table 4A) (Figure 5A-E). In the Risso's dolphin (GGL1) and in the striped dolphin (RT1), the higher dose of PAHs even caused the death of all cells. The fibroblast vitality was assessed with trypan blue, a quality control test to check the cell preparation. In the event of cell damage, trypan blue penetrates the cell membrane, and dead or damaged cells appear blue [36]. CYP2B also, showed induction in the three species treated with PAHs (Table 4B). The striped dolphin (RT1) showed the same trend for the two cytochromes (Table 4A-B).

A CYP1A1	*Acetone 0.1%*	*Dose C*	*Dose B*	*Dose A*
MBE3 (Bryde's whale)	100	31.6	106.2	55.8
PM6 (sperm whale)	100	111.5	69.6	68.2
GGL1 (Risso's dolphin)	100	107.7	90.2	D.C.
RT1 (striped dolphin)	100	279.8	77.3	D.C.
MDC12 (long-beaked common dolphin)	100	155.6	279.7	123.2
B CYP2B	*DMSO 0.1%*	*0.01 µg/ml*	*0. 1 µg/ml*	*1 µg/ml*
MBE3 (Bryde's whale)	N.C.	N.C.	N.C.	N.C.
PM6 (sperm whale)	100	125.0	64.2	101.6
GGL1 (Risso's dolphin)	N.C.	N.C.	N.C.	N.C.
RT1 (striped dolphin)	100	108.6	25.3	D.C.
MDC12 (long-beaked common dolphin)	100	218.6	65.6	64.3

Table 4. (A-B): Mean values of immunofluorescence of CYP1A1 (A) and CYP2B (B) revealed in cultured fibroblasts of different species treated with PAHs (Dose C = 0.5µM BaP + 10µM BnF, Dose B = 2.5µM BaP + 50µM BnF and Dose A = 12.5µM BaP + 250µM BnF). The immunofluorescence is expressed as index numbers respect to solvent control. Different colour of box is related to different increase of these proteins. N.C. = no cells. D.C. = death cells.

6.5. CYP1A1 and CYP2B in different species after treatment with BPA

In Table 5A-B we reported the results of the mean levels of immunofluorescence of CYP1A1 (A) and CYP2B (B), revealed in cultured fibroblasts of killer whale (MOO12) treated with BPA, expressed as index numbers.

A CYP1A1	*Ethanol 0.1%*	*0.1 µg/ml*	*1 µg/ml*	*10 µg/ml*	*100 µg/ml*
MOO12 (killer whale)	100	800.7	886.3	747.3	D.C.
B CYP2B	*Ethanol 0.1%*	*0.1 µg/ml*	*1 µg/ml*	*10 µg/ml*	*100 µg/ml*
MOO12 (killer whale)	100	48.3	40.4	49.2	D.C.

Table 5. (A-B): Mean values of immunofluorescence of CYP1A1 (A) and CYP2B (B) revealed in cultured fibroblasts of different species treated with BPA. The immunofluorescence is expressed as index numbers respect to solvent control. Different colour of box is related to different increase of these proteins. D.C. = death cells.

Only one species, the killer whale, was treated with this potent estrogenic compound since the study to evaluate the cetacean susceptibility to BPA has just started in our laboratories. However, this preliminary experiment confirms that BPA is a potent CYP2B inhibitor [35] while it gives information on CYP1A1 inductive capacity (Figure 6 A-B). To be highlighted that the highest used dose (100 μg/ml) is 100% lethal for fibroblast cells.

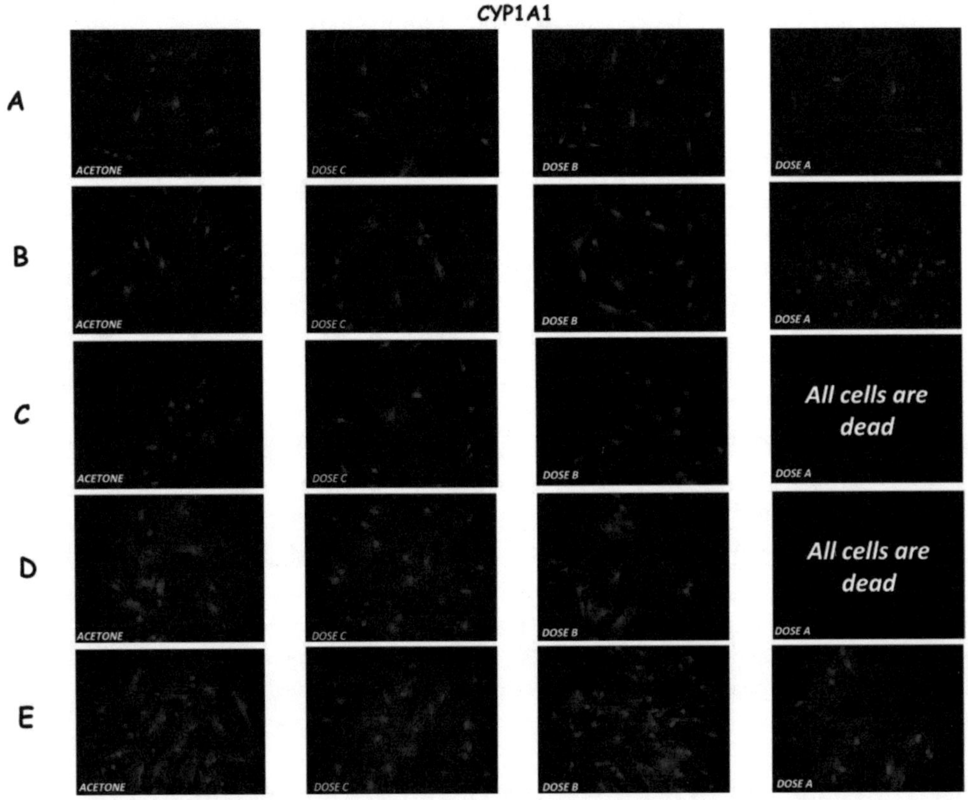

Figure 5. A-E: Immunofluorescence (AUF/nucleus) of CYP1A1 (A-E) in fibroblast cells of Bryde's whale (MBE3) (A), sperm whale (PM6) (B), Risso's dolphin (GGL1) (C), Striped dolphin (RT1) (D) and long-beaked common dolphin (MDC12) (E) treated with PAHs. DAPI and Alexa Fluor 594 (Intensity 200ms) images of Acetone and the three PAH treatments.

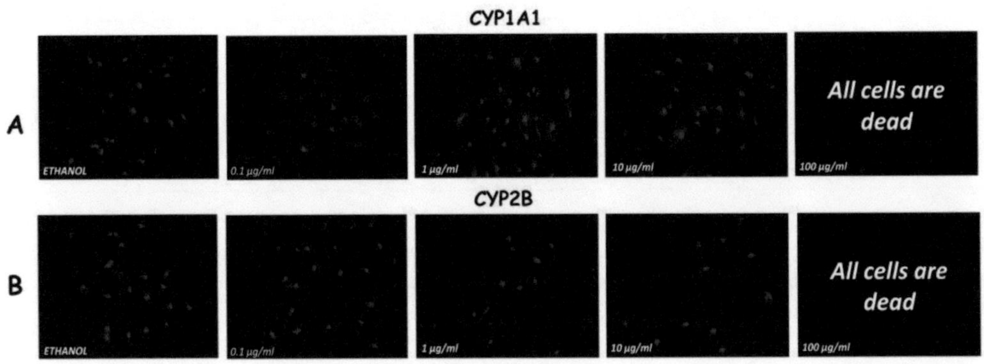

Figure 6. A-B: Immunofluorescence (AUF/nucleus) of CYP1A1 (A-B) and CYP2B (C-D) in fibroblast cells of killer whale (MOO12) (A, B) treated with BPA. DAPI and Alexa Fluor 594 (Intensity 200ms) images of Ethanol and the four BPA treatments.

7. Qualitative and quantitative mica protein expression as toxicological stress marker of the immune system

Cetacean morbilliviruses and papillomaviruses as well as *Brucella* spp. and *Toxoplasma gondii* are thought to reduce population abundance by inducing high mortalities, lowering reproductive success or by synergistically increasing the virulence of other diseases. Severe cases of lobomycosis and lobomycosis-like disease (LLD) may contribute to the death of some dolphins [37]. Environmental contamination seems to play a role in these diseases because many pollutants are known immunosuppressants and can markedly affect the immune status of the cetacean specimens [38, 39, 40, 41, 42]. As already mentioned, the cetacean skin is an important tissue of the immune system and the MICA protein, used in this study as toxicological stress marker of the immune system, is expressed in fibroblasts. In fact, the aim of this study was to evaluate the MICA protein expression in fibroblast cell cultures of cetaceans (skin biopsies of free-ranging animals and skin samples of stranded cetaceans dead within 2-12 h). Here we present the immunofluorescence technique in cultured fibroblasts used for qualitative and quantitative evaluation of MICA expression, induced by treatment with OC mixture, flame retardants, PAHs and BPA, as toxicological stress marker of the immune system of different species of odontocetes (sperm whale (PMAS1), killer whale (OO12), striped dolphin (RT23), long-beaked common dolphin (MDC12)) and mysticetes (fin whale (RT25), Bryde's whale (MBE3)).

7.1. Basal levels of MICA in different species

The basal level of MICA, evaluated with immunofluorescence technique in the fibroblasts of different cetacean species before treatment with different mixtures, is the first important result of this research step as it provides an indication of the immune status of these marine mammals. The results, expressed as immunofluorescence for cell (AUF/nucleus) mean values are presented in the Table 6 and, as histograms, in the Figure 7.

	MICA (UAF/nucleus)
Bryde's whale (MBE3)	50482.0
Long-beaked common dolphin (MDC12)	94751.0
Killer whale (MOO12)	195378.0
Sperm whale (PMAS1)	12647.0
Bottlenose dolphin (TTA1)	31473.0
Striped dolphin (RT23)	18511.0
Fin whale (RT25)	31325.0

Table 6. Basal levels of immunofluorescence (AUF/nucleus) of MICA in fibroblast cells of Bryde's whale, long-beaked common dolphin, killer whale, sperm whale, bottlenose dolphin, striped dolphin and fin whale.

We can highlight that the three specimens belonging to the three species sampled in the Sea of Cortez (Bryde's whale, long-beaked common dolphin and killer whale) showed higher basal activity of MICA with respect to all Mediterranean specimens, regardless of the

Figure 7. Basal levels of immunofluorescence (AUF/nucleus) of MICA in fibroblast cells of Bryde's whale (B.e), long-beaked common dolphin (D.c.), killer whale (O.o.), sperm whale (P.m.), bottlenose dolphin (T.t.), striped dolphin (S.c.) and fin whale (B.p.).

species and the fact that they were sampled free-ranging or found stranded alive and then died. Moreover, the basal activity of MICA in the three Mexican species seems to be related to their different diet, with an increasing activity with the increase of the trophic level. Bryde's whale is not strictly planktophagus as is the fin whale in the Mediterranean, feeding mainly on blue fish. So this species in the food chain is definitely closer to long-beaked common dolphin and the other toothed whales in this study than to the other mysticete species. The free-ranging specimen of sperm whale sampled in the surrounding water of Asinara Island (Mediterranean Sea) showed a basal activity very similar to that showed by the striped dolphin, but lower than the other species sampled in the Mediterranean Sea (bottlenose dolphin and fin whale).

This sharp distinction between the activity of MICA found in the Sea of Cortez and Mediterranean Sea specimens is probably the most important result to be highlighted: it seems that the environment in which specimens live and, therefore, the anthropogenic stress to which they are subjected are determinant in the response of this protein. In the light of this result we can hazard the conclusion that the lower the anthropic stress of the specimens, the higher the basal activity of MICA. Regarding the Mediterranean species, the two stranded specimens (striped dolphin RT23 and fin whale RT25), were both affected by *morbillivirus*; it would be very interesting to know the basal activity in the same species sampled free-ranging, to understand whether in case of immunosuppression the activity of MICA increases or decreases. To assess whether MICA increases or decreases, because of the presence of an inducer or a repressor of the immune system, we treated fibroblast cell cultures with cyclosporine A (CsA), a drug that belongs to the category of immunosuppressants, and with β-glucan, a polysaccharide known to increase the response of the immune system. The results of each specimen whose cells were treated with the two compounds are showed in Table 7.

Killer whale							
	n° cells	Mean	Median	Minimum	Maximum	S.D.	S.E.
BA	98	195378	151675	42646	592187	162020	46176
Inducer	151	189172	202628	50932	350323	109320	30320
Repressor 0.8μg/ml	135	106656	86680	46865	248599	61607	19481
Repressor 80μg/ml	79	279350	250709	20631	636284	174779	50454.4
Bottlenose dolphin							
	n° cells	Mean	Median	Minimum	Maximum	S.D.	S.E.
BA	87	31473	30832	26302	39107	4196.4	1713.2
Inducer	72	19230	19319	9871	44949	10153	3210
Repressor 0.8μg/ml	66	21673	19480	19637	55989	13259	4193
Repressor 80μg/ml	71	20647	17264	10492	56499	13536	4280
Striped dolphin							
	n° cells	Mean	Median	Minimum	Maximum	S.D.	S.E.
BA	75	18511	17896	8825	42083	9254	3055
Inducer	76	17606	18690	8556	26951	5977	1890
Repressor 0.8μg/ml	69	20677	124525	41662	329043	13440	4250
Repressor 80μg/ml	All cell death						
Fin whale							
	n° cells	Mean	Median	Minimum	Maximum	S.D.	S.E.
BA	114	31325	25172	18459	63324	16085	5086
Inducer	244	17409	17867	9673	23938	4270	1350
Repressor 0.8μg/ml	78	28304	29231	13275	37051	69853	21061
Repressor 80μg/ml	35	19031	18254	7299	56686	14102	4252

Table 7. Descriptive statistics of immunofluorescence per cell (AUF/nucleus) of MICA revealed in cultured fibroblasts of Bryde's whale (MBE3), striped dolphin (RT23), bottlenose dolphin (TTA1), and killer whale (OO12) treated with the inducer (β-glucan) and the repressor (Cyclosporine A). BA represents the blank (cultured fibroblasts treated only with primary and secondary antibodies).

In all cases, with the exception of fibroblasts of killer whale treated with the highest dose of repressor and fibroblasts of striped dolphin treated with the lower dose of repressor (highlighted in dark grey), we can see a decrease of the response of the MICA compared to the basal activity. After the treatment with these doses, it therefore remains difficult to understand the response of MICA which we must expected from a toxicological stress. Probably the choice of the two compounds, known to have such capabilities in relation to the immune system but not specifically in respect to MICA, should be reassessed.

7.2. MICA in different species after treatment with OC mixture

From Table 8 it appears evident that the carrier of OC mixture (DMSO) confounds the evaluation of the responses of the cells to the treatments, as previously demonstrated in other studies [15, 43].

MICA	*DMSO 0.05%*	*0.01 µg/ml*	*0.1 µg/ml*	*1 µg/ml*	*5 µg/ml*	*25 µg/ml*
MBE3 (Bryde's whale)	100	100	174	62	/	/
MDC12 (long-beaked common dolphin)	100	104	98	29	/	/
MOO12 (killer whale)	100	31	25	26	47	22
PMAS1 (sperm whale)	100	137	94	291	2523	11195
RT23 (striped dolphin)	100	53	N.C.	59	54	N.C.
RT25 (fin whale)	100	119	96	100	120	69

Table 8. Mean values of immunofluorescence of MICA revealed in cultured fibroblasts of different species treated with OC mixture. The immunofluorescence is expressed in index numbers respect to solvent control. Different colour of box is related to different increase of this protein.

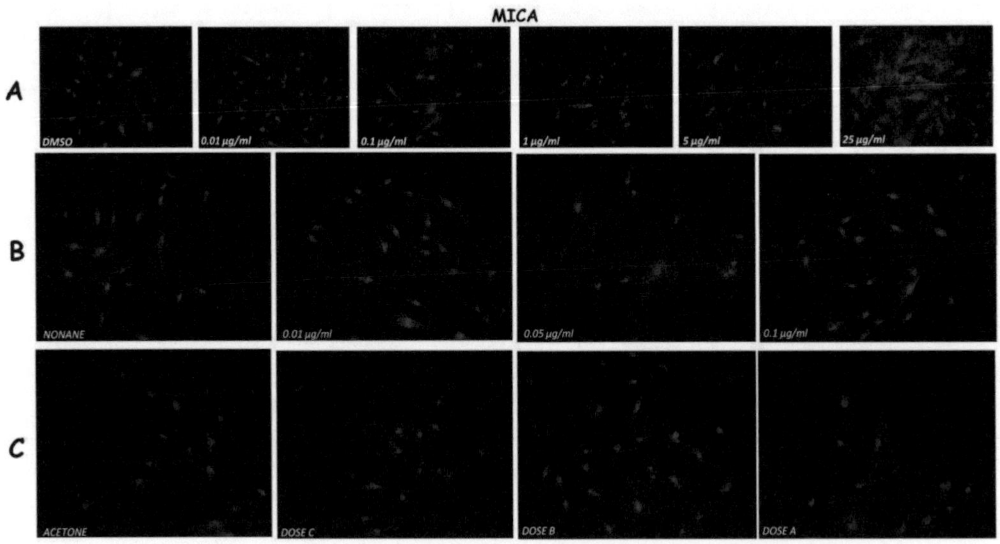

Figure 8. A-C: Immunofluorescence (AUF/nucleus) of MICA in fibroblast cells of sperm whale (PMAS1).DAPI and Alexa Fluor 594 (Intensity 200ms) images of DMSO and the five OC mixture treatments (A); Nonane and three flame retardant treatments (B) and Acetone and three PAH treatments (C).

Only the sperm whale demonstrated a clear inductive effect on MICA due to OCs, compared to DMSO (Figure 8A). In fibroblasts of killer whale (MOO12) and striped dolphin (RT23) no dose caused an increase of the MICA with respect to the carrier (DMSO). A dose dependent induction of MICA was present in sperm whale (PMAS1), a bell-shaped response was present in Bryde's whale (MBE3) and in long-beaked common dolphin (MDC12) while discontinuous induction response was showed by fin whale (RT25).

7.3. MICA in different species after treatment with Flame Retardants

In the Table 9 are reported the results of the mean levels of immunofluorescence of MICA, revealed in cultured fibroblasts of different species treated with the flame retardants, expressed as index numbers.

MICA	*Nonane 0.01 µg/ml*	*0.01 µg/ml*	*0.05 µg/ml*	*0.1 µg/ml*
MBE3 (Bryde's whale)	100	49	35	55
MDC12 (long-beaked common dolphin)	100	69	96	209
MOO12 (killer whale)	100	66	54	96
PMAS1 (sperm whale)	100	102	74	102
RT23 (striped dolphin)	100	103	132	107
RT25 (fin whale)	100	101	134	120

Table 9. Mean values of immunofluorescence of MICA revealed in cultured fibroblasts of different species treated with flame retardants. The immunofluorescence is expressed in index numbers respect to solvent control. Different colour of box is related to different increase of this protein.

In striped dolphin (RT23) and fin whale (RT25) the highest response of MICA was related to flame retardants, with a bell-shaped response in both species. Discontinuous induction response is shown by sperm whale (PMAS1) (Figure 8B) and long-beaked common dolphin (MDC12), while Bryde's whale (MBE3) and killer whale (MOO12) showed no induction response.

7.4. MICA in different species after treatment with PAHs

In Table 10 are reported the results of the mean levels of immunofluorescence of MICA, revealed in cultured fibroblasts of different species treated with the PAHs, expressed as index numbers.

MICA	*Acetone 0.1%*	*Dose C*	*Dose B*	*Dose A*
MDC12 (long-beaked common dolphin)	100	68	93	55
PMAS1 (sperm whale)	100	101	181	189

Table 10. Mean values of immunofluorescence of MICA revealed in cultured fibroblasts of different species treated with PAHs. The immunofluorescence is expressed in index numbers respect to solvent control. Different colour of box is related to different increase of this protein.

Only two species, the long-beaked common dolphin (MDC12) and the sperm whale (PMAS1) were treated with PAHs. However, even with this treatment, there is a significant dose/response type increase in the level of MICA in the sperm whale (PMAS1) (Figure 8C). Long-beaked common dolphin (MDC12) showed no induction response.

7.5. MICA in different species after treatment with BPA

Table 11 shows the results of the mean levels of immunofluorescence of MICA, revealed in cultured fibroblasts of different species treated with the BPA, expressed as index numbers.

MICA	*Ethanol 0.1%*	*0.1 μg/ml*	*1 μg/ml*	*10 μg/ml*	*100 μg/ml*
RT23 (striped dolphin)	100	84	99	85	D.C.
RT25 (fin whale)	100	62	68	153	D.C.

Table 11. Mean values of immunofluorescence of MICA revealed in cultured fibroblasts of different species treated with BPA. The immunofluorescence is expressed in index numbers respect to solvent control. Different colour of box is related to different increase of this protein.

The results showed that the higher dose caused the death of all cells, thus focusing all its toxicity. Only fin whale (RT25) fibroblasts had an inductive phenomenon with respect to solvent control.

8. Presence of DNA damage by comet assay as genotoxicity biomarker

We treated fibroblast cells with three different doses of a mixture of benzo(a)pyrene and beta-naphthoflavone for 4 h. The viability of fibroblast cells, assessed by the trypan blue test, was very high in control (>95%) and slightly decreased following the different treatments. We analysed DNA damage by the Comet assay. The principle of the Comet assay is that smaller DNA molecules migrate faster in an electric field than larger molecules. The treated cells are encapsulated in gel and lysed by alkali, which also denatures the DNA. Subsequent electrophoresis causes migration of the DNA. While the undamaged DNA appears as a "head", fragmented DNA move faster, giving the characteristic appearance of a comet tail. Figure 9 shows four normal cells at increasing degree of damaged DNA.

The cells were processed at different conditions of electrophoresis, to evaluate different types of strand breaks. Varying the pH during lysis and electrophoresis effects the type of strand breaks expressed. When cells are lysed and subjected to electrophoresis under neutral conditions (pH 8) only double strand breaks were detected. Under pH 12.1 conditions the double and single strand breaks were detected while under pH > 13 conditions the double strand breaks, single strand breaks and alkali labile lesions were detected [44].

The results of Comet assay evaluated in cells processed at pH 12.1 are shown in Figure 10. An increase of DNA migration was observed in fibroblasts exposed for 4 h at doses C and B, while a slight decrease was observed at dose A in comparison to the control. The acetone, used as carrier for benzo(a)pyrene and beta-naphthoflavone exposure, showed DNA fragmentation values very similar to the control.

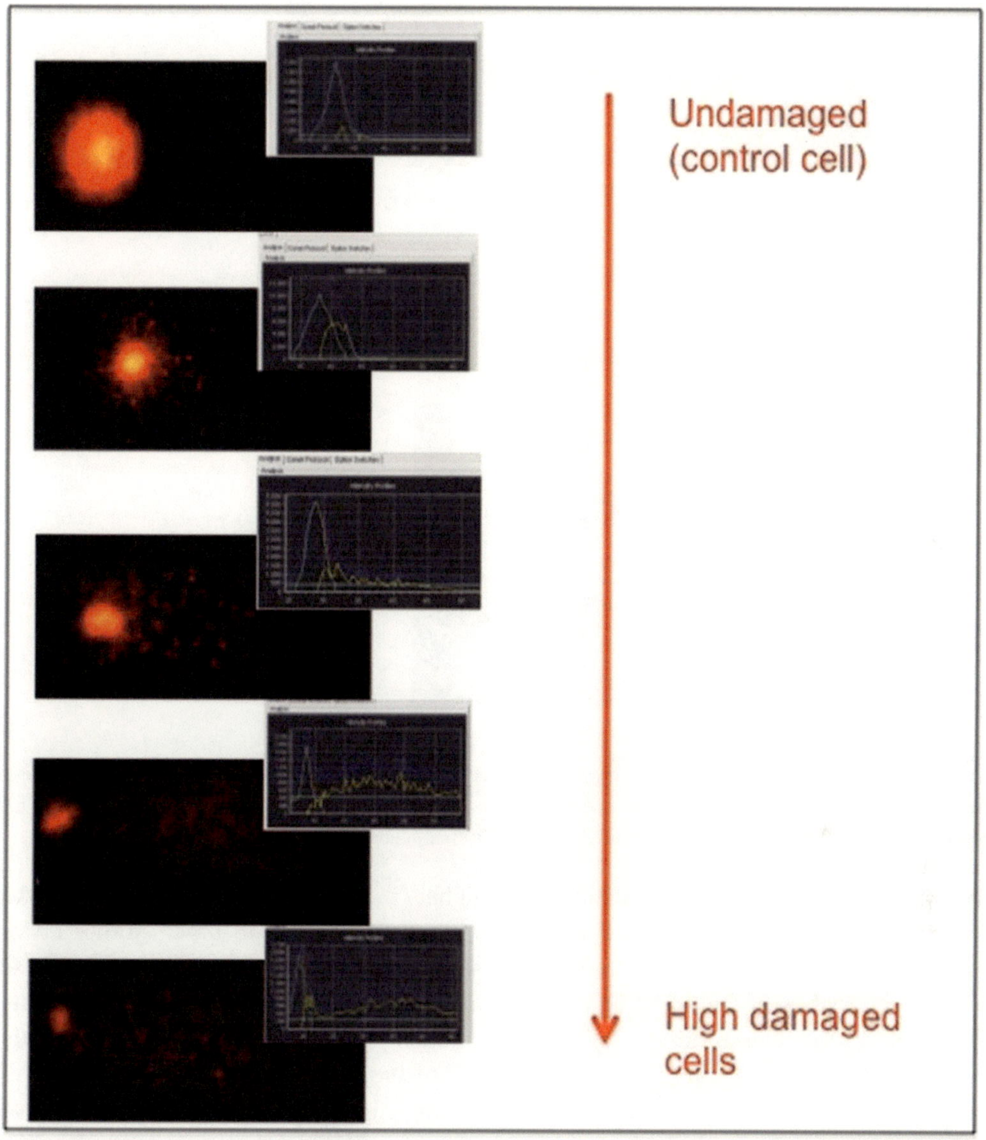

Figure 9. A photomicrograph of fibroblast cells of striped dolphin processed for the DNA comet assay. Four normal cells at increasing degree of damaged DNA: normal DNA, low-level DNA damage, DNA with a long tail, DNA almost completely fragmented.

The highest DNA fragmentation was observed at dose C and decreasing DNA tail values were observed from dose C to dose A. The results of the Comet assay evaluated in striped dolphin fibroblast cells processed at three different pHs showed a similar trend (Table 12), although the trend was more evident for the pH 12.1. Our results, although preliminary, suggest that alkaline Comet assay (pH 12.1) is the optimal version capable of detecting the DNA damage in fibroblast cells for future analysis.

Electrophoresis condition	Tail DNA %		
	Dose C	Dose B	Dose A
pH 8	30.91	19.68	24.8
pH 12.1	30.98	24.88	17.25
pH 13	35.09	27.54	25.71

Table 12. DNA migration evaluated in fibroblast cells of striped dolphin at three different pH conditions.

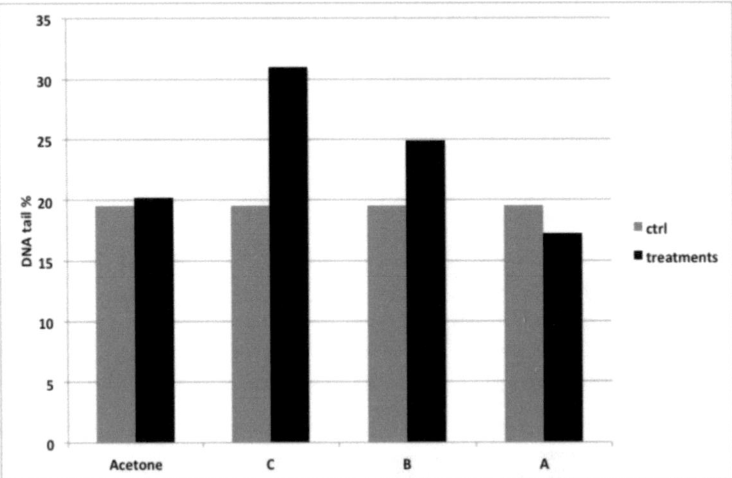

Figure 10. Effects of different doses of the mixture benzo(a)pyrene and beta-naphthoflavone at three doses (C = (0.5μM BaP + 10μM BnF), B = (2.5μM BaP + 50μM BnF) and A = (12.5μM BaP + 250μM BnF)) exposure on the DNA integrity of striped dolphin fibroblast cells after 4 h exposure.

The induction of DNA fragmentation was higher at the lowest dose (Dose C), while decreased at higher doses (Dose B and A), a result in contrast to other studies on cell mammal cultures [45, 46]. These investigations have demonstrated that an increase in the percentage of DNA in the tail region of the comets occurred in a concentration-dependent manner after exposure to different classes of genotoxic compounds, such as PAHs, methyl methanesulfonate (MMS) and H_2O_2. Our earlier Comet assay data on dolphin fibroblast cells exposed to benzo(a)pyrene are in agreement with the bibliography data. Thus, the decrease observed in our present data could be probably due to the action of the beta-naphthoflavone. This hypothesis is consistent with Gravato *et al.* [47] who demonstrated a decrease of DNA damage after exposure of specimens of *Anguilla anguilla* to beta-naphthoflavone. However, further studies are needed to confirm the genotoxic potential of mixture of PAHs for cetacean fibroblasts and investigate the potential genotoxicity of other classes of contaminants.

9. Conclusion

The aim of the present study was to propose cetacean fibroblast cell cultures as an "in vitro" method, called "Test Tube Cetaceans", to investigate effects of environmental contaminants in these marine mammals. The data reported in this chapter confirm that the use of Test Tube Cetaceans is a good non destructive surrogate of "in vivo" cetacean test (killing) to evaluate the different hazards of cetaceans to pollution.

Regarding the toxicological susceptibility to some xenobiotic compounds and to PAHs, the main results showed that the basal level of CYP1A1 and CYP2B of different cetacean species is very dissimilar and this seems to be especially species-specific rather than related to the geographic range, diet, toxicological status, etc. in which the specimens were found. All pollutants, at different level depending on the species and of the dose of treatment, showed an inductive capacity of these cytochromes. At times the response was dose dependent, other was bell-shaped response and other was a discontinuous induction response.

The qualitative and quantitative MICA protein expression as toxicological stress marker of the immune system showed that the three species sampled in the Sea of Cortez (Bryde's whale, long-beaked common dolphin and killer whale) showed higher basal level of MICA in respect to all Mediterranean specimens, regardless of the species and the fact that they were sampled free-ranging or found stranded alive and then died. This sharp distinction between the activity of MICA found in the Sea of Cortez and Mediterranean Sea specimens is probably the most important result to be highlighted: it seems that the environment in which specimens live and, therefore, the anthropogenic stress to which they are subjected, are determinant in the response of this protein of the immune system. In the light of this result we can hazard the conclusion that the lower the anthropic stress of the specimens, the higher is the basal level of MICA.

The Comet assay proved to be a very useful tool for assessing the potential genotoxicity of PAHs in cetacean fibroblast cell cultures. Future investigations will be conducted to investigate the genotoxic effects of different classes of contaminants in striped dolphin and other cetacean species. This technique led to the evaluation of possible DNA damage in species never studied before in this field, in order to investigate the different susceptibility to various contaminants, using fibroblast cell cultures.

In conclusion, "Test Tube Cetaceans" can be proposed to the scientific community as the "in vitro" method used to replace the "scientific whaling" to study the toxicological threats of different species of cetaceans, primarily of endangered species such as fin whale and Mediterranean common dolphin, since the research priorities is the conservation for the maintenance of cetacean biodiversity.

Author details

Letizia Marsili*, Silvia Maltese, Daniele Coppola, Ilaria Caliani, Laura Carletti,
Matteo Giannetti, Tommaso Campani, Matteo Baini, Cristina Panti, Silvia Casini
and M. Cristina Fossi
Department of Environmental Sciences, University of Siena, Siena, Italy

* Corresponding Author

Acknowledgement

The authors particularly thank the precious collaborators in the cetacean sampling: for free-ranging cetaceans Jorge Urban Ramirez, Lorenzo Rojas-Bracho, Claudia Diaz, Carlos A. Nino Torres, Lorena Viloria, Carlos Alberto (Mexico); for stranded cetaceans Fabrizio Cancelli, Tommaso Renieri, Fabrizio Serena, Cecilia Mancusi, Sandro Mazzariol (Italy).

10. References

[1] Schipper J, Chanson JS, Chiozza F, Cox NA, Hoffmann M, Katariya V, Lamoreux J, Rodrigues AS, Stuart SN, Temple HJ, Baillie J, Boitani L, Lacher TE Jr, Mittermeier RA, Smith AT, Absolon D, Aguiar JM, Amori G, Bakkour N, Baldi R, Berridge RJ, Bielby J, Black PA, Blanc JJ, Brooks TM, Burton JA, Butynski TM, Catullo G, Chapman R, Cokeliss Z, Collen B, Conroy J, Cooke JG, da Fonseca GA, Derocher AE, Dublin HT, Duckworth JW, Emmons L, Emslie RH, Festa-Bianchet M, Foster M, Foster S, Garshelis DL, Gates C, Gimenez-Dixon M, Gonzalez S, Gonzalez-Maya JF, Good TC, Hammerson G, Hammond PS, Happold D, Happold M, Hare J, Harris RB, Hawkins CE, Haywood M, Heaney LR, Hedges S, Helgen KM, Hilton-Taylor C, Hussain SA, Ishii N, Jefferson TA, Jenkins RK, Johnston CH, Keith M, Kingdon J, Knox DH, Kovacs KM, Langhammer P, Leus K, Lewison R, Lichtenstein G, Lowry LF, Macavoy Z, Mace GM, Mallon DP, Masi M, McKnight MW, Medellín RA, Medici P, Mills G, Moehlman PD, Molur S, Mora A, Nowell K, Oates JF, Olech W, Oliver WR, Oprea M, Patterson BD, Perrin WF, Polidoro BA, Pollock C, Powel A, Protas Y, Racey P, Ragle J, Ramani P, Rathbun G, Reeves RR, Reilly SB, Reynolds JE 3rd, Rondinini C, Rosell-Ambal RG, Rulli M, Rylands AB, Savini S, Schank CJ, Sechrest W, Self-Sullivan C, Shoemaker A, Sillero-Zubiri C, De Silva N, Smith DE, Srinivasulu C, Stephenson PJ, van Strien N, Talukdar BK, Taylor BL, Timmins R, Tirira DG, Tognelli MF, Tsytsulina K, Veiga LM, Vié JC, Williamson EA, Wyatt SA, Xie Y, Young BE. The status of the world's land and marine mammals: diversity, threat, and knowledge. Science 2008;322(5899) 225-230.

[2] European Commission. Our life insurance, our natural capital: an EU biodiversity strategy to 2020, May 3, 2011, Brussels, Belgium. COM 2011;244 1-16.

[3] Fossi MC, Marsili L, Neri G, Casini S, Bearzi G, Politi E, Zanardelli M, Panigada S. Skin biopsy of Mediterranean cetaceans for the investigation of interspecies susceptibility to xenobiotic contaminants. Marine Environmental Research 2000;50(1-5) 517-521.

[4] Fossi MC, Marsili L. Multi-Trial Ecotoxicological Diagnostic Tool in Cetacean Skin Biopsies. In: Uday Khopkar (ed.) Skin Biopsy – Perspectives. Rijeka: InTech; 2011. p317-333. Available from http://www.intechopen.com/articles/show/title/multi-trial-ecotoxicological-diagnostic-tool-in-cetacean-skin-biopsies (accessed 21 February 2012).

[5] Di Guardo G, Marruchella G, Agrim, U, Kennedy S. *Morbillivirus* infections in aquatic mammals: a brief overview. Journal of Veterinary Medicine Series A-Physiology Pathology Clinical Medic 2005;52(2) 88-93.

[6] Fossi MC, Marsili L. Endocrine disrupters in aquatic mammals. Pure and Applied Chemistry 2003;75(11-12) 2235-2247.

[7] Alaee M, Arias P, Sjodin A, Bergman Å. An overview of commercially used brominated flame retardants, their applications, their use patterns in different countries/regions and possible modes of release. Environment International 2003;29(6) 683-689.

[8] Yang M, Park MS, Lee HS. Endocrine disrupting chemicals: human exposure and health risks. Journal of Environmental Science and Health - Part C: Environmental Carcinogenic & Ecotoxicology Reviews 2007;24(2)183-224.

[9] Hammond JA, Hall AJ, Dyrynda L. Comparison of polychlorinated biphenyl (PCB) induced effects on innate immune functions in harbour and grey seals. Aquatic Toxicology 2005;74(2) 126-138.

[10] Canesi L, Borghi C, Ciacci C, Fabbri R, Lorusso LC, Vergani L, Marcomini A, Poiana G. Short-term effects of environmentally relevant concentrations of EDC mixtures on *Mytilus galloprovincialis* digestive gland. Aquatic Toxicology 2008;87(4) 272-279.

[11] Marsili L, Fossi MC, Casini S, Savelli C, Jimenez B, Junin M, Castello H. Fingerprint of polycyclic aromatic hydrocarbons in two populations of southern sea lions (*Otaria flavescens*). Chemosphere 1997;34(4) 759-770.

[12] Villeneuve DL, Khim JS, Kannan K, Giesy JP. Relative potencies of individual polycyclic aromatic hydrocarbons to induce dioxinlike and estrogenic responses in three cell lines. Environmental Toxicology 2002;17 128-137.

[13] Caliani I, Porcelloni S, Mori G, Frenzilli G, Ferraro M, Marsili L, Casini S, Fossi MC. Genotoxic effects of produced waters in mosquito fish (*Gambusia affinis*). Ecotoxicology 2009;18(1) 75-80.

[14] Marsili L, Fossi MC, Neri G, Casini S, Gardi C, Palmeri S, Tarquini E, Panigada S. Skin biopsies for cell cultures from Mediterranean free-ranging cetaceans. Marine Environmental Research 2000;50(1-5) 523-526.

[15] Marsili L, Casini S, Bucalossi D, Porcelloni S, Maltese S, Fossi MC. Use of immunofluorescence technique in cultured fibroblasts from Mediterranean cetaceans as new "in vitro" tool to investigate effects of environmental contaminants. Marine Environmental Research 2008;66(1) 151-153.

[16] Goldstone HMH, Stegeman JJ. A revised evolutionary history of the CYP1A subfamily: gene duplication, gene conversion, and positive selection. Journal of Molecular Evolution 2006;62 708–717.

[17] Ngui JS, Bandiera SM. Induction of hepatic CYP2B is a more sensitive indicator of exposure to aroclor 1260 than CYP1A in male rats. Toxicology and Applied Pharmacology 1999;161(2) 160-170.

[18] Dehn PF, Allen-Mocherie S, Karek J, Thenappan A. Organochlorine insecticides: impacts on human HepG2 cytochrome P4501A, 2B activities and glutathione levels. Toxicology In Vitro 2005;19(2) 261-273.

[19] Marsili L, Maltese S, Carletti L, Coppola D, Casini S, Fossi MC. MICA expression as toxicological stress marker in fibroblast cell cultures of cetaceans. Comparative

Biochemistry and Physiology - Part A: Molecular & Integrative Physiology 2010;157(1) S24-S25.

[20] Meyer T, Stockfleth E, Christophers E. Immune response profiles in human skin. British Journal of Dermatology 2007;157(2) 1-7.

[21] Frenzilli G, Scarcelli V, Del Barga I, Nigro M, Forlin L, Bolognesi C, Sturve J. DNA damage in eelpout (*Zoarces viviparus*) from Goteborg Harbour. Mutation Research 2004;552 187-195.

[22] Frenzilli G, Nigro M, Lyons BP. The Comet assay for the evaluation of genotoxic impact in aquatic environments. Mutation Research 2009;681(1) 80-92.

[23] Dhawan A, Bajpayee M, Parmar D. Comet assay: a reliable tool for the assessment of DNA damage in different models. Cell Biology and Toxicology 2009;299(25) 5-32.

[24] Berubè M, Palsboll P. Identification of sex in cetaceans by multiplexing with three ZFX and ZFY specific primers. Molecular Ecology 1996;5 283-287.

[25] Frame GM, Cochran JW, Bowadt SS. Complete PCB congener distributions for 17 aroclor mixtures determined by 3 HRGC systems optimized for comprehensive quantitative, congener-specificus analysis. Journal of High Resolution Chromatography 1996;19 657-668.

[26] Pelkonen O, Bremier DD. Role of environmental factors in the pharmacokinetics of drugs: considerations with respect to animal models, P450 enzymes, and probe drugs. In: Welling PG, Balant LP (ed.) Handbook of Experimental Pharmacology 10. Berlin: Sprinter Verlag; 1994. p298-332.

[27] Stegeman JJ, Hahn ME. Biochemistry and molecular biology of monooxygenases: Current perspectives on forms, function, and regulation of cytochrome P450 in aquatic species. In: Malins DC, Ostrander GK (ed.) Aquatic Toxicology: Molecular, biochemical and cellular perspectives. USA: Lewis Publishers; 1994. p87-206.

[28] Pelkonen O, Raunio H. Metabolic activation of toxins: tissue-specific expression and metabolism in target organs. Environmental Health Perspectives 1997;104(4) 767-774.

[29] Kim JH, Stansbury KH, Walzer NJ, Trush MA, Strickland PT, Setter TR. Metabolism of benzo(a)pyrene and benzo(a)pyrene-7,8-diol by human cytochrome P450 1B1. Carcinogenesis 1998;19 1847-1853.

[30] Lewis DFV, Ioannides C, Parke DV. Cytochromes P450 and species differences in xenobiotici metabolism and activation of carcinogenesis. Environmental Health Perspectives 1998;106 633-641.

[31] Nims RW, Lubet RA. Induction of cytochrome P-450 in the Norway rat, *Rattus norvegicus,* following exposure to potential environment contaminants. Journal of Toxicology and Environmental Health 1995;46 271-292.

[32] You L, Chan SK, Bruce JM, Archibeque-Engle S, Casanova M, Corton JC, Heck H. Modulation of testosterone-metabolizing hepatic cytochrome P-450 enzymes in developing Sprague-Dawley rats following in utero exposure to p,p'-DDE. Toxicology and Applied Pharmacology 1999;158 197-205.

[33] Sanders JM, Burka LT, Smith CS, Black W, James R, Cunningham ML. Differential expression of CYP1A, 2B, and 3A genes in the F344 rat following exposure to a

polybrominated diphenyl ether mixture or individual components. Toxicological Sciences 2005;88 127-133.

[34] Takeshita A, Koibuchi N, Oka J, Taguchi M, Shishiba Y, Ozawa Y. Bisphenol-A, an environmental estrogen, activates the human orphan nuclear receptor, steroid and xenobiotici receptor-mediated transcription. European Journal of Endocrinology 2001;145(4) 513-517.

[35] Hanioka N, Jinno H, Tanaka-Kagawa T, Nishimura T, Ando M. Interaction of bisphenol a with rat hepatic cytochrome P450 enzymes. Chemosphere 2000;41(7) 973-978.

[36] Dammacco F, editor. Diagnostica immunologica. Padova: Piccin-Nuova libraria; 1995.

[37] Van Bressem MF, Raga JA, Di Guardo G, Jepson PD, Duignan PJ, Siebert U, Barrett T, Santos MC, Moreno IB, Siciliano S, Aguilar A, Van Waerebeek K. Emerging infectious diseases in cetaceans worldwide and the possible role of environmental stressors. Diseases of Aquatic Organisms 2009;86(2) 143-157.

[38] Fowles JR, Fairbrother A, Baecher-Steppan L, Kerkvliet NI. Immunologic and endocrine effects of the flame-retardant pentabromodiphenyl ether (DE-71) in C57BL/6J mice. Toxicology 1994;86(1-2) 49-61.

[39] Ndeble K, Tchounwou PB, McMurray RW. Coumestrol, bisphenol-A, DDT, and TCDD modulation of interleukin-2 expression in activated CD+4 Jurkat T cells. International Journal of Environmental Research and Public Health 2004;1(1) 3-11.

[40] Ohshima Y, Yamada A, Tokuriki S, Yasutomi M, Omata N, Mayumi M. Transmaternal exposure to bisphenol a modulates the development of oral tolerance. Pediatric Research 2007;62(1) 60-64.

[41] Yoshitake J, Kato K, Yoshioka D, Sueishi Y, Sawa T, Akaike T, Yoshimura T. Suppression of NO production and 8-nitroguanosine formation by phenol-containing endocrine-disrupting chemicals in LPS-stimulated macrophages: involvement of estrogen receptor-dependent or –independent pathways. Nitric Oxide 2008;18(3) 223-228.

[42] Schwacke LH, Zolman ES, Balmer BC, De Guise S, George RC, Hoguet J, Hohn AA, Kucklick JR, Lamb S, Levin M, Litz JA, McFee WE, Place NJ, Townsend FI, Wells RS, Rowles TK. Anaemia, hypothyroidism and immune suppression associated with polychlorinated biphenyl exposure in bottlenose dolphins (*Tursiops truncatus*). Proceedings of the Royal Society - Biological Sciences 2012;279(1726) 48-57.

[43] Spinsanti G, Panti C, Bucalossi D, Marsili L, Casini S, Frati F, Fossi MC. Selection of reliable reference genes for qRT-PCR studies on cetacean fibroblast cultures exposed to OCs, PBDEs, and 17beta-estradiol. Aquatic Toxicology 2008;87(3) 178-186.

[44] Lee RF, Steiner S. Use of the single cell gel electrophoresis/comet assay for detecting DNA damage in aquatic (marine and freshwater) animals. Mutation Research 2003;544 43–64.

[45] Stang A, Witte I. The ability of the high-through put comet assay to measure the sensitivity of five cell lines toward methyl methane sulfonate, hydrogen peroxide, and pentachlorophenol. Mutation Research 2010; 701 103–106.

[46] Platt KL, Aderhold S, Kulpe K, Fickler M. Unexpected DNA damage caused by polycyclic aromatic hydrocarbons under standard laboratory conditions. Mutation Research 2008;650 96–103.

[47] Gravato C, Teles M, Oliveira M, Santos MA. Oxidative stress, liver biotransformation and genotoxic effects induced by copper in *Anguilla anguilla* L. – the influence of pre-exposure to b-naphthoflavone. Chemosphere 2006;65 1821–1830.

Chapter 4

Assessing Biomagnification and Trophic Transport of Persistent Organic Pollutants in the Food Chain of the Galapagos Sea Lion (*Zalophus wollebaeki*): Conservation and Management Implications

Juan José Alava and Frank A.P.C. Gobas

Additional information is available at the end of the chapter

http://dx.doi.org/10.5772/51725

1. Introduction

Bioaccumulation of persistent organic pollutants (POPs) represents a risk to the marine environment and wildlife, including marine mammals and birds [1-4]. Biomagnification is a special case of bioaccumulation and is defined as the process by which concentrations of contaminants or chemical substances (i.e. thermodynamic activities of chemical substances often measured by the lipid normalized concentration) in consumer and higher trophic level organisms exceed those concentrations in the diet or organism's prey [5-7]. This process can occur at each step in a food chain, potentially producing very high and toxic concentrations in upper-trophic-level species [7].

Bioaccumulation and biomagnification are important considerations in the categorization and risk assessment of chemical compounds under the treaty of the Stockholm Convention for POPs and regulatory and management efforts in several nations such as the Canadian Environmental Protection Act Canada (CEPA [8]), the Toxic Substances Control Act (TSCA [9]) in the United States and the Registration, Evaluation, Authorisation and Restriction of Chemicals program (REACH) in the European countries [10]. Due to the long-range atmospheric transport and global fractioning of POPs northward from low or mid latitudes [11, 12], the Arctic and northern hemisphere have remained as active regions of research to study biomagnification of POPs in trophic chains and food webs [2, 13-15]. However, very little is known about the bioaccumulative behaviour and fate of these substances in tropical zones of the planet.

There are several measures that have been used to express the degree of biomagnification. The simplest measure is the Biomagnification Factor (BMF), which is described as the ratio of the chemical concentrations in the organism (C_B) and the diet of the organism (C_D), i.e., BMF = C_B/C_D, where the chemical are usually expressed in units of mass of chemical per kg of the organism (in wet weight or in a lipid basis) and mass chemical per kg of food (in wet weight or in a lipid basis) [6]. Biomagnification of organic contaminants and foraging preferences in aquatic and marine food webs can also be investigated using stable nitrogen isotope as biomarkers of trophic level [15-20]. Stable isotope analysis (SIA) has emerged as a tool in foraging ecology/habitat use, physiology and ecotoxicology, and is applied widely to study marine mammal ecology [21]. Stable nitrogen isotope analysis is a known well established technique for assessing predator–prey interactions and organism trophic levels (TL) in food webs [22-25]. Specifically,$\delta^{15}N$, the concentration ratio of $^{15}N/^{14}N$, expressed relative to a standard (i.e., atmospheric N_2), has been shown to increase with increasing trophic level due to the preferential excretion of the lighter nitrogen isotope [26]. Likewise, carbon isotope signatures (δ^{13}C) provide information on habitat use and general sources of diet of organisms, i.e., marine/freshwater, coastal/oceanic, pelagic/benthic [27].

Studies of the biomagnification and food web transport of POPs in tropical systems such as remote islands around the equatorial Pacific Ocean are lacking. Due to the remoteness and isolation of the Galapagos Islands relative to other better studied geographical areas, the Galapagos Island food web offers a unique opportunity to undertake research related to the transport, bioaccumulative nature and biomagnification of globally distributed contaminants in tropical environments at the ecosystem level. The low population levels and generally good environmental control and management practices on the islands ensures that local pollutant sources are in most cases insignificant compared to global sources. These conditions provide a unique mesocosm to study the behaviour of global pollutants in marine mammalian food-chains.

The Galapagos sea lion(*Zalophus wollebaeki*) is an endemic marine mammal residing year round in the islands and exhibiting a high degree of dietary plasticity, consuming several groups of fish prey (99% of the diet). The Galapagos sea lion diet includes Cupleidae (thread herrings and sardines), Engraulidae (anchovies), Carangidae (bigeye scads), Serranidae (groupers, whitespotted sand bass or camotillo), Myctophidae (lantern fishes), Mugilidae (mullets) and Chlorophtalmidae fishes, and a low proportion of squid, as reported in the existing literature [28-31]. Although the information about diet and trophic level is limited for sea lions at several rookeries in the Galapagos Islands, it is known that the dietary preferences of Galapagos sea lions are also a function of the local variation in prey availability and regional climate-oceanic variability such as the El Niño events, when sea lions can switch their diet composition to more abundant fish items [30, 32, 33]. The Galapagos sea lion has been recognized as a key species for the functioning and health of the marine ecosystem of the islands under the environmental management action plan of the Galapagos Marine Reserve (GMR) [33]. Because of its high trophic position, relative abundance in the islands and non-migratory behaviour, Galapagos sea lions can serve as local sentinels of food web contamination [33-35]. Concentrations of polychlorinated biphenyls (PCBs) and dichlorodiphenyltrichloroethane (DDT) were recently detected in this species, underlying the health risk due to the toxicity and bioaccumulation potential of these contaminants in the Galapagos food web [34, 35]. Thus, equivalent to the role of killer whales as global sentinels of pollution in the Northeastern Pacific [1], the Galapagos sea lion can be used as an eco-marker of environmental pollution and a key indicator of not only the coastal marine health, but the public health in the region.

With the aim to contribute to the understanding of the behaviour and fate of POPs in marine food webs of tropical regions, this chapter provides an advanced primer on biomagnification assessment of POPs in the Galapagos Islands based on the existing literature on baseline levels of DDT detected in Galapagos sea lions [35] and recent unpublished data on organochlorine pesticides (i.e. mirex, dieldrin, chlordanes, β-HCH) and PCBs in Galapagos sea lions and fish preys. To accomplish this work, we made use of concentration data measured in Galapagos sea lions and their fish prey and determination of predator-prey biomagnification factors to assess biomagnification in this tropical system. Insights on the impact of biomagnification and conservation and management implications at the ecosystem level in the Galapagos are discussed.

2. Materials and methods

2.1. Tissue collection from Galapagos sea lion pups

In a recent study [35], blubber biopsy and hair samples of 20 Galapagos sea lion pups of 2–12 months of age were obtained from four rookeries in the Galapagos Archipelago (3°N–4°S, 87°–94°W) between 24-29March 2008. Briefly, pups were sampled at Isabela (Loberia Chica, n = 5), Floreana, (Loberia, n =6) and Santa Cristobal (Puerto Baquerizo, n = 4; Isla Lobos, n = 5) islands. Pups were captured with hoop nets and manually restrained. Age was estimated by visual observation of both the size and weight of the animal. In all circumstances, capture stress and holding time were minimized (< 10-15 min). Hair samples were obtained using a sterile scissor to trim or a scalpel to shave the region to be used prior to the biopsy collection and deposited into labelled zipper bags. Biopsies (100 mg; 6mm–Miltex biopsy punch) were collected from an area 10-20 cm lateral to the spinal column and anterior to the pelvis. The biopsy site was pre-cleaned with alcohol and betadine. Biopsies were wrapped in hexane-rinsed aluminum foil and placed in a cooler with wet ice and transferred into cryovals placed in a cryoship (-20°C) during the field sampling, and, afterwards stored at -80°C in the laboratory until chemical analysis.

Pups were chosen because (a) the animals are readily accessible and relatively easy to capture in most of the rookeries of the Galapagos Islands year round; (b) the animals are of similar age (3-10 months), minimizing the influence of life history parameters on contaminant concentrations; (c) because they are nursed by adult reproductive females they have a high trophic position because they are feeding on mother's milk, ingesting energy and pollutants and analogous to a predator–prey relationship [35]. The rationale of the study design to justify the use of pups as ecosystem based sentinels of biomagnification is also explained in Figure 1.

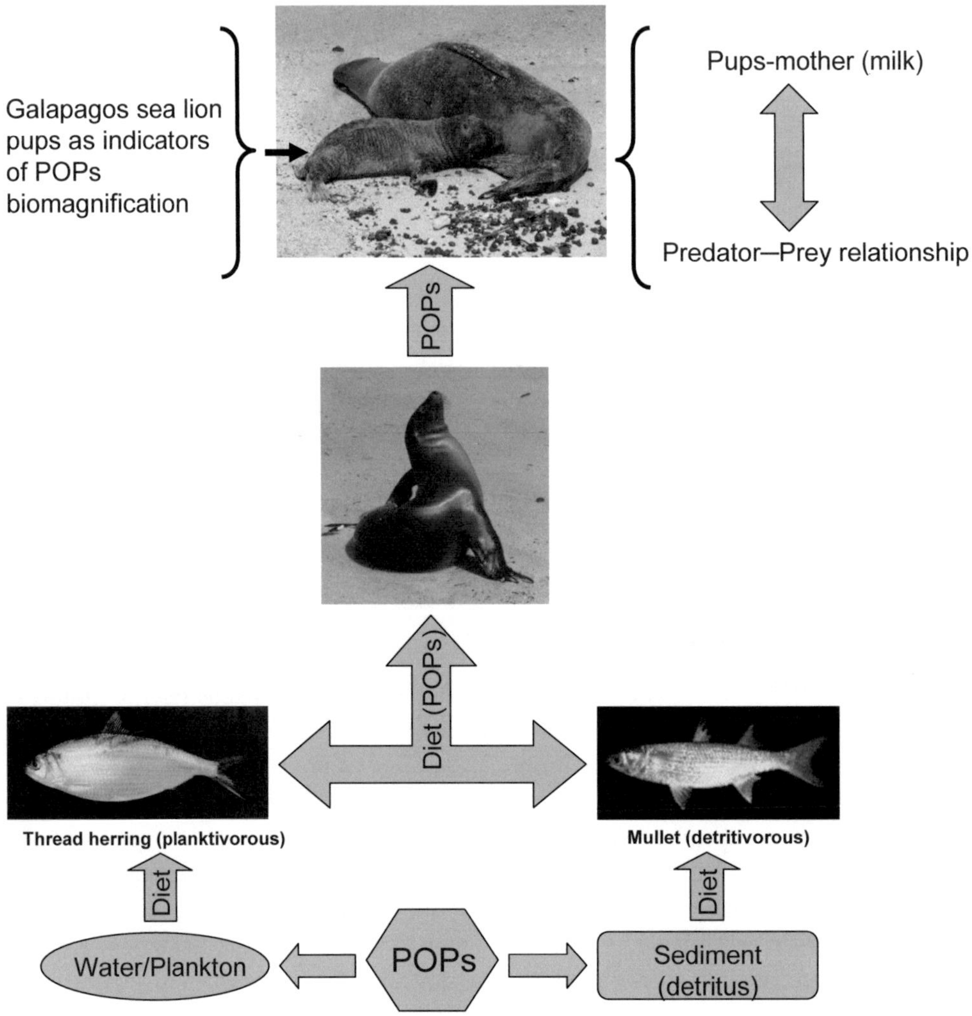

Figure 1. Conceptual model illustrating the bioaccumulation process in a representative, food chain of the Galapagos sea lion. Piscivorous Galapagos sea lions can be exposed to persistent organic pollutants (POPs), mainly through dietary ingestion. Low trophic level prey fish can absorb POPs from water and plankton (planktivorous fish), as well as from sediments (detritivorous fish). Nursing pups can bioaccumulate POPs from adult females by nursing and thus occupy a higher tropic level relative to their mothers because$\delta^{15}N$ isotopic enrichment.

2.2. Fish collection and homogenization

Two species of fish (mullets, *Mugil curema*; n = 11; and, Galapagos thread herrings, *Ophistonema berlangai*; n = 4), which for the purpose of this study were assumed to be major prey items of Galapagos sea lions, were collected from Galapagos waters by fishers during

March-April 2008. Mullets are coastal fish, inhabiting nearshore habitats, and demersal-benthic feeders (detritivorous), grazing on detritus and bottom sediments and digesting the nutritive matter (iliophagous foraging), while Galapagos thread herrings are endemic, pelagic and schooling fishes that filter-feed (planktivorous) mainly on tiny planktonic organisms (e.g., phytoplankton) in open waters [36].

After field collection, fish specimens were frozen until further transportation to the lab, where they were stored at -80ºC. Each fish was measured, weighed and sexed. Muscle biopsies were extracted from the dorsal, lateral muscle of each fish, using a 6mm–biopsy punch (Accuderm, USA), and saved in vials for stable isotope analysis.

Each individual fish was homogenized using a clean, hexane-acetone rinsed meat grinder (Omcam Inc., Italy). The ground fish was then further homogenized in a homogenizer (Omni, USA and/or Polytron, Kinematica, GmbH, Switzerland) at dial position 5-6 for ≈1 min until material was well mixed and homogenous in appearance. Homogenized samples and subsamples were transferred to clean glass jars and stored at -80 ºC until further chemical analysis.

2.3. Sample preparation for Stable Isotopes Analysis (SIA)

Each set of hair samples collected from Galapagos sea lion pups was cleaned for lipid and particle removal by washing the hair three times with a chloroform:methanol 2:1 v/v solution using a clean Pasteur glass pipette. Samples were transferred into labelled scintillation vials and desiccated overnight, and, then, lyophilized using a freeze drier (Free Zone ® Plus 4.5 Liter Cascade; Labconco, Kansas City, MO) for 24 hr (Vacuum pressure set point: 0.01 mBar).

Fish biopsy samples were freeze dried overnight (Vacuum pressure set point: 0.01 mBar). Biopsy samples were weighed and freeze dried again to determine if there were differences in weights after the second freeze drying. Once the sample weight was constant (i.e., no remaining moisture), one set of freeze dried samples was stored in the desiccator until further analysis for $\delta^{15}N$. The set of freeze dried replicates underwent an extraction protocol to remove lipids to be used for $\delta^{13}C$ analysis. First, freeze dried samples were pulverized using a mortar and transferred into a glass tube for lipid extraction by adding 5ml of chloroform:methanol 2:1 v/v; and, then vortex mixed for 30 seconds. Solids were dispersed with sonification in bath sonicator for 10 min. Samples were allowed to settle for 30 min at room temperature, followed by an additional 30 second vortex and sonification. Samples were centrifuged for 5 minutes at 1000 rpm (model GS6R, Beckman, USA) to enhance pellet formation. The solvent was carefully removed with glass Pasteur pipette (pipette was changed for each sample), without transferring any particulate matter, and the solvent was disposed in the waste bottle. A second extraction was repeated. The supernatant was carefully removed with pipette and the residue was left at -20ºC overnight. Samples were dried under Nitrogen and transferred to a clean, amber vial for analysis of stable isotopes of carbon and nitrogen.

2.4. Stable Isotopes Analysis (SIA)

Carbon and nitrogen isotopic analyses on fish biopsies and Galapagos sea lion hair were accomplished by continuous flow, isotopic ratio mass spectrometry (CF-IRMS) using a GV-Instruments® IsoPrime attached to a peripheral, temperature-controlled, EuroVector® elemental analyzer (EA) (University of Winnipeg Isotope Laboratory, UWIL). One-mg samples were loaded into tin capsules and placed in the EA auto-sampler along with internally calibrated carbon/nitrogen standards. Nitrogen and carbon isotope results are expressed using standard delta (δ) notation in units of per mil (‰).The delta values of carbon ($\delta^{13}C$) and nitrogen ($\delta^{15}N$) represent deviations from a standard. $\delta^{15}N$ isotope ratios (‰) were determined using the following equation [21,26]:

$$\delta^{15}N = [(^{15}N/^{14}N_{SAMPLE}/^{15}N/^{14}N_{STANDARD}) - 1] \times 1000$$

where $^{15}N/^{14}N_{SAMPLE}$ is the isotope ratio of the tissue sample analyzed; and, $^{15}N/^{14}N_{STANDARD}$ represents the ratio of the international standard of atmospheric N_2 (air), IAEA-N-1 (IAEA, Vienna), for $\delta^{15}N$. The equivalent equation for $\delta^{13}C$ isotope ratios (‰) is:

$$\delta^{13}C = [(^{13}C/^{12}C_{SAMPLE}/^{13}C/^{12}C_{STANDARD}) - 1] \times 1000$$

The standard used for carbon isotopic analyses was the Vienna PeeDee Belemnite (VPDB). Analytical precision, determined from the analysis of duplicate samples, was ±0.13‰ for $\delta^{13}C$ and ±0.6‰ for $\delta^{15}N$. The analytical precision based on standards, which are more isotopically homogeneous than samples, was ± 0.19‰ for $\delta^{13}C$ and ±0.24 for $\delta^{15}N$.

2.5. Trophic level estimations

The trophic positions ($TP_{CONSUMER}$) of the prey species (i.e. fish) and the predator (Galapagos sea lion) were determined relative to the baseline $\delta^{15}N$ (assumed to occupy a trophic level 2), using the following algorithm [37, 38]:

$$TP_{CONSUMER} = \frac{\left(\delta^{15}N_{CONSUMER} - \delta^{15}N_{BASELINE}\right)}{3.4} + 2$$

Where $\delta^{15}N_{CONSUMER}$ is the average $\delta^{15}N$ signature value of the predator; $\delta^{15}N_{BASELINE}$ is the $\delta^{15}N$ signature at the base of the food web; and 3.4‰ is the isotopic, trophic level enrichment factor ($\Delta^{15}N$), recommended to be used for constructing food webs when a priori knowledge of $\Delta^{15}N$ is unavailable [39]. The $\delta^{15}N_{BASELINE}$ was established as the $\delta^{15}N$ signature of the particulate organic matter (POM) of bottom sediments in the eastern equatorial Pacific Ocean (250 km south of the islands) with a value of 5.5‰ [31, 40], which is relatively close to the $\delta^{15}N$ value of 7.3‰, reported recently for phytoplankton in the Galapagos [30]. The rationale for using this signature is supported by the fact that the assimilation of nitrogen (i.e., NO_3^-) up taken from near surface marine waters by phytoplankton is reflected by $\delta^{15}N$ values of POM, which is also a major component of the carbon flux and sediments [40].

Although pups instead of adult individual sea lions were sampled in this study, the $\delta^{15}N$ signature in the pup is expected to reflect the isotopic nitrogen signature of the mother, as pups feed only on mothers' tissue (i.e., milk proteins) analogous to a predator-prey relationship, resulting in a $\delta^{15}N$ isototipc enrichment of 2.1‰ and 0.9‰ $\delta^{13}C$ enrichment in relation to adult females [41, 42]. Because of lactation, pups can be at a higher trophic level than their mothers (Figure 1). However, the $\delta^{15}N$ signature in the pups can provide useful information about the foraging habits (i.e., diet) of adult female animals [43].

2.6. Sample preparation for chemical analysis

Contaminant analyses were conducted in the Regional Dioxin Laboratory (RDL) at the Institute of Ocean Sciences (IOS), Fisheries and Ocean Canada (DFO), based on analytical methodologies described elsewhere [44]. In brief, the muscle-blubber biopsy samples of Galapagos sea lion pups (0.053 to 0.212 g wet weight) and subsamples of fish homogenate (9.23 to 10.5 g) were spiked with a mixture of surrogate internal standards which contained all fifteen $^{13}C_{12}$-labeled PCBs, and a mixture of labelled organochlorine pesticides (OCPs): D_3 1,2,4-Trichlorobenzene, $^{13}C_6$ 1,2,3,4 Tetrachlorobenzene, $^{13}C_6$ Hexachlorobenzene, $^{13}C_6\gamma$-HCH, $^{13}C_6\gamma$-HCH, $^{13}C_{10}$ trans Nonachlor, $^{13}C_{12}$ TeCB-47, $^{13}C_{12}$*p,p'*-DDE, $^{13}C_{12}$ Dieldrin, $^{13}C_{12}$*o,p*-DDD, $^{13}C_{12}$*p,p'*-DDD, $^{13}C_{12}$*o,p*-DDT, $^{13}C_{12}$*p,p'*-DDT, $^{13}C_{10}$ Mirex. All surrogate internal standards were purchased from Cambridge Isotope Laboratories (Andover, MA). The spiked samples were homogenized with Na_2SO_4 in a mortar, transferred quantitatively into an extraction column, and extracted with DCM/hexane (1:1 v/v). For some of the samples the extract formed two layers/phases, a waxy-precipitate layer and the solvent layer. The solvent layer was transferred to a clean flask and the waxy precipitate was treated with several aliquots of hexane and DCM. Each of these was transferred to the flask that contained the solvent layer of the extract. Despite the treatment with additional volumes of hexane and DCM, vortexing and pulverization, the waxy precipitate (for sea lions) did not dissolved in the solvents used and as a result it was not included in the corresponding sample extract that was used for lipid and contaminants determinations.The DCM:Hexane sample extracts were evaporated to dryness and the residue was weighted in order to determine the total lipid in the samples. Subsequently the residue was re-suspended in 1:1 DCM/Hexane and divided quantitatively into two aliquots. The larger aliquot (75% of the extract) was subjected to sample-cleanup for PCBs determinations. The remaining (25% of the extract) was used for OCP determinations.

2.7. PCB and OC pesticides analyses

Sample extracts were analyzed for PCB congeners and target OCPs by gas chromatography/high-resolution mass spectrometry (GC/HRMS). The specific methodology and protocols for the quantification and analytical methods to determine PCB congeners and OCPs have previously been reported in prior published papers (34, 35).

2.8. Quality assurance/quality control measures

The mass spectrometry conditions used for all the analyses, the composition of the linearity calibration solutions, the criteria used for congener identification and quantification and the quality assurance – quality control procedures used for the quantification of PCBs and OCPs followed those described in detail elsewhere [34, 35, 44].

2.9. Bioaccumulation parameter

In general, the biomagnification of contaminants is basically quantified as the biomagnification factor in terms of the concentration of a given chemical in the consumer or predator relative to the concentration in the diet or prey (i.e. $BMF = C_B/C_D$, where C_B is the chemical concentration in the organism and C_D is the chemical concentration of the diet). To quantify biomagnification in the Galapagos sea lions relative to prey items (i.e., thread herring and mullet) and to explore the effect of the magnitude of trophic level differences on the BMF measures, the predator-prey biomagnification factor (BMF_{TL}) was used for data interpretation in this study.The criterion applied to indicate the capability of the chemical to biomagnify was a BMF > 1. A BMF statistically greater than 1 indicates that the chemical is a probable bioaccumulative substance [7].

2.9.1. Predator-prey Biomagnification Factor (BMF_{TL})

Following this approach, the mean lipid normalized concentration of each contaminant measured in Galapagos sea lion pups was divided by the mean lipid adjusted concentration in the prey. Then, the biomagnification factor can be adjusted to represent exactly one trophic level in difference using the trophic level estimated from $\delta^{15}N$. Therefore, the field based predator-prey biomagnification factor normalized to trophic position or $BMF_{TROPHIC\ LEVEL}$ (BMF_{TL}) is calculated using the following equation [15]:

$$BMF_{TL} = \frac{(C_{PREDATOR} / C_{PREY})}{TL_{PREDATOR} - TL_{PREY}}$$

Where $C_{predator}$ and C_{prey} are chemical concentrations in the predator and prey, expressed in units of mass of chemical (μg) per kg of the predator and mass chemical (μg) per kg of prey in a lipid normalized basis (i.e. $BMF_{LIPID\ WEIGHT}$),and $TL_{predator}$ and TL_{prey} are the trophic levels of the predator and prey. The BMF_{TL} values were used to measure biomagnification in the tropical food chain between two adjacent trophic levels (i.e., the difference in TL between predator and prey is small), assuming steady state in contaminant concentrations between predator and prey. Since BMF_{TL} can be related to the trophic magnification factor (TMF), which describes the increase of contaminants from one trophic level to the other (derived from the slope, b, of the relationship between an organism's log lipid normalized chemical concentration), it can also be expressed as BMF_{TL}* as proposed by Conder *et al.* [45]:

$$BMF_{TL}* = 10^{\left[\frac{\log_{10}([C_{PREDATOR}]/[C_{PREY}])}{TL_{PREDATOR}-TL_{PREY}}\right]}$$

Where $C_{predator}$ and C_{prey} are appropriately normalized (e.g., lipid normalized) chemical concentrations in the predator and prey, and $TL_{predator}$ and TL_{prey} are the trophic levels of the predator and prey. In essence, the BMF_{TL} is the biomagnification factor normalized to a single trophic level increase in the food-web [45].The use of trophic magnification factors (TMFs) is currently an emerging approach to better assess the biomagnification of POPs in marine food webs [16]. An important number of studies in the northern hemisphere have relied on the use of the TMF for this purpose [15, 16, 18]. Thus, the use of TMF coupled with stable isotope analysis (SIA) to track the amplification and transport of POPs in food webs is a recommended methodology in eco-toxicology to study the biomagnification of POPs. The lack of prey samples and minimal trophic levels required (≥ 3) precluded to undertaking a trophic magnification factor (TMF) assessment in this study.

2.10. Data treatment and supporting statistical analysis

Concentrations of all detected POPs were blank corrected using the method detection limit (MDL), defined as the mean response of the levels measured in three procedural blanks used plus three times the standard deviation (SD) of the blanks ($MDL = Mean_{blanks} + 3*SD_{blanks}$). Following this methodology, the concentration of each PCB congener and OC pesticide was determined based on concentrations above the MDL only. Only PCBs detected in 100% of samples and above the MDL were used for data analysis and calculations of BMFs. Contaminant concentration data were log-transformed to fit the assumption of normality criterion before statistical analysis. ∑PCB concentrations were calculated as the sum of PCB-52, PCB 74, PCB 95, PCB-99, PCB-101, PCB-105, PCB-118, PCB 128, PCB - 138/163/164, PCB-146, PCB 153, PCB 156, PCB 174, PCB 180, PCB 183, PCB 187, PCB 201 and PCB 202. ∑DDTs were defined as the sum of *o, p'*-DDE, *p, p'*-DDE, *o, p'*-DDD, *p, p'*-DDD, *o, p'*-DDT and *p, p'*-DDT, and ∑chlordanes as the sum of *trans*-chlordane, *cis*-chlordane, *trans*-nonachlor and *cis*-nonachlor.

To further support the analysis of biomagnification of POPs in the tropical food chain of the Galapagos, statistical comparisons between the concentrations of selected PCBs (e.g., PCBs 153, 180), ∑DDTs, *p,p'*-DDE and other organochlorine pesticides measured in the Galapagos sea lion and those detected in diet items (i.e., mullet and thread herring) were conducted. These comparisons were conducted using analyses of variance (ANOVA) if variances were homoscedastic (i.e., equal variances) or Welch's analyses of variance if variances or standard deviations were heteroscedastic (i.e., unequal variances as tested by Levene's test or Bartlett test, $p< 0.05$), and a Tukey-Kramer honestly significant difference (HSD) test, which is a post-hoc method recommended to test differences between pairs of means among groups that contain unequal sample sizes [46]. Inter-site comparisons among rookeries samples followed the same statistical methods. Statistical comparison tests were conducted at a level of significance of $p< 0.05$ ($\alpha = 0.05$).

Principal Components Analyses (PCAs) were conducted on the fractions of PCBs and organochlorine pesticides relative to total concentrations by contaminant group (i.e., contaminants expressed as a fraction of total) for each sample to visualize spatial differences in patterns in sea lion pups from different sites within the Galapagos Archipelago and elucidate potential sources (i.e., local versus global-atmospheric). First, samples with undetectable values were replaced by a random number between the lowest and the highest concentration that were detectable (> MDL) to account for uncertainty before PCA (i.e., *trans*-chlordane and PCB 110 showed zero values in blanks in three and two samples out of 20, respectively; therefore; there was not possible to calculate MDLs), or otherwise removed from the PCAs. Secondly, samples were normalized to the concentration total before PCA to remove artefacts related to concentrations differences between samples. Finally, the centered log ratio transformation (division by the geometric mean of the concentration-normalized sample followed by log transformation) was then applied to this compositional data set to produce a data set that was unaffected by negative bias or closure [47]. Regressions, statistical comparisons and PCAs were run using JMP 7.0 (SAS Institute Inc.; Cary, NC, USA).

3. Results and discussion

3.1. Stable isotope profiles and trophic levels

The values of $\delta^{15}N$ and $\delta^{13}C$ (mean ± standard deviation) found here are consistent to those reported in Galapagos sea lion pups (i.e., 13.1‰ ± 0.5‰ for $\delta^{15}N$, and -14.5‰ ± 0.5‰ for $\delta^{13}C$) in a recent study [31].No significant relationship was observed between isotopic values and length of the pups ($\delta^{15}N$: $r = 0.005$, $p =0.7594$; $\delta^{13}C$:$r = 0.18$, $p = 0.0626$) or weight ($\delta^{15}N$:$r = 0.0001$, $p =0.9645$; $\delta^{13}C$: $r = 0.18$, $p = 0.0752$). Although female pups appeared to exhibit higher values of $\delta^{15}N$ compared to male pups (*t*-test = 2.3767, $p = 0.0288$), $\delta^{13}C$ values between males and females were similar (*t*-test = -0.3326, $p = 0.7433$). In addition, no significant inter-site differences in $\delta^{15}N$ (ANOVA, $p = 0.4235$) and $\delta^{13}C$ (ANOVA, $p = 0.8378$) values were found among rookeries. This indicates that site or foraging location had minimal influence on the isotope ratios. The lack of differences was further minimized by sampling similar ontogenetic stages (i.e., pups of similar age, development and size), and a metabolically inactive tissue (i.e., fur hair), which is corroborated by the fact that hair is an inert tissue containing physiological and dietary information (isotopic signals) [48].

Based on the $\delta^{15}N$values, the trophic level (TL) measured here for the Galapagos sea lion ($\delta^{15}N = 13.0$; TL = 4.2) fall within the range of those recently reported (i.e., $\delta^{15}N$ =12.6–13.4; TL = 4.1–4.4) elsewhere [30, 31, 43]. The $\delta^{15}N$ values for thread herrings and mullets were 9.4‰ ± 1.77‰ (TL = 3.1), and 12.7‰ ± 1.10‰ (TL = 4.1), respectively, while the $\delta^{13}C$ values for thread herrings and mullets were -17.0 ±0.70 and -9.34 ±0.80.

3.2. POP concentrations in animals and inter-site comparisons

3.2.1. *Galapagos sea lions*

Observed concentrations of selected POPs in Galapagos sea lion and two of its main prey items are summarized in Table 1. Galapagos sea lions represented the largest number of organisms sampled in this study (n = 41) and exhibited the highest concentrations of PCBs and OC pesticides. The multi-comparison post hoc analysis, including sea lions and prey fish, showed that no significant differences in OC pesticides and PCB congener concentrations were observed between male and female pups. Fish prey commonly exhibited significantly lower concentrations than Galapagos sea lion pups (ANOVA and multi-comparisons Tukey-Kramer (HSD) post-hoc test, $p < 0.05$) (Table 1, Figure 2).

Concentrations of ∑DDTs in Galapagos sea lions ranged from 16.0 to 1700 μg/kg lipid and ∑DDTs were the predominant OC pesticide in Galapagos sea lion pups, as previously reported [35]. ∑Chlordanes were the second most abundant group of contaminants present. *Trans*-nonachlor represented 68% of ∑chlordanes, followed by *cis*-chlordane, *cis*-nonachlor and *trans*-chlordane (Table 1), a pattern comparable to that reported in pups of southern elephant seals (*Mirounga leonina*) [49] and Weddell seals (*Leptonychotes weddellii*) [50]. This indicates that *trans*-nonachlor is a predominant chlordane compound in pinnipeds.

Within the hexachlorocyclohexanes (HCHs), β-HCH was the only isomer detectable in all pups (>MDL). β-HCH was the dominant HCH isomer in blubber samples of California sea lions (*Zalophus californianus*) from Baja California [51] and in toothed cetaceans from tropical and temperate waters of the Indian and North Pacific oceans [52] due to the greater biomagnification of the most bioaccumulative β-HCH versus γ-HCH [3, 20]. Interestingly, the mean β-HCH concentration in Galapagos sea lions was higher than the mean ∑HCH concentrations measured in spinner dolphins (*Stenella longirostris*) (21.3 μg/kg lipid) captured in a marine area of the Eastern Tropical Pacific [52] in offshore waters north of the Galapagos.

Both dieldrin and mirex were detected in all pups with concentrations ranging from 0.85 to 24 μg/kg lipid for mirex and from 9.00 to 83.0 μg/kg lipid for dieldrin. Concentrations of ∑PCBs (i.e., sum of 20 PCB congeners) ranged between 16.0 and 380 (μg/kg lipid) in pups and from 1.0 to 140 (μg/kg lipid) in fish preys (Table 1).

3.2.2. *Fish prey*

OC pesticides, including ∑DDTs, chlordanes, β-HCH, dieldrin and mirex, and individual PCB congeners detected in Galapagos sea lion pups were also detected (> MDL) in all sampled thread herring and mullet prey samples. Significantly lower concentrations of OC pesticides and PCBs were found in thread herrings and mullets than in Galapagos sea lion pups (ANOVA and multi-comparisons Tukey-Kramer (HSD) post-hoc test, $p < 0.05$; Table 1). PCB 202 was the only congener exhibiting similar concentrations in sea lions and fish (ANOVA, $p > 0.05$), suggesting a lack of its bioaccumulation in the food chain. Although thread herring and mullet showed differences in $\delta^{15}N$ values or trophic levels and foraging strategies, concentrations of POPs in these two species were similar (Figure 2) with the exception of mirex and *cis*-nonachlor, which were higher in planktivorous thread herrings than in mullets. Endosulfan sulphate was detected in all mullet samples ranging from 0.07 to 0.22 μg/kg lipid, with an arithmetic mean of 0.16 μg/kg lipid. Only two thread herring samples exhibited detectable concentration of this pesticide (0.002–0.05 μg/kg lipid). Endosulfan sulphate was not detected in any of the biopsy samples of pups.

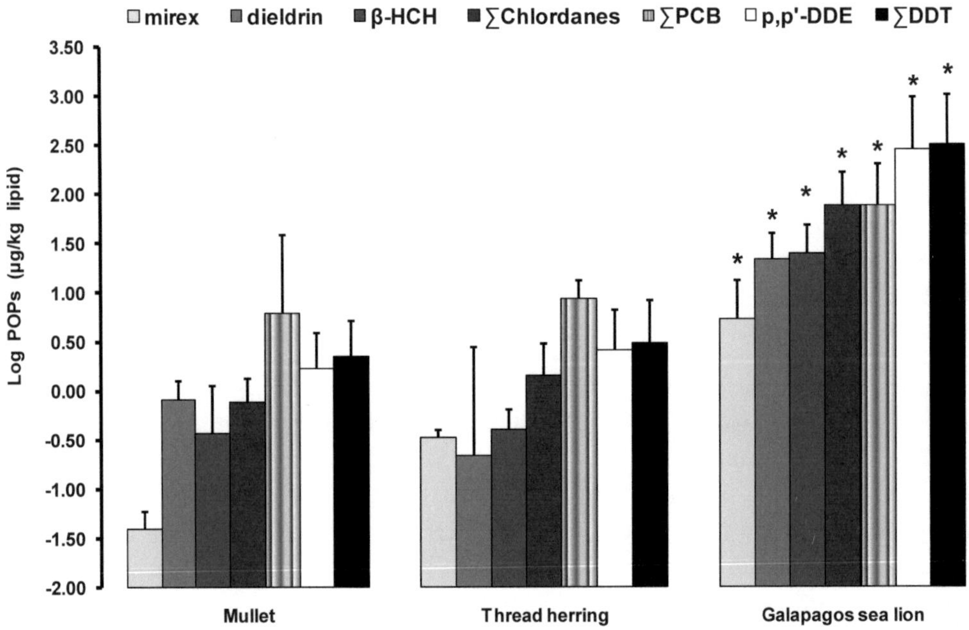

Figure 2. Inter-species comparisons of ∑PCB and organochlorine pesticide (mirex, dieldrin, β-HCH, ∑Chlordanes, *p,p*-DDE, ∑DDT) concentrations. Asterisks indicate that concentration in the Galapagos sea lion were significantly higher ($p < 0.05$) than those found in mullets and thread herrings. Error bars are standard deviations.

The PCB composition in prey showed a different composition of PCB congeners compared to that of sea lions pups (Figure 3). Higher chlorinated PCBs, i.e., Hepta, Octa and Nona-chlorinated biphenyls (PCBs 180–201) were more abundant in thread herrings and mullets than in Galapagos sea lion pups. This indicates the possible role of biotransformation, reduced uptake of PCBs, or a natural placental barrier for heavier PCBs in sea lions. Lower chlorinated PCB congeners, ranging from PCB 43/44 to PCB 118 (Tetra to Penta- chlorinated biphenyls), make up an important contribution (≈ 37% ± 7.25%) to the total PCB concentrations suggesting a lighter PCB signature ("equatorial fingerprint") in the Galapagos sea lion, mullet and thread herring compared to that observed in many arctic biota.

	Galapagos sea lion (predator)		Fish (prey)		*p*-value
	Female pups	Male pups	Thread herring	Mullet	
	(n = 10)	(n = 10)	(n = 4)	(n = 6)	
Lipid (%)	75.9 ± 3.50	77.8 ± 2.45	1.22 ± 0.86	2.86 ± 2.00	
***p,p'*-DDE**	480 ± 120 **A**	505 ± 180 **A**	3.30 ± 1.00 **B**	2.22 ± 0.700 **B**	<0.05*
	(65.4–1183)	(13.6–1650)	(0.669–5.00)	(0.620–5.20)	
***p,p'*-DDT**	13.0 ± 2.85 **A**	8.60 ± 1.08 **A**	0.070 ± 0.046 **B**	0.130 ± 0.051 **B**	<0.05*
	(1.70–29.0)	(0.974–12.0)	(ND–0.195)	ND–0.300	
***p,p'*-DDD**	20.0 ± 4.73 **A**	17.0 ± 4.60 **A**	0.440 ± 0.140 **B**	0.550 ± 0.170 **B**	<0.05*
	(1.88–44.0)	(0.965–54.0)	(0.036–0.70)	(0.155–1.30)	
∑DDT	516 ± 125 **A**	533 ± 183 **A**	4.00 ± 1.26 **B**	3.00 ± 0.910 **B**	<0.05*
	(71.2–1230)	(16.3–1666)	(0.705–6.05)	(0.820–6.80)	
Mirex	8.60 ± 1.76 **A**	6.40 ± 2.20 **A**	0.330 ± 0.030 **B**	0.040 ± 0.008 **C**	<0.05**
	(2.50–21.0)	(0.850–24.0)	(0.250–0.400)	(0.028–0.080)	
Dieldrin	31.0 ± 7.26 **A**	22.0 ± 4.80 **A**	0.600 ± 0.204 **B**	0.880 ± 0.128 **B**	<0.05**
	(9.00–83.0)	(9.00–63.0)	(0.005–0.90)	(0.400–1.30)	
***β*-HCH**	34.2 ± 4.00 **A**	26.0 ±7.05 **A**	0.440 ± 0.090 **B**	0.495 ± 0.095 **B**	<0.05**
	(18.3–52.0)	(7.75–78.0)	(0.229–0.620)	(0.041–0.650)	
***trans*-chlordane**	0.410 ± 0.100 **A**	0.65 ± 0.10 **A**	0.070 ± 0.027 **B**	0.040 ± 0.015 **B**	<0.05**
	(ND–0.840)	(0.273–1.03)	(ND–0.130)	(ND–0.110)	
***cis*-chlordane**	17.2 ± 2.67 **A**	15.0 ± 2.75 **A**	0.455 ± 0.140 **B**	0.250 ± 0.053 **B**	<0.05*
	(6.800–34.0)	(3.60–31.0)	(0.049–0.670)	(0.120–0.482)	
***trans*-nonachlor**	73.0 ± 12.0 **A**	65.0 ± 22.0 **A**	0.860 ± 0.191 **B**	0.40 ± 0.072 **B**	<0.05**
	(37.0–146)	(11.0–214)	(0.430–1.30)	(0.160–0.570)	
***cis*-nonachlor**	16.0 ± 3.20 **A**	10.0 ± 2.10 **A**	0.300 ± 0.109 **B**	0.195 ± 0.050 **C**	<0.05*
	(3.7–31.8)	(3.56–25.8)	(ND–0.510)	(0.075–0.380)	
∑Chlordanes	107 ±15.0 **A**	90.5 ± 25.2 **A**	1.70 ± 0.445 **B**	0.870 ± 0.175 **B**	<0.05*
	(48.1–180)	(18.8–255)	(0.481–2.50)	(0.372–1.50)	
PCB 52	3.20 ± 0.530 **A**	2.10 ± 0.610 **A**	0.210 ± 0.030 **B**	2.20 ± 1.85 **B**	<0.05**
	(1.13–5.60)	(0.332–7.05)	(0.136–0.270)	(0.055–11.0)	
PCB 74	2.60 ± 0.410 **A**	2.00 ± 0.510 **A**	0.100 ± 0.009 **B**	0.280 ± 0.220 **B**	<0.05**
	(1.40–5.10)	(0.340–4.40)	(0.050–0.085)	(0.012–1.40)	
PCB-95	2.80 ± 0.303 **A**	2.20 ± 0.320 **A**	0.300 ± 0.090 **B**	2.02 ± 1.70 **B**	0.05**
	(1.63–4.83)	(0.873–3.75)	(0.018–0.413)	(0.026–10.4)	

PCB-99	11.0 ± 2.07 **A**	8.30 ± 2.70 **A**	0.570 ± 0.073 **B**	2.62 ± 2.14 **B**	<0.05**
	(4.99–27.0)	(1.30–23.0)	(0.390–0.740)	(0.090–13.0)	
PCB-101	8.70 ± 1.38 **A**	4.30 ±1.38 **A**	0.630 ± 0.186 **B**	3.35 ± 2.70 **B**	<0.05**
	(4.36–18.3)	(1.79– 16.4)	(0.115–0.980)	(0.090–17.0)	
PCB-105	2.05 ± 0.630 **A**	1.30 ± 0.445 **A**	0.205 ± 0.070 **B**	0.760 ± 0.600 **B**	<0.05**
	(0.715–7.40)	(0.140–4.10)	(0.062–0.374)	(0.020–3.70)	
PCB-118	14.0 ± 3.50 **A**	9.70 ± 3.40 **A**	1.00 ± 0.170 **B**	3.80 ± 3.00 **B**	<0.05**
	(5.70–43.0)	(1.26–32.0)	(0.710–1.46)	(0.118–19.0)	
PCB 128	2.50 ± 0.750 **A**	1.60 ± 0.570 **A**	0.180 ±0.060 **B**	0.560 ± 0.450 **B**	<0.05**
	(0.740–8.76)	(0.201–5.25)	(0.071–0.350)	(0.015–2.80)	
PCB 138/163/164	24.0 ± 6.70 **A**	15.50 ± 5.60 **A**	1.30 ± 0.360 **B**	3.30 ± 2.60 **B**	<0.05*
	(7.80–80.0)	(2.080–50.0)	(0.690–2.20)	(0.150–16.0)	
PCB 146	6.00 ± 1.40 **A**	2.80 ± 1.10 **A,B**	0.40 ± 0.078**B,C**	0.600 ± 0.460 **C**	<0.05**
	(2.10–16.0)	(0.620–11.5)	(0.210–0.570)	(0.030–3.00)	
PCB 153	35.0 ± 8.90 **A**	25.0 ± 9.80 **A**	1.60 ± 0.580 **B**	3.80 ± 3.00 **B**	<0.05*
	(11.3–99.3)	(2.60–95.4)	(0.601–3.10)	(0.180–19.0)	
PCB-156	0.610 ± 0.137 **A**	0.40 ± 0.110 **A**	0.17 ± 0.035**A,B**	0.400 ± 0.320 **B**	<0.05**
	(0.170–1.60)	(0.090–1.07)	(0.075–0.240)	(0.012–1.96)	
PCB-174	0.680 ± 0.110 **A**	0.420 ± 0.096 **A**	0.090 ± 0.050 **B**	0.370 ± 0.300 **B**	<0.05**
	(0.140–1.30)	(0.100–0.860)	(0.025–0.230)	(0.014–1.80)	
PCB 180	16.0 ± 4.24 **A**	12.0 ± 4.40 **A**	1.66 ± 0.420 **B**	1.90 ± 1.50 **B**	<0.05*
	(3.90–44.0)	(1.00–44.0)	(0.600–2.60)	(0.130–9.10)	
PCB-183	2.20 ± 0.669 **A**	1.40 ± 0.536 **A**	0.215 ± 0.072 **B**	0.440 ± 0.350 **B**	<0.05*
	(0.516–7.45)	(0.170–5.26)	0.008–0.330	0.030–2.20	
PCB 187	3.40 ± 0.812 **A**	1.45 ± 0.43 **A,B**	0.620 ± 0.130 **B**	0.930 ± 0.680 **B**	<0.05*
	(0.965–9.50)	(0.470–4.55)	(0.230–0.840)	(0.080–4.32)	
PCB 201	1.20 ± 0.515 **A**	0.60 ± 0.20 **A,B**	0.140 ± 0.04**A,B**	0.370 ± 0.280 **B**	<0.05*
	(0.140–5.60)	(0.050–2.00)	(0.060–0.240)	(0.030–1.80)	
PCB 202	0.355 ± 0.180 **A**	0.160 ± 0.050 **A**	0.070 ±0.020 **A**	0.120 ± 0.090 **A**	>0.05*
	(0.022–1.90)	(0.008–0.470)	0.033–0.126	0.010–0.600	
∑PCBs	136 ± 32 **A**	91.0 ± 30.0 **A**	9.35 ± 1.90 **B**	28.0 ± 22.0 **B**	<0.05**
	(50.2–384)	(16.0–282)	(5.40–14.0)	(1.20–138)	

*Homocedastic: Welch's analysis of variances not used; **Heteroscedastic: Welch's analysis of variances used; ND = non-detectable concentration

Table 1. POP concentrations (μg/kg lipid) in Galapagos sea lion, thread herring and mullet sampled in 2008. Lipid contents are arithmetic mean ± standard deviations (SD). Concentrations are mean ± standard error (SE), and the range is indicated between brackets. Different letters (i.e. A, B, and C) indicate significant differences among sea lion pups and fish species (ANOVA and multi-comparisons Tukey-Kramer (HSD) post-hoc test)

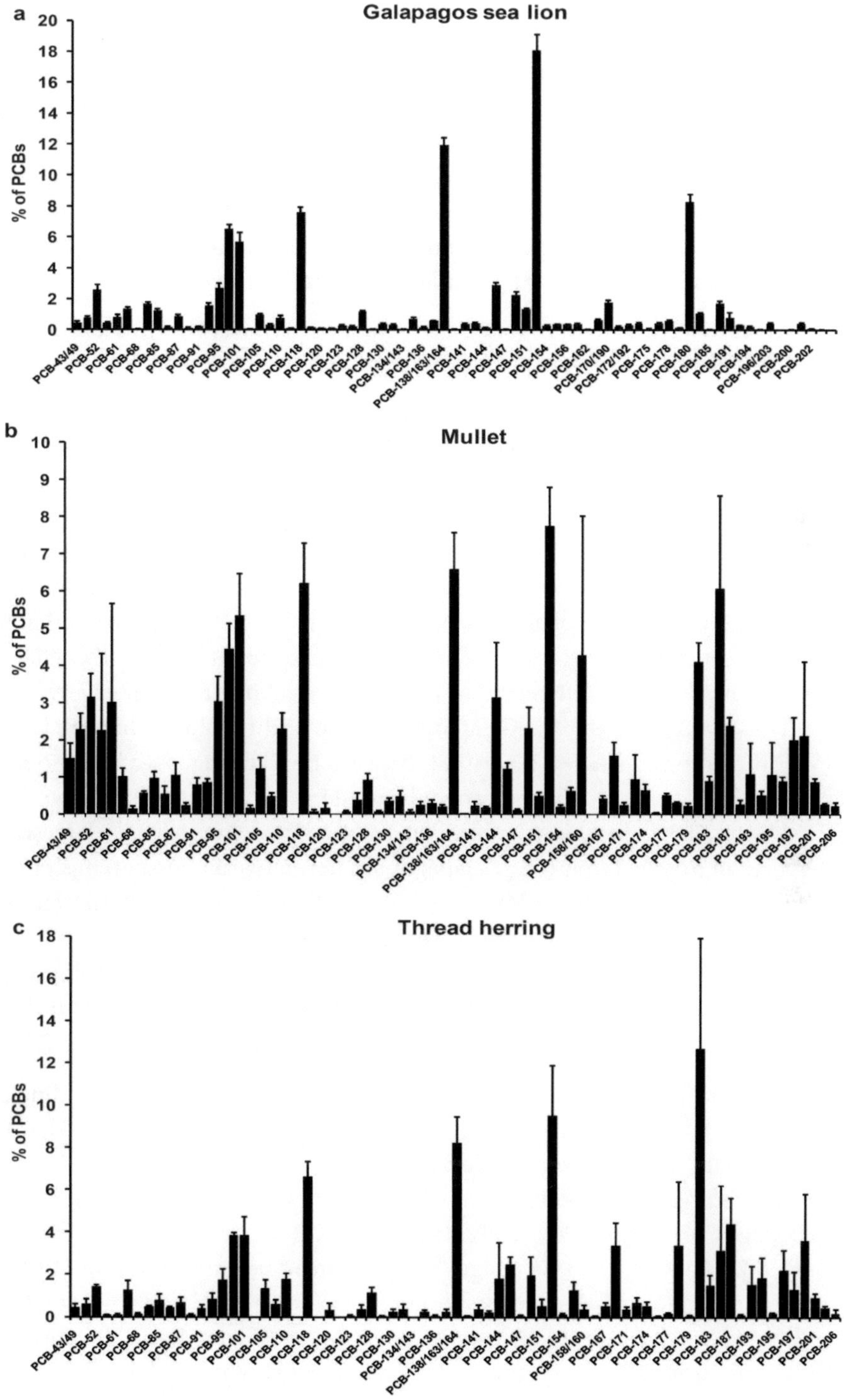

Figure 3. Composition of PCB congeners in Galapagos sea lion pups (a), mullet (b) and thread herring (c). Error bars are standard errors.

3.2.3. Intersite comparisons

The relative concentrations of contaminants observed in all sites exhibited a general common pattern, ∑DDT > ∑Chlordane > ∑PCBs >*β*-HCH> dieldrin > mirex, which was dominated by ∑DDTs, followed by chlordanes and PCBs, and secondly by *β*-HCH, dieldrin and mirex. Concentrations of ∑PCBs and OC pesticides detected in Galapagos sea lion pups showed no significant differences among rookeries (ANOVA for all comparisons, $p> 0.05$). This might suggest a common, global source of contamination delivering POPs to the animals, and that localized sources play a little role in contributions of POPs.

3.3. Biomagnification factors

The interpretation of the data resulting from the use of biomagnification factors are focused on BMF_{TL} as the BMF and BMF_{TL}* was used in this study as an optional approach for evaluation of BMF methods. When the BMF is calculated for the Galapagos sea lion/thread herring case, the BMF values were consistent among the methodologies used (Table 2). In contrast, the three methods differed markedly from 9 to $9.5x10^{18}$ orders of magnitude higher for OC pesticides and from 4.8 to 1.9 $x10^{7}$ orders of magnitude higher for PCBs when the predator-prey BMF_{TL} approaches versus the conventional $C_{PREDATOR}/C_{PREY}$ ratio in the Galapagos sea lion/mullet relationship are compared. These fluctuations appear to be driven by the effect of the magnitude resulting from the differences in trophic levels. While the trophic level difference ($TL_{predator} - TL_{prey} = 1.1$) between the Galapagos sea lion and the thread herring is large, the trophic level difference ($TL_{predator} - TL_{prey} = 0.11$) between the Galapagos sea lion and the mullet is statistically insignificant ($p >0.05$) and cannot be used in the calculation of the predator-prey BMF_{TL} .Thus, the predator-prey biomagnification factor methodologies (BMF_{TL}) are sensitive to small differences in trophic levels (i.e., Galapagos sea lion-mullet). Based on this observation, the best way of expressing the BMF is the calculation of the BMF calculated as the $C_{PREDATOR}/C_{PREY}$ ratio, which was similar between the Galapagos sea lion/herring and Galapagos sea lion/mullet cases.The use of different biomagnification factor measures showed that BMF_{TL} and BMF_{TL}* are more appropriate to assess biomagnification if differences in trophic levels of predator/prey relationships are large (i.e. >1), as depicted in Table 2.

Calculated biomagnification factors of OC pesticides and PCB congeners, including octanol-water (K_{OW}) and octanol-air partition coefficients (K_{OA}), are shown in Table 2. The BMF_{TL} of OC pesticides ranged from 7.3 (*trans*-chlordane) to 140 (*p,p*'-DDT) kg/kg lipid in Galapagos sea lion/thread herring and from 130 (*trans*-chlordane) to as high as 2000 (*p,p*'-DDE) kg/kg lipid in Galapagos sea lion/mullet, while BMF_{TL} for PCB congeners ranged from 2.7 (PCB 156) to 30 (PCB 74) kg/kg lipid in Galapagos sea lion/thread herring, and from 11 (PCB 52) to 72 (PCB 153) kg/kg lipid in Galapagos sea lion/mullet (Table 3). No significant correlations were found between the BMF_{TL} of OC pesticides and K_{OW} (Figure 4b,d). Yet, BMF_{TL} values decrease for some pesticides (e. g., mirex; trans-chlordane) when a K_{OW} of $10^{5.5}$ or $10^{6.0}$ is exceeded. As a function of the octanol-air partition coefficient (K_{OA}), the BMF_{TL} for OC pesticides increased markedly as the K_{OA} increased from $10^{7.5}$ to 10^{9}, and then dropped for the rest of pesticides as K_{OA} exceeds $10^{9.5}$ (Figure 4a,c).

Compound	Log K_{OW} 25-26 ºC	Log K_{OA} 37 ºC	BMF sea lion/ thread herring	BMF sea lion/ mullet	BMF$_{TL}$se a lion/ thread herring	BMF$_{TL}$ sea lion/ mullet	BMF$_{TL}$ * sea lion/ thread herring	BMF$_{TL}$ * sea lion/ mullet
p,p'-DDE	6.93	9.44	150	220	140	2000	100	2.10×10^{21}
p,p'-DDT	6.39	10.7	150	84.0	140	760	106	3.00×10^{17}
p,p'-DDD	6.30	10.3	41.0	33.0	38.0	300	31.0	6.60×10^{13}
∑DDT	6.41	10.7	132	180	122	1630	92.0	3.10×10^{20}
β-HCH	3.81	10.5	68.5	60.7	60.0	550	50.0	1.60×10^{16}
trans-chlordane	6.27	10.1	7.90	14.0	7.00	130	6.80	2.67×10^{10}
cis-chlordane	6.20	10.1	35.0	65.0	33.0	590	27.0	2.86×10^{16}
trans-nonachlor	6.35	10.0	80.0	177	74.0	1610	57.5	2.70×10^{20}
cis-nonachlor	6.08	8.38	44.0	68.0	40.0	615	33.0	4.30×10^{16}
∑Chlordanes			58.0	113	54.0	1030	43.0	4.70×10^{18}
Mirex	7.50	7.96	22.0	176	21.0	1600	18.0	2.50×10^{20}
Dieldrin	5.48	8.73	45.0	30.0	41.0	270	34.0	2.70×10^{13}
PCB-52	5.90	8.39	12.5	1.21	12.0	11.0	10.0	5.80×10^{0}
PCB 74	7.70	8.41	32.0	7.87	30.0	72.0	25.0	1.40×10^{8}
PCB 95	7.30	8.98	8.78	1.25	8.10	11.0	7.50	7.50×10^{0}
PCB-99	6.60	9.36	16.7	3.64	15.5	33.1	13.5	1.30×10^{5}
PCB-101	6.30	9.11	10.3	1.90	9.53	18.0	8.66	4.20×10^{2}
PCB-105	6.80	9.56	8.10	2.20	7.50	20.0	6.95	1.28×10^{3}
PCB-118	6.70	8.24	12.0	3.17	11.0	29.0	10.0	3.60×10^{4}
PCB 128	7.00	9.16	11.4	3.60	10.5	33.0	9.50	1.10×10^{5}
PCB -138/163/164	7.20	10.0	15.0	5.90	14.0	54.0	12.0	1.10×10^{7}
PCB-146	7.30	9.22	11.8	7.33	11.0	67.0	9.80	7.30×10^{7}
PCB 153	6.90	9.79	19.0	7.90	18.0	72.0	15.0	1.50×10^{8}
PCB 156	7.40	9.74	2.95	1.28	2.70	12.0	2.72	9.15×10^{0}
PCB 174	7.00	9.62	6.05	1.50	5.60	14.0	5.30	3.90×10^{1}
PCB 180	7.20	9.83	8.30	7.40	7.70	67.0	7.10	7.60×10^{7}
PCB 183	7.00	9.88	8.50	4.10	7.90	38.0	7.30	4.00×10^{5}
PCB 187	7.25	9.71	3.95	2.60	3.70	24.0	3.60	6.90×10^{3}
PCB 201	7.10	10.3	6.26	2.35	5.80	21.0	5.50	2.40×10^{3}
PCB 202	7.10	NR	3.80	2.10	3.55	19.0	3.50	9.40×10^{2}
∑PCBs			12.0	4.10	11.0	37.0	10.0	3.65×10^{5}

NR= non reported;
Values for log K_{OW} and log K_{OA} were obtained from Kelly *et al.* [2] and Mackay *et al.* [56].

Table 2. Biomagnification factors (BMF), Predator-prey Biomagnification factors (BMF$_{TL}$) and Log Predator-prey Biomagnification factors (BMF$_{TL}$*) in units of kg/kg lipid for organochlorine pesticides (OCP) and PCB congeners in the Galapagos sea lion. The logarithmic values of the octanol-water (K_{OW}) and octanol-air (K_{OA}) partition coefficients for each contaminant are also reported as supporting indicators of bioaccumulation.

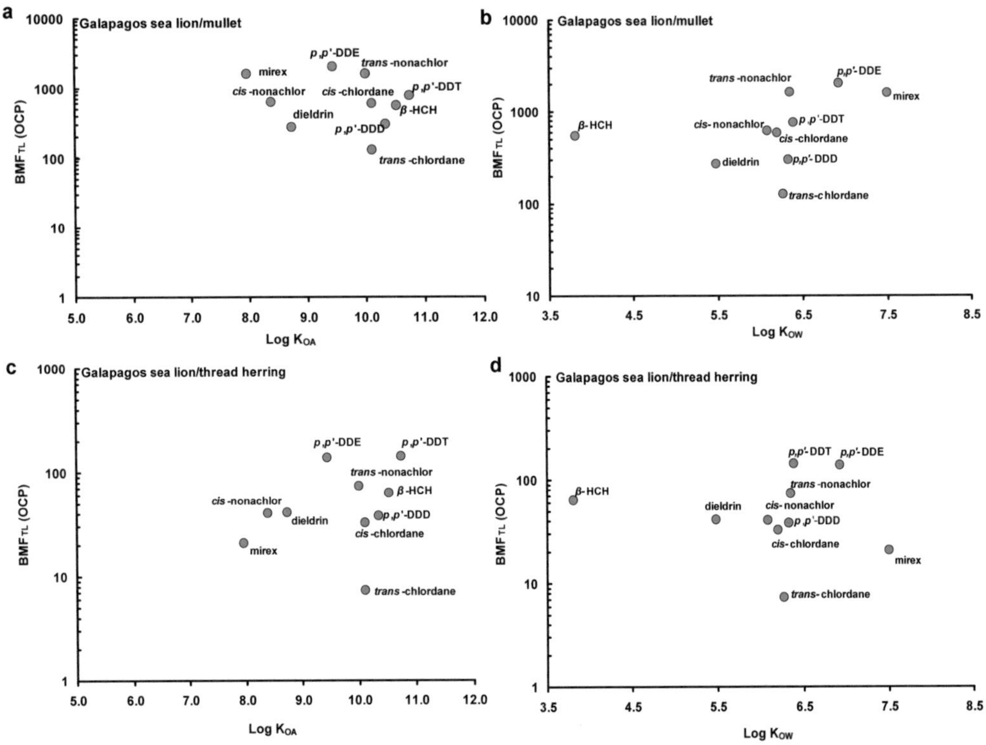

Figure 4. Predator-prey biomagnification factors (BMF_{TL}) in the Galapagos sea lion as expressed by the OC pesticide concentration ratios sea lion/ mullet (a, b) and sea lion/ thread herring (c, d) as a function of log K_{OA} (a, c) and log K_{OW} (b, d).The figure illustrates that while the Stockholm Convention for POPs uses a log K_{OW}> 5 as a criterion to identify bioaccumulative substances, substances including β-HCH with a log K_{OW}< 5 can biomagnify in marine mammals. Log K_{OA} appears to be a better predictor of substances that have the potential to biomagnify in marine mammals. Values for log K_{OW} and log K_{OA} were obtained from Kelly *et al.* [2] and Mackay *et al.* [56].

The BMF_{TL} of PCBs showed different trends when looking a different prey items in terms of K_{OA}. While no correlation was found between the BMF_{TL} of PCBs and log K_{OA} in the Galapagos sea lion/ mullet relationship (Figure 5a), BMF_{TL} for PCBs increased as the K_{OA} increased from $10^{7.6}$ to $10^{8.4}$ and then appeared to decrease gradually with increasing log K_{OA} in the Galapagos sea lion/thread herring relationship (Figures 5c). No correlation was found between the BMF_{TL} of PCBs and log K_{OW} for the Galapagos sea lion/thread herring or Galapagos sea lion/mullet feeding relationship (Figure 5b, d).

These observations demonstrate that these halogenated substances biomagnify and achieve concentrations in Galapagos sea lions that exceed those in their prey, although physiological processes and biotransformation may limit the biomagnification of some contaminants. When comparing the plots of BMF_{TL} of PCBs versus log K_{OW} or versus log K_{OA} similar patterns were observed for both Galapagos sea lion/thread herring and Galapagos sea lion/mullet feeding relationships (Figure 5a,d and Figure 5b,d, respectively). This is explained by the strong correlation usually observed between log K_{OA} and log K_{OW} of PCBs [53].

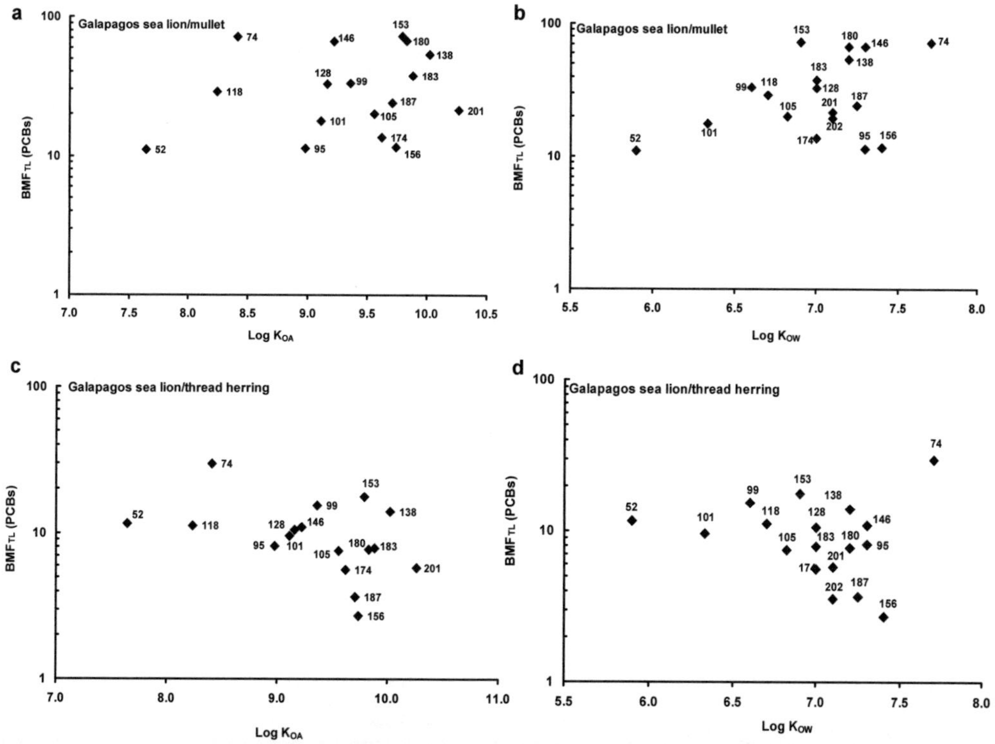

Figure 5. Predator-prey biomagnification factors (BMFTL) in the Galapagos sea lion as expressed by the PCB congeners' concentration ratios sea lion/mullet (a, b) and sea lion/thread herring (c, d) as a function of log KOA (a, c) and log KOW (b, d). For PCBs, log KOW appears to be an adequate predictor of the bioaccumulative potential of PCBs in marine mammals because all PCBs tested have a high log KOA> 6. Values for log KOW and log KOA were obtained from Kelly *et al.* [2] and Mackay *et al.* [56].

The BMFTL for organochlorine pesticides expressed by the concentration ratios sea lion/thread herring and sea lion/mullet of the Galapagos sea lion are higher than those reported for harp seals (*Pagophilus groenlandicus*) from the contaminated Barents Sea [15], (Table 3). However, the BMFTL for PCBs of the Galapagos sea lion are lower than those reported for harp seals. This indicates the biomagnification predominance of organochlorine pesticides in tropical-equatorial regions versus the predominant biomagnification of PCBs in Arctic regions. To further explore these comparisons, the ratio of the BMFTL for *p,p'*-DDE (the DDT dominant metabolite) to the BMFTL for PCB 153 (used here as the most recalcitrant PCB congener) was calculated for both species of pinnipeds and then compared. As shown in Table 3, the ratio *p,p'*-DDE BMFTL/PCB 153 BMFTL was much higher in the Galapagos compared to that of the Barents Sea, which is driven by the predominance of *p,p'*-DDE biomagnification in the Galapagos. Vapor pressures of organic contaminants are expected to be higher in tropical systems due to warmer/higher temperature in comparisons to cold/lower temperature in the Arctic; and, therefore, higher thermodynamic gradients and increase in concentrations are likely to occur during the trophic transfer of contaminant mass from prey to predator, resulting in a high biomagnification factor.

	Galápagos Islands (Ecuador) Galapagos sea lion	Barents Sea Harp seal[a]
	BMF_{TL}	BMF_{TL}
p,p'-DDE	139–2014	319
p,p'-DDT	142–760	NR
∑DDT	122–1631	NR
β-HCH	63.0–552	4.1
cis-chlordane	32.7–587	NR
trans-chlordane	7.34–128	NR
trans-nonachlor	73.7–1609	141.7
∑Chlordanes	54.1–1029	NR
PCB 52	11.0–11.6	NR
PCB 99	15.5–33.1	147.0
PCB 101	9.53–17.7	NR
PCB 105	7.51–20.0	18.1
PCB 118	11.2–28.8	41.6
PCB 138	13.9–53.9	327.7
PCB 153	17.7–72.2	416
PCB 180	7.72–66.9	NR
∑PCBs	11.2–37.2	NR
Ratio BMF_{TL} *p,p*′-DDE to BMF_{TL} PCB 153	7.85–27.9	0.77

NR= non reported

[a] Borga *et al.* [15].

Table 3. Comparison of BMF_{TL} for remote marine food chains between the Galapagos Islands and an Arctic reion for selected organochlorine pesticides and PCBs. The BMF_{TL} for Galapagos sea lions are expressed as the range of concentration ratios of both sea lion/thread herring and sea lion/mullet feeding relationships.

3.4. Biomagnification behaviour of POPs in the Galapagos food-chain

It is well recognized that the increase in organic chemical concentrations in lipids of organisms with increasing trophic level in food-webs originates from the magnification of the chemical concentration in the gastro-intestinal tract caused by food digestion and absorption [5,14]. In this study, the biomagnification capacity of organochlorine contaminants in the tropical food chain of the Galapagos sea lion is established (i.e. $C_{PREDATOR}>C_{PREY}$, BMF > 1).

However, a range of various factors directly or indirectly affect magnification process in predators, including animal ecologies and physiologies, feeding preferences, life history parameters (sex, age, body size and corporal condition), reproduction, geographic locations and stochastic-climatic events. Furthermore, the composition of contaminants can be shaped through toxicokinetics processes (i.e., uptake, metabolism, respiration and excretion), influencing the persistence and food-web biomagnification of POPs. Due to these factors, it is complex to elucidate whether a wild predator is at a steady state with its diet; therefore, calculated BMFs may not always reflect actual biomagnification [54]. As shown in this study, predator-prey BMFs revealed the biomagnification capacity of POPs in the food chain of the Galapagos sea lions, which is an apex predator possessing flexible feeding preferences (dietary plasticity).

Efficient uptake and dietary assimilation and slow depuration/excretion rates of these compounds (PCBs with K_{OW} ranging 10^5–10^7, and OC pesticides K_{OW} ranging $10^{3.8}$–$10^{7.0}$) explain the high degree of biomagnification in the Galapagos marine food chain. Dietary absorption efficiencies of Penta and Hexachlorobiphenyls are typically between 50-80% in fish and 90-100% in mammals [55] and chemical half-lives ($t_{1/2}$) for recalcitrant PCBs such as PCB 153 in organisms exceed 1000 days [56]. The analysis of BMF_{TL} estimates of PCBs and OC pesticides (Figures 4-5) indicates that OC pesticides and PCBs are accumulated by fish and sea lions and also biomagnify in the food chain. Based on contaminants' predator-prey BMFs, the DDT metabolites, *p,p'*-DDT and *p,p'*-DDE, followed by *trans*-nonachlor (Figure 4), are the most bioaccumulative pesticides, while PCB 74 and 153 are the most bioaccumulative PCB congeners in the Galapagos sea lion (Figures 5). The less bioaccumulative compounds are *trans*-chlordane and PCB 156.

Of particular importance is the biomagnification behaviour of *β*-HCH with a $K_{OW} < 10^4$ ($K_{OW} = 10^{3.8}$; Figure 4b,d), but with a K_{OA} of $10^{8.9}$–$10^{10.5}$ (Figure 4a,c), contrasting with the regulatory criteria and current management policies (i.e. Stockholm Convention; CEPA) for POPs that consider only chemicals with K_{OW} values $>10^5$ as bioaccumulative substances [7]. The predator-prey biomagnification factors (BMF_{TL} = 63–552) of *β*-HCH in Galapagos sea lions exceed equivalent biomagnification factors of PCB 153 (BMF_{TL} =18.0–72.2) and PCB 74 (BMF_{TL} =30.0–72.0), as shown in Table 2. This portrays that *β*-HCH, a relatively hydrophilic and nonmetabolizable chemical, biomagnifies in the tropical marine mammalian food chain of an air breathing organism (the Galapagos sea lions), which is explained by the relatively high K_{OA} of *β*-HCH ($K_{OA} > 10^{7.0}$) and its negligible respiratory elimination. Biomagnification of *β*-HCH was evident in the lichen-caribou-wolf terrestrial food chain, in the maritime and interior grizzly bears' food chains, and in a marine mammalian food web (including water-respiring and air-breathing organisms) from temperate regions of Canada and the Canadian Arctic [2,14,19].

3.5. Environmental transport of contaminants

Lack of significant differences and consistent uniformity of PCBs and OC pesticides, particularly for PCBs, among sites might indicate common sources of contamination. Concentrations of PCBs were also similar among rookeries in an earlier baseline study [34], although DDT concentrations were found to be significantly different [35]. Furthermore, principal components analysis represented a more comprehensive approach for exploring spatial differences and behaviour of POPs. The two first principal components (i.e., PC 1 and PC2) accounted for 55.2% of the total variation in Galapagos sea lion pups. PCA score plot results for the 2008 data further revealed that contaminants follow a similar trend, aggregated near to the centre of the axes, among sites, showing lack of discrimination and differentiation in contaminant patterns (Figure 6a). The first principal component (i.e., loading plots, PC1: 40.1% of the total variance) segregated in a significant degree the heavier PCB congeners (upper and lower left quadrants) from the lighter PCBs (upper and lower right quadrants; as seen in Figure 6b). A high positive PC1 score was correlated with higher percentages of low chlorinated PCBs (e.g., PCBs 43/49, 47/48/49, 52, 60, 61, 66, 74, 85, 86/97,87, 92, 95, 101, 110, 123, 132, 135, 136, 141, 144, 149) and *p,p′*-DDD, *p,p′*-DDT, dieldrin, *cis*-nonachlor, *trans*-chlordane, *cis*-chlordane and β-HCH, while a high negative score in PC 1 (upper and lower left quadrant) was correlated with a lower proportion of heavily and several, more persistent chlorinated PCBs (e. g. PCBs 118, 138/163/164, 137, 153, 158/160, 171, 177, 180, 183, 170/190, 172/192, 193, 194, 195, 196/203, 201, 202), as well as the semi-volatile and more bioaccumulative *p,p′*-DDE. These patterns show that PC1 appeared to be related to vapour pressure (Henry's Law constant or H) due to a high contribution of more volatile halogenated contaminants (pesticides) and less chlorinated (lighter) PCB congeners. A significant correlation was also observed between the log of the Henry's law constant (Log H) for the PCBs and PC1 (the variable loadings of the first principal component;$p < 0.05$, $r = 0.27$; Figure 7), suggesting that log H represented an important factor influencing the transport pathways and partitioning of PCB mixtures in remote environments; and, therefore, affecting the ultimate composition pattern observed in Galapagos sea lions. The Henry's law constant for each PCB is a fundamental parameter that represents the air-water equilibrium partitioning between surface waters and the atmosphere [57]. This indicates that local sources of exposure for high chlorinated PCBs are minimal in the Galapagos and that most of the contamination by POPs is coming from common atmospheric or continental sources.

Dieldrin is a metabolite of aldrin, which was used for agriculture and public health purposes at beginning of the 1950s until its production was cancelled in 1989 in North America, but as with other pesticides, it continues to enter the environment via erosion of soils contaminated in the past and atmospheric deposition [58]. Mirex is a very unreactive and hydrophobic insecticide that was used in North America to control fire ants and as a fire retardant, persisting in the environment because of chronic small inputs from the atmosphere [59]. The presence of this compound in these blubber samples might be related to the past use of mirex in continental Ecuador [60] because of the possible use as insecticide (bait) to control invasive ants in the Galapagos and continental Ecuador. β-HCH is a major constituent of technical HCHs, which is likely one of the sources of this residue. Another potential source of β-HCH can be lindane (i.e., γ-HCH) since this pesticide is currently being used in several countries in the southern hemisphere as evidenced by its detection in blubber samples of southern elephant seals and minke whales (*Balaenoptera acutorostrata*) from the Antarctic Ocean [49, 61]. At the continental coast of Ecuador, lindane has recently been detected in sediments and aquatic organisms from the Taura River in the Gulf of Guayaquil [62]. The atmospheric influx of HCHs source formulations used in the Asian and South American tropics (i.e., lindane) and North America (i.e. technical HCH) might explain the incidence of β-HCH in these samples. Uncertain records of use of legacy OC pesticides exist for the Galapagos, although anecdotic suggested the use of CUP for agriculture (Dr. Alan Tye, former Head Scientist, Department of Plant and Invertebrate Science, Charles Darwin Foundation, Galapagos Islands), and the widespread use of DDT to eliminate introduced rats in the Galapagos by the US Armed Forces during the 1940s and 1950s [35].

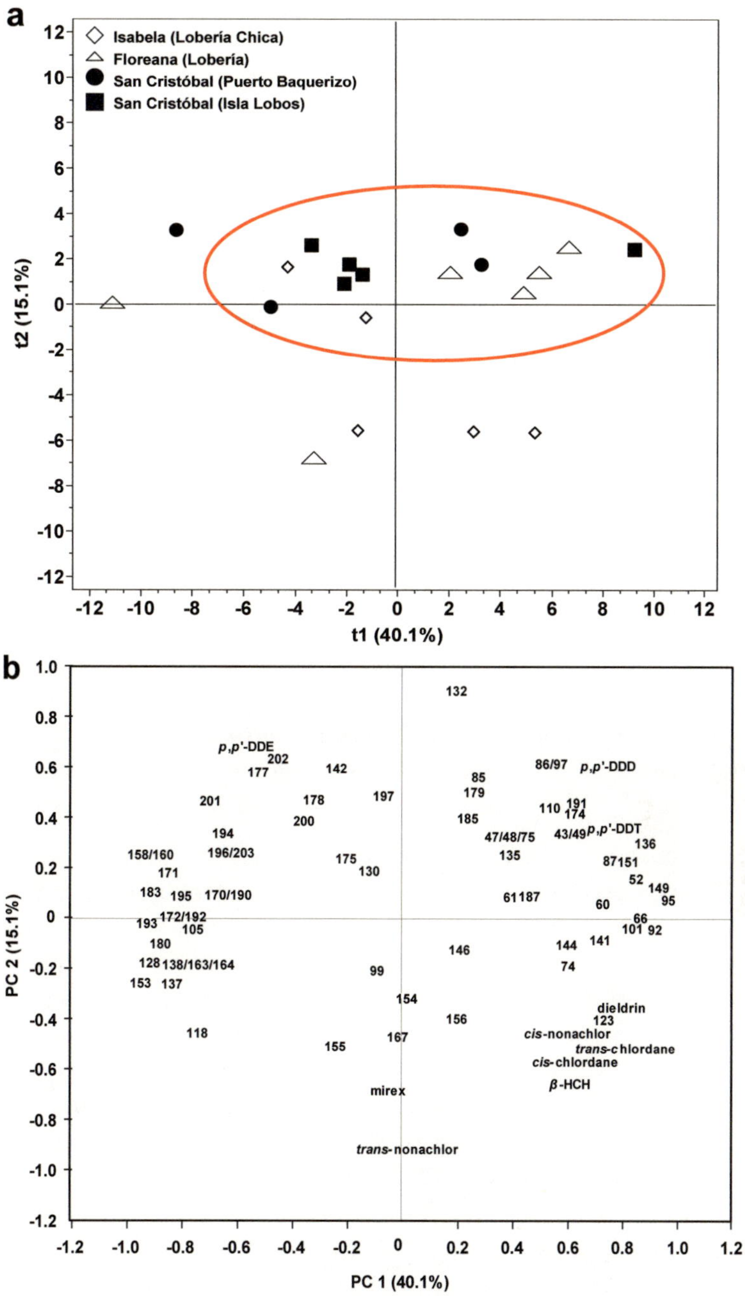

Figure 6. Principal components analysis where the variance accounted for by each principal component is shown in parentheses after the axis label: (a) score plots for patterns of POPs for the first two principal components shows that most of the pups from different rookeries have a similar contaminant pattern, as demonstrated here by the sample scores plot (t1 and t2) of 20 individuals; (b) loadings plots (PC1 and PC2) showing values of individual PCB congeners and pesticides in Galapagos sea lion pups, where numbers are PCB congeners based on the IUPAC system.

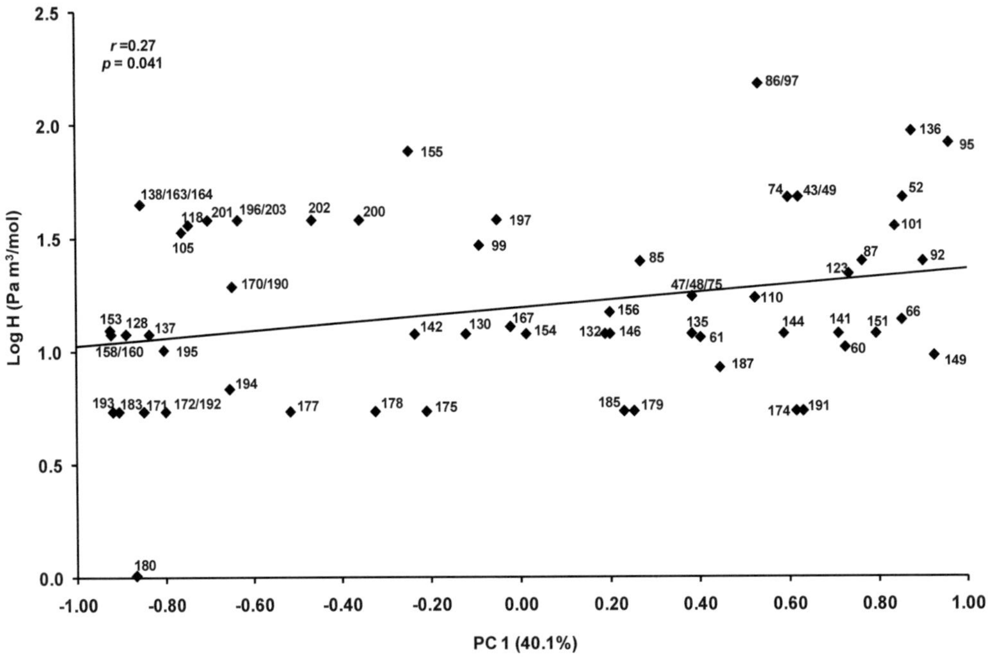

Figure 7. Relationship between the Henry's law constant (Log H) for polychlorinated biphenyl (PCB) congeners and the first principal component (PC1). PC1 is significantly correlated with Log H for PCB congeners, suggesting that Galapagos sea lions from the remote Galapagos Islands are more exposed to light PCB mixtures, consistent with atmospheric signals. Numbers are PCB congeners based on the IUPAC system.

The long range atmospheric transport coupled with global fractionation have usually been described as the major mechanism delivering POPs from lower or mid latitudes to the polar regions [11, 63, 64], but it is likely that a similar mechanism or redistribution from mid latitudes may be also expanding or delivering volatile or semi-volatile pesticides such as HCHs and DDTs to isolated islands around the equator (i.e., the Galapagos Archipelago). These observations suggest that the contamination by organochlorine pesticides might be coming from both local and continental sources because pesticides were used in the recent past in countries in the southern hemisphere [49, 65]. Trans-Pacific air pollution of contaminants from tropical Asia to the eastern Pacific [63, 66] cannot be ruled out as a global and common pathway of POPs of atmospheric origin.

3.6. Health risk assessment

The health risk of POP biomagnification in Galapagos sea lions is of serious concern in the long term, as we have previously reported that 1% of the male pups exceeded the *p,p'*-DDE toxic effect concentration associated with potent anti-androgenic effects [35]. DDT concentrations in Galapagos sea lion pups are near levels expected to be associated with impacts on the immune systems, and in minor degree on the endocrine systems in males. Adult male Galapagos sea lions can be expected to exhibit DDT concentrations that are higher than those in pups as DDTs accumulate throughout the animal's life because they are unable to offload contaminants during reproduction [35].While concentrations of DDTs pose protracted health risk because of lifetime exposure, the ∑PCB concentrations in Galapagos sea lion pups were lower than the new toxicity reference value of1,300 μg/kg lipid for risk of immunotoxicity and endocrine disruption in harbor seals [67]. Other POPs with a similar mode of toxicity such as polybrominated diphenyl ether (PBDEs) flame retardants, which were also detected recently in these animals [34], can further exacerbate the immune and endocrine response. A compromised immune and endocrine system impairs the ability of animals to combat disease and to successfully reproduce.

4. Conservation implications and future research

The Galapagos is one of the last evolutionary biology labs to preserve biodiversity. Yet, it has already been declared a UNESCO-Heritage site at risk because of invasive species, escalating human population growth and burgeoning tourism [68]. This study corroborated that POPs biomagnify to a significant degree in the tropical marine food chain of the Galapagos' marine ecosystem. This has important implications for management and control of organochlorine pesticides and conservation of marine ecosystems in tropical regions since pollution in the Galapagos has been categorized as an aesthetic issue rather than a chronic problem.

Recently, the World Health Organization (WHO) has reactivated the use of the malaria mosquito-fighting pesticide DDT in tropical countries because of increasing malaria cases [69]. While the concentrations of DDT and associated health risks in wildlife are generally believed to be declining, this may no longer be the case in tropical countries where DDT is increasingly used and can biomagnify in food chains. A renewed use of DDT to combat malaria is likely to increase DDT concentrations in the Southern Hemisphere and in particular put bird and marine mammal populations at greater risk because of the biomagnification of these substances in their food webs.

Since the ratification of the UN Stockholm Convention on POPs by Ecuador in 2004, the National Plan for the Inventory and Management of POPs was undertaken [70, 71]. DDT is included on Schedule 2 of the Stockholm Convention because of its damaging health effects in human and wildlife populations. Continuation of this initiative will help to control DDT contamination in the Galapagos. While DDT can save human lives, it can also adversely affect wildlife, local food production and opportunities for ecotourism. DDT use requires that tradeoffs need to be made between the conservation of valued, sensitive wildlife (e.g. Galapagos sea lions), fragile ecosystems and public health programs to control malaria.

Additional research and field sampling efforts may include other organisms integrating the trophic guilds of the Galapagos sea lion food web by measuring legacy and emerging POPs, stable isotopes and subsequent estimations of trophic levels. This will allow assessing in a higher degree the food web amplification of pollutants through the use of TMFs and food web bioaccumulation models in marine ecosystem of the remote Galapagos Islands.

Our findings provide sound scientific information on food chain contamination and potential ecological impacts in the Galapagos that can be used for conservation plans at the ecosystem level, and portrays the implications for environmental management and control of bioaccumulative, persistent and toxic contaminants (e. g. DDT). Finally, this study serves as a reference point against which possible future impact of DDT use in tropical marine ecosystems can be measured, underlying the use of more environmental friendly substances to control pests and vectors in developing countries.

Author details

Juan José Alava*
School of Resource & Environmental Management, Faculty of Environment,
Simon Fraser University, Burnaby, British Columbia, Canada
Fundación Ecuatoriana para el Estudio Estudio de Mamíferos Marinos (FEMM), Guayaquil, Ecuador
Charles Darwin Foundation, Puerto Ayora, Santa Cruz, Galápagos Islands, Ecuador

Frank A.P.C. Gobas
School of Resource & Environmental Management, Faculty of Environment,
Simon Fraser University, Burnaby, British Columbia, Canada

Acknowledgement

The author thanks P. Ross, J. Elliott, A. Harestad and L. Bendell for his valuable insights to improve this manuscript. Special thanks to the Charles Darwin Foundation for the Galapagos Islands, the Galapagos National Park and the Santa Barbara Marine Mammal Centre for their support and logistic field assistance.We are indebt with S. Salazar, D. Páez-Rosas, G. Jimenez-Uzcátegui, M. Cruz, P. Martinez, G. Merlen, J. Geraci, P. Howorth and the volunteers from the Marine Mammal Center in Santa Barbara (E. Stetson, C. Powell, D. Noble, N. Stebor, D. Storz and S. Crane) for their assistance in the field work and live capture of pups. Official permits for carrying out this research and exporting of samples were given by the Galapagos National Park.

5. References

[1] Ross PS, Ellis GM, Ikonomou MG, Barrett-Lennard LG, Addison RF. High PCB Concentrations in free-ranging Pacific killer whales, *Orcinus orca*: Effects of age, sex and dietary preference.Marine Pollution Bulletin2000; 40(6): 504-515.

[2] Kelly BC, Ikonomou MG, Blair JD, Morin AE, Gobas FAPC. Food Web Specific Biomagnification of Persistent Organic Pollutants.Science2007; 317(5835): 236-239.

* Corresponding Author

[3] Rattner BA., Scheuhammer AM.,Elliott JE. History of Wildlife Toxicology and the Interpretation of Contaminant Concentrations in Tissues.in:Beyer WN., Meador JP. (eds.) Environmental Contaminants in Biota: Interpreting Tissue Concentrations, 2nd ed. Boca Raton, FL, USA: CRC Press; 2011. p9-44.

[4] Loseto LL., Ross, PS. Organic Contaminants in Marine Mammals: Concepts in Exposure,Toxcity and Management. In: Beyer WN., Meador JP. (eds.) Environmental Contaminants in Biota: Interpreting Tissue Concentrations, 2nd ed. Boca Raton, FL, USA: CRC Press; 2011. p349-375.

[5] Gobas FAPC, Zhang X, Wells R. Gastrointestinal magnifications: The mechanism of biomagnification and foodchain accumulation of organic chemicals. Environmental Science and Technology 1993;27(13): 2855-2863.

[6] Gobas FAPC, Morrison HA. Bioconcentration and Bioaccumulation in the Aquatic Environment. in: Boethling R., Mackay D. (eds.) Handbook of Property Estimation methods for chemicals: Environmental and Health Sciences. Boca Raton, FL, USA: CRC Press LLC; 2000. p189-231.

[7] Gobas FAPC, de Wolf W, Verbruggen E, Plotzke K, Burkhard L. Revisiting Bioaccumulation Criteria for POPs and PBT Assessments. Integrated Environmental Assessment and Management2009; 5(4): 624-637.

[8] Government of Canada. Canadian Environmental Protection Act. Canada Gazette Part III. 22. Canada, Ottawa, Ontario. Canada: Public Works and Government Services; 1999. 249p.

[9] US Environmental Protection Agency [USEPA]. Toxic Substances Control Act (1976). Washington DC: USEPA; 1976.

[10] Council of the European Union. Regulation (EC) No 1907/2006 of the European Parliament and of the Council of 18 December 2006 concerning the Registration, Evaluation, Authorisation and Restriction of Chemicals (REACH), establishing a European Chemicals Agency; 2006.

[11] Wania F, Mackay D. Global fractioning and cold condensation of low volatility organochlorine compounds in polar regions. Ambio 1993; 22(1): 10-18.

[12] Guglielmo F, Lammel G, Maier-Reimer E. Global environmental cycling of gamma-HCH and DDT in the 1980s-a study using a coupled atmosphere and ocean general circulation model. Chemosphere 2009; 76(11): 1509-1517.

[13] Muir DCG, Savinova T, Savinov V, Alexeeva L, Potelov V, Svetochev V. Bioaccumulation of PCBs and chlorinated pesticides in seals, fishes and invertebrates from the White Sea, Russia.Science of the Total Environment 2003; 306(1-3): 111-131.

[14] Kelly BC, GobasFAPC. An Arctic Terrestrial Food-Chain Bioaccumulation Model for Persistent Organic Pollutants.Environmental Science and Technology 2003; 37(13): 2966-2974.

[15] Borgå K, Fisk AT, Hoekstra PF, Muir DCG. Biological and chemical factors of importance in the bioaccumulation and trophic transfer of persistent organochlorine contaminants in Arctic marine food webs.Environmental Toxicology and Chemistry 2004; 23 (10): 2367-2385.

[16] Borgå K, Kidd KA, Muir DCG, Berglund O, Conder JM, Gobas FAPC, Kucklick J, Malm O, Powell DE. Trophic Magnification Factors: Considerations of Ecology, Ecosystems, and Study Design. Integrated Environmental Assessment and Management2011; 8(1):64-84

[17] Kidd KA, Bootsma HA, Hesslein RH, Muir DCG, Hecky RE. Biomagnification of DDT through the Benthic and Pelagic Food Webs of Lake Malawi, East Africa: Importance of trophic level and carbon source. Environmental Science and Technology 2001; 35(1):14-20.

[18] Fisk AT, Hobson KA, Norstrom RJ. Influence of Chemical and Biological Factors on Trophic Transfer of Persistent Organic Pollutants in the Northwater Polynya food web.Environmental Science and Techonology 2001; 35(4): 732-738.

[19] Christensen JR, Macduffee M, Macdonald RW, Whiticar M, Ross PS. Persistent organic pollutants in British Columbia Grizzly Bears: Consequence of Divergent Diets.Environmental Science and Technology 2005; 39(18): 6952-6960.

[20] Cullon DL, Yunker MB, Alleyne C, Dangerfield N, O'NeilS, Whiticar MJ, Ross PS. Persistent organic pollutants (POPs) in Chinook salmon (*Oncorhyncus tshawytscha*): Implications for Northeastern Pacific Resident Killer Whales.Environmental Toxicology and Chemistry 2009; 28(1):148-161.

[21] Newsome SD, Clementz MT, Koch PL. Using Stable Isotope Biogeochemistry to Study Marine Mammal Ecology. Marine Mammal Science 2010; 26(3): 509-572.

[22] Peterson B, Fry B. Stable Isotopes in Ecosystem Studies. Annual Review of Ecology and Systematics 1987;18: 293-320.

[23] Hobson KA, Welch HE. Determination of Trophic Relationships within a High Arctic Marine Food Web using d13C and d15N Analysis. Marine Ecology Progress Series1992; 84: 9-18.

[24] Hanson S, Hobbie J, Elmgren R, Larsson U, Fry B, Johansson S. The stable nitrogen isotope ratio as a marker of food web interactions and fish migration.Ecology 1997; 78(7): 2249-2257.

[25] Hobson KA, Fisk A, Karnovsky N, Holst M, Gagnon JM, Fortier M.A Stable Isotope ($\delta^{13}C$ and $\delta^{15}N$) Model for the North Water Food Web: Implications for Evaluating Trophodynamics and the Flow of Energy and Contaminants. Deep-Sea Research Part II-Topical Studies in Oceanography2002; 49(22-23): 5131-5150.

[26] DeNiro MJ, Epstein S. Influence of Diet on the Distribution of Nitrogen Isotopes in Animals. Geochimica Cosmochimica Acta 1981; 45(3): 341-353.

[27] Burton RK, Koch PL. Isotopic Tracking of Foraging and Long Distance Migration in Northeastern Pacific Pinnipeds. Oecologia 1999; 119(4): 578-585.

[28] Dellinger T, Trillmich F. Fish Prey of the Sympatric Galápagos Fur Seals and Sea Lions: Seasonal Variation and Niche Separation. Canadian Journal of Zoology 1999; 77(8): 1204-1216.

[29] SalazarSK. Variación temporal y espacial del espectro trófico del lobo marino de Galápagos. MSc thesis. Centro Interdisciplinario de Ciencias Marinas, Instituto Politécnico Nacional, La Paz, Baja California Sur, México; 2005.

[30] Páez-Rosas D,Aurioles-Gamboa D, Alava JJ, Palacios DM. Stable Isotopes indicate Differing Foraging Strategies in two Sympatric Otariids of the Galapagos Islands. Journal of Experimental Marine Biology and Ecology 2012; 424-425:44-52.

[31] Aurioles-Gamboa D, Newsome SD, Salazar-Pico S, Koch PL. Stable isotope differences between sea lions (*Zalophus*) from the Gulf of California and Galapagos Islands.Journal of Mammalogy 2009; 90(6): 1410-1420.

[32] Salazar S, Bustamante RH. Effects of the 1997-98 El Niño on population size and diet of the Galápagos sea lions (*Zalophus wollebaeki*). Noticias de Galápagos (Galapagos Newsletter) 2003; 62: 40-45.

[33] Alava JJ., Salazar S.Status and Conservation of Otariids in Ecuador and the Galapagos Islands.In: TritesAW., Atkinson SK., DeMaster DP., Fritz LW, Gelatt TS., Rea L.D., Wynne KM. (eds.)Sea lions of the World. Fairbanks, USA: Alaska Sea Grant College Program, University of Alaska; 2006.p495-519.

[34] Alava JJ, Ikonomou MG, Ross PS, Costa D, Salazar S, Gobas FAPC. Polychlorinated Biphenyls (PCBs) and Polybrominated Diphenyl Ethers (PBDEs) in Galapagos sea lions (*Zalophus wollebaeki*). Environmental Toxicology and Chemistry 2009; 28(11): 2271-2282.

[35] Alava JJ,Ross PS, Ikonomou MG, Cruz M, Jimenez-Uzcategui G, Salazar S, Costa DP, Villegas-Amtmann S, Howorth P, Gobas, FAPC. DDT in endangered Galapagos Sea Lions (*Zalophus wollebaeki*). Marine Pollution Bulletin 2011; 62(4): 660-671.

[36] Grove JS, Lavenberg RJ, The fishes of the Galapagos Islands. Stanford, CA: Standford University Press; 1997.

[37] Vander Zanden MJ, Cabana G, Rasmussen JB. Comparing the trophic position of littoral fish estimated using stable nitrogen isotopes (δ^{15}N) and dietary data. Canadian Journal of Fishery and Aquatic Sciences 1997; 54(5): 1142-1158.

[38] Vander Zanden MJ, Rasmussen JB. Primary consumer δ^{15}Nand δ^{13}C and the trophic position of aquatic consumers. Ecology1999; 80(4): 1395-1404.

[39] Jardine TD, Kidd KA, Fisk AT. Applications, Considerations, and Sources of Uncertainty when using Stable Isotope Analysis in Ecotoxicology. Environmental Science and Technology 2006; 40(24): 7501-7511.

[40] Farrell JW, Pederson TF, Calvert SE, Nielsen B. Glacial-interglacial changes in nutrient utilization in equatorial Pacific Ocean.Nature1995; 377(6549): 514-517.

[41] Fogel ML, Tuross N, Owsley DW. Nitrogen isotope tracers of human lactation in modern and archeological populations. In: Carnegie Institution 1988-1989, Annual Report Geophysical Laboratory. Washington, D.C.: Geophysical Laboratory, Carnegie Institution; 1989.p111-117.

[42] Porras-Peters H, Aurioles-Gamboa D, Cruz-Escalona VH, Koch PL. Trophic level and overlap of sea lions (*Zalophus californianus*) in the Gulf of California, Mexico. Marine Mammal Science 2008;24(3): 554-576.

[43] Páez-Rosas D, Aurioles-Gamboa D. Alimentary niche partitioning in the Galápagos sea lion, *Zalophus wollebaeki*. Marine Biology2010;157(12): 2769-2781.

[44] Ikonomou MG, Fraser TL, Crewe NF, Fischer MB, Rogers IH, HeT, Sather PJ, Lamb RF. A Comprehensive Multiresidue Ultra-Trace Analytical Method, Based on HRGC/HRMS, for the Determination of PCDDs, PCDFs, PCBs, PBDEs, PCDEs, and

Organochlorine Pesticides in Six Different Environmental Matrices. Sidney, BC, Canada: Fisheries and Oceans Canada, Canadian Technical Report of Fisheries and Aquatic Sciences No. 2389; 2001.

[45] Conder JM, Gobas FAPC, Borgå K, Muir DCG, Powell DE. Use of Trophic Magnification Factors and Related Measures to Characterize Bioaccumulation Potential of Chemicals.Integrated Environmental Assessment and Management2011; 8(1):85-97.

[46] ZarJH.Biostatistical analysis. 4th ed. Upper Saddle River, New Jersey,USA: Prentice Hall; 1999.

[47] Ross PS, Jeffries SJ, Yunker MB, Addison RF, Ikonomou MG, Calambokidi JC. Harbor seals (*Phoca vitulina*) in British Columbia, Canada, and Washington State, USA, Reveal a Combination of Local and Global Polychlorinated Biphenyl, Dioxin, and Furan Signals. Environmental Toxicology and Chemistry 2004; 23(1):157-165.

[48] Darimont CT, Reimchen TE. Intra-hair Stable Isotope Analysis Implies Seasonal Shift to Salmon in Gray Wolf Diet. Canadian Journal of Zoology 2002;80(9): 1638-1642.

[49] Miranda–Filho KC, Metcalfe TL, Metcalfe CD, Robaldo RB, Muelbert MMC, Colares EP, Martinez PE, Bianchini A. Residues of Persistent Organochlorine Contaminants in Southern Elephant Seals (*Mirounga leonina*) from Elephant Island, Antarctica. Environmental Science and Technology 2007;41(11): 3829-3835.

[50] Kawano M, Inoue T, Wada T, Hidaka H, Tatsukawa R. Bioconcentration and Residue Patterns of Chlordane Compounds in Marine Animals: Invertebrates, Fish, Mammals, and Seabirds.Environmental Science and Technology1988; 22(7): 792-797.

[51] Del ToroL, Heckel G, Camacho-Ibar VF, Schramm Y. California sea lions (*Zalophus californianus californianus*) have Lower Chlorinated Hydrocarbon Contents in Northern Baja California, Mexico, than in California, USA.Environmental Pollution2006; 142(1): 83-92.

[52] Prudente M, Tanabe S, Watanabe M, Subramnian A, Miyazki N, Suarez P, Tatsukawa R. Organochlorine Contamination in some Odontoceti Species from the North Pacific and Indian Ocean.Marine Environmental Research, vol. 44, no. 4, pp. 415–427, 1997.

[53] Gobas FAPC, Kelly BC, Arnot JA. Quantitative Structure Activity Relationships for Predicting the Bioaccumulation of POPs in Terrestrial Food-webs.QSAR & Combinatorial Science 2003; 22(3): 346-351.

[54] Christensen JR, Letcher RJ, Ross PS. Persistent or not Persistent? Polychlorinated Biphenyls are readily Depurated by Grizzly Bears (*Ursus arctos horribilis*).Environmental Toxicology and Chemistry2009; 28(10): 2206-2215.

[55] Kelly BC, Gobas FAPC, McLachlan MS. Intestinal Absorption and Biomagnification of Organic Contaminants in Fish, Wildlife and Humans. Environmental Toxicology and Chemistry 2004; 23(23): 2324–2336.

[56] Mackay D, Shui WY, Ma KC. Illustrated Handbook of Physical–Chemical Properties and Environmental Fate of Organic Chemicals. Boca Raton, FL: Lewis Publishers; 1992.

[57] Fang F, Chu S, Hong C.Air-water Henry's Law constants for PCB congeners: Experimental Determination and Modeling of Structure-property Relationship. Analytical Chemistry 2006; 78(15): 5412-5418.

[58] ATSDR. Toxicological Profile for Aldrin/Dieldrin (Update). Department of Health and Human Services, Public Health Services.Atlanta, GA: Agency for Toxic Substances and Disease Registry, Division of Toxicology/Toxicology Information Branch;2002.
[59] Sergeant DB, Munawar M, Hodson PV, Bennies DT, Huestis SY. Mirex in the North America Great Lakes: New Detections and their Confirmation.Journal of the Great Lakes Research 1993;19(1): 145-157.
[60] Solórzano L. Status of Coastal Water Quality in Ecuador.in:Olsen S., Arriaga L. (eds.)A Sustainable Shrimp Mariculture Industry for Ecuador. Technical Report Series TR-E-6. Narragansett, RI, USA: Coastal Resources Center, University of Rhode Island; 1989.p163-177
[61] Aono S, Tanabe S, Fujise Y, Kato H, Tatsukawa R. Persistent Organochlorines in Minke Whale (*Balaenoptera acutorostrata*) and their Prey Species from the Antarctic and the North Pacific, "Environmental Pollution1997;98(1): 81-89.
[62] MontañoM, ResabalaC. Pesticidas en Sedimentos, Aguas, y Organismos de la Cuenca del Rio Taura.Revista de Ciencias Naturales y Ambientales 2005;1: 93-98.
[63] Iwata H, Tanabe S, Sakai N, Tatsukawa R. Distribution of Persistent Organochlorines in the Oceanic Air and Surface Seawater and the Role of Ocean on their Global Transport and Fate.Environmental Science and Technology1993; 27(6):1080–1098.
[64] IwataH, Tanabe S, SakaiN, NishimuraA, Tatsukawa R. Geographical Distribution of Persistent Organochlorines in Air, Water and Sediments from Asia and Oceania, and their Implications for Global Redistribution from lower Latitudes.Environmental Pollution1994; 85(1): 15–33.
[65] Blus LJ. Organochlorine pesticides.in:Hoffman DJ., Rattner BA., Burton GA., Cairns J. (eds.) Handbook of Ecotoxicology. Boca Raton, FL, USA: CRC Press; 2003.p313-339
[66] Wilkening KE, Barrie LA, Engle M. Trans-Pacific Air Pollution. Science 2000; 290(5489): 65-67.
[67] Mos L, Cameron M, Jeffries SJ, Koop B, Ross PS. 2010. Risk-based analysis of PCB toxicity in harbour seals. Integrated Environmental Assessment and Management 2010; 6(4): 631–640.
[68] Watkins G, Cruz F.Galapagos at Risk: A Socioeconomic Analysis of the Situation in the Archipelago. Puerto Ayora, Province of Galápagos, Ecuador: Charles Darwin Foundation; 2007.
[69] World Health Organization. WHO: WHO gives Indoor Use of DDT, a Clean Bill of Health for Controlling Malaria: WHO Promotes Indoor Residual Spraying with Insecticides as One of Three Main Interventions to fight Malaria. Washington, DC:World Health Organization; 2006.
http://www.who.int/mediacentre/news/releases/2006/pr50/en/index.html. (accessed 19 October 2008).
[70] Ministerio del Ambiente. Inventario de Plaguicidas COPs en elEcuador. Informe Técnico Final. Proyecto GEF/2732-02-4456. Global Environmental Facility (GEF). Programa Nacional Integrado para la GestiónRacional de las Sustancias Químicas, Ministerio del ambiente del Ecuador: Escuela Superior Politécnica del Ecuador (ESPOL)-Instituto de Ciencias Químicas (ICQ); 2004.

[71] Ministerio del Ambiente. Plan Nacional de Implementación para la Gestión de los Contaminantes Orgánicos Persistentes en el Ecuador," GEF/2732-02-4456. Global Environmental Facility (GEF), Convenio de Estocolmo sobreContaminantesOrgánicos Persistentes (COPs), PNUMA, Ministerio delAmbiente del Ecuador; 2006.

Chapter 5

Skin Biopsy Applications in Free Ranging Marine Mammals: A Case Study of Whale Skin Biopsies as a Valuable and Essential Tool for Studying Marine Mammal Toxicology and Conservation

Catherine F. Wise, John Pierce Wise, Jr.,
Sandra S. Wise and John Pierce Wise, Sr

Additional information is available at the end of the chapter

http://dx.doi.org/10.5772/54212

1. Introduction

The need to study, evaluate and understand the impacts of marine pollution on marine life is real and urgent. We depend on the ocean for food, transportation, economic gain, leisure and to enhance the quality of our lives. The residence times of pollutants in ocean water are short. The pollutants either settle to the bottom and attach to sediments or enter the food chain and accumulate in marine organisms. These outcomes mean that the only effective ways to assess marine pollution are to study the concentrations of pollutants in sediments or to study them in marine organisms. The average depth of the world ocean is 3,790 meters making it technically impractical to assess pollutants in sediments worldwide because of the great ocean depths under extreme pressures and vast amount of area. Thus, the best approach to assessing ocean pollution is to study it in marine organisms and, because of their relationship to humans (both biologically, culturally and inspirationally) and their ability to integrate air, water and prey, the best marine organisms to focus on are marine mammals.

There are many marine mammal species. Many of them are listed as endangered or threatened. A species that is considered to be one of the most endangered in the world is the North Atlantic right whale. Their population numbers only about 400 individuals [1]. This species suffers detrimental losses to their population through boat strikes and entanglements in fishing gear [1]. Regulations are being implemented to prevent extinction; however, boat strikes and entanglements may not be the only reason the population numbers remain so low. Other factors, perhaps pollutants, might be affecting the overall survival and reproductive ability of these animals [2].

Ocean pollution is a threat to marine mammals. Some pollutants cause obvious and direct harm to the animals such as plastics and other debris. Others can be less evident, such as, agricultural runoff and industrial wastes. Industrial wastes can include air pollution that the animals breathe in and pollutants that can be found at different levels of the water column. These types of pollutants can have long term and persistent exposures.

Whales are exposed to all environmental pollutants that reach the ocean. They spend time at the surface and travel throughout the water column so they are exposed to pollutants that remain at the surface and those that disperse through the water column. They experience dermal exposure through pollutants in the water. They feed at different depths so they get pollutants through ingesting animals that may have accumulated them. They all breathe air and, thus, are exposed to air pollutants. Consequently, they make excellent models to use for studying the threats and consequences of ocean pollution. The challenge, of course, is to develop an approach for studying the toxicology of marine pollution in marine mammals that provides species specific data along with an individual and population context.

Marine mammal research is difficult and expensive. There are laws that have been implemented to specifically protect marine mammals. Some species not only fall under this protection but also are protected by the Endangered Species Act. Thus, research with these animals is strictly enforced and regulated through various permitting agencies (eg, National Marine Fisheries Service (NMFS) and United States Fish and Wildlife Service, USFWS). These permits limit the amount, time and ways they can be studied. In addition to the permits, properly trained personnel and proper equipment are required. Whales in particular are difficult to study because they are typically found far from shore requiring extensive travel and vessel time. There is specialized equipment to aid in finding the whales while they are underwater but most of the search requires visual sightings of animals for the short period of time they are at the surface, requiring trained crew to be on watch during all daytime hours in a variety of environmental conditions. Because of these factors, researching whales is a particularly challenging task and one that requires careful use of resources and the ability to extract as much information as possible.

A unique and effective way to study marine mammals is through skin biopsies. The skin can be used to determine the genetics of an individual which can allow for gender identification as well as determining intraspecific relationships, genetic diversity of different subpopulations and tracking individual animals over time. Biopsies can provide important information about environmental pollutants and their effects. For example, they can be used to measure the levels of metals because many metals are known to accumulate in the skin. Through a skin biopsy, blubber is also collected which can be measured for a variety of organic pollutants. In addition, the interface between the skin and blubber can be used to create living cell lines. These cell lines can further be used to determine the toxicity of a pollutant by measuring levels of toxicity and DNA damage.

In this chapter we use chromium as an example for how skin biopsies of free ranging whales can be used to evaluate the environmental impact of a particular ocean contaminant. We chose to use chromium because it is a known carcinogen and a known reproductive and developmental toxicant that is understudied in the marine environment [3]. Our approach is straightforward: 1) Determine if Cr exposure has occurred, 2) Determine if Cr is cytotoxic and genotoxic to cultured whale cells and 3) Compare data in cultured whale cells to levels in whale tissue to gain a toxic context. We exemplify this using biopsy samples from both sperm whales and North Atlantic right whales.

2. Materials and methods

2.1. Determining if Cr exposure has occurred in whales

2.1.1. Biopsy collection

Whale skin biopsies were collected using a specialized biopsy dart and crossbow according to standard methods [4]. The biopsy dart has a stainless steel tip that collects a skin sample that is about 25 mm long or less and 7 mm in diameter. A buoyant stopper located behind the tip prevents the biopsy dart from penetrating beyond the depth of the tip. The stopper causes the dart to bounce off and float for an easy retrieval with a net. Tips are stored in 70% ethanol until use. Upon retrieval, the tissue sample is removed from the tip using Teflon forceps and placed into a glass Petri dish.

2.1.2. Biopsy processing

The skin and blubber were separated leaving the interface to be used for skin fibroblast cell growth. The blubber and skin were used for analysis of genetics, levels of metals and organics. The interface, once isolated, was immersed in a tissue buffer (PBS with 20% penn/strep and 2% gentamicin) for 30 minutes to get rid of any bacteria that may have been present on the skin. Tissue was then placed in a Petri dish and cut into approximately 1 mm pieces. These pieces were transferred into two T-25 flask with 1 ml of medium (DMEM-F12, cosmic calf serum, L-glutamine, penicillin, streptomycin, sodium pyruvate) and placed upside down in a 33°C humidified incubator with 5% CO_2. After 24 h, 5 ml of medium was added and the flask was gently turned right side up and monitored for cell growth. Living cells typically plated out on the flask directly from the tissue explants within one week.

2.1.3. Measuring Levels in the Skin from the Biopsy

The whale skin biopsies were analyzed using inductively coupled plasma mass spectroscopy to determine the total chromium in the tissue according to published methods using a Perkin-Elmer/Sciex ELAN ICPMS. Samples were rinsed with deionized water and allowed to air dry in a laminar flow hood to minimize contamination. Approximately 0.1 g of tissue was placed in a digestion vessel, 2 ml of Optima grade nitric acid was added, the vessel placed in a hot block, and refluxed at 95°C for 4 h. The sample was cooled, 2 ml Optima grade hydrogen and deionized water (3:2 v/v) was added, heated until the effervescence subsided, cooled, and brought up to a final volume of 20 ml. Standard quality assurance procedures were employed (Table 1) and include the analysis of standard reference materials, a duplicate sample and a pre-digestion spike. Instrument response was evaluated initially and after 10 samples, using commercially available calibration verification standards and a blank. All calibration verifications were within the acceptance criterion of 85-115% recovery and the preparation blank values were below 3x the limit of detection. Standard reference materials were used to assess method performance, where applicable. Interference check solutions were analyzed with all sample runs to check for matrix effects which might be interfering with sample analysis. The mean limit of detection (LOD) was the lowest analyte concentration likely to be reliably distinguished from the blank and at which detection is feasible. The LOD was previously determined by utilizing both the measured blank and test replicates of a matrix matched sample known to contain a low concentration of analyte. All samples were diluted 2x for analysis by ICP-MS. All data are presented as ug/g wet weight.

2.2. Determining if Cr is cytotoxic and genotoxic to cultured whale cells

2.2.1. *Chemical treatments*

There are two major biologically relevant valence states for chromium, hexavalent and trivalent, with the hexavalent forms considered more potent than the soluble forms. This study focused on the hexavalent form because the marine environment favors the hexavalent form [5]. Moreover, the total Cr levels in the whales were found to be high and considering that Cr(III) is poorly absorbed by mammals [6], for the whales to accumulate these levels of total Cr, the original exposure to Cr would almost certainly have been to Cr(VI).

We treated whale cells with both a water soluble (sodium chromate) and a water-insoluble particulate (lead chromate) to determine its ability to induce DNA damage. The concentration units for the two chemicals are different (uM for sodium chromate and ug/cm^2 for lead chromate) because sodium chromate dissolves fully in water and thus is a solution, while lead chromate does not and instead forms a slurry of particles in water. The cells were treated with this slurry of intact particles. Since all the lead chromate is not dissolved to express it in units of molarity would overestimate the dose. Thus, its units are weight per surface area.

2.2.2. *Cytotoxicity assay*

Cells were seeded into a 6 well plate and allowed to grow for 48 h. The cells were treated for 24 h with 1-25 uM of sodium chromate or 0.05-10 ug/cm^2 lead chromate. After 24 h we harvested the cells using a standard protocol [7]. Briefly, cells were harvested, counted and re-plated into 100 mm dishes at colony forming density, allowed to grow approximately 2 weeks, then stained and counted. Treated dishes were compared to the untreated control.

2.2.3. *Chromosome damage assay*

Cells were seeded into 100 mm dishes and allowed to grow for 48 h and treated as they were in the cytotoxicity assay described above. Cells were harvested and analyzed using published protocols [7]. Briefly, cells were harvested, treated with potassium chloride to swell the cells, fixed and dropped onto slides. Slides were stained with Giemsa, coverslipped and 100 metaphases were analyzed per treatment concentration.

3. Results

3.1. Determining if exposure has occurred

Once a skin biopsy was collected from a free ranging whale the skin was used to measure the levels of chromium present. The different valence states of chromium cannot accurately be determined, so we measured the total level of chromium in the tissue (Figure 1). For North Atlantic right whales in the Bay of Fundy, 7 biopsies were collected. The total chromium levels in right whale skin (wet weight) ranged from 4.9 to 10 ug Cr/g tissue with an average of 7.0 ug Cr/g tissue. (Figure 1A). For sperm whales, 331 biopsies were collected from 17 different regions around the globe. In sperm whales, the total chromium ranged from 0.9 to 122.6 ug Cr/g tissue with an average of 9.3 ug/g w.w (Figure 1B). The highest mean levels by region were found in the Bahamas with an average of 81.9 ug/g w.w for 2 animals. The average was slightly higher in sperm whale tissues than North Atlantic right whale tissues (Figure 1C).

3.2. Determining if Cr is cytotoxic and genotoxic to cultured whale cells

3.2.1. Cytotoxicity

Both soluble and particulate forms of chromium are cytotoxic to sperm whale and Northern right whale skin cells. The data show that 1, 2.5, 5, 10 and 25 uM sodium chromate (the soluble form of chromate) induced 78, 60, 49, 11, 1 percent survival, respectively, in right whale cells; and 80, 51, 13, 2 and 0 percent survival, respectively, in sperm whale cells (Figure 2A). Doses of 0.1, 0.5, 1, 5 and 10 ug/cm^2 lead chromate induced 126, 64, 49, 11 and 1 percent cell survival in right whale cells and doses of 0.05, 0.1, 0.5, 1, 5 and 10 ug/cm^2 lead chromate induced 100, 91, 68, 63, 36 and 7 percent cell survival, respectively (Figure 2B).

3.2.2. Genotoxicity

Both soluble and particulate forms of chromium induced chromosome damage in sperm whale and North Atlantic right whale skin cells. We measured genotoxicity by induced chromosome damage in two ways: Percent of metaphases with damage and total damaged chromosomes in 100 metaphases. In sperm whale cells 1, 2.5, 5, 10 and 25 uM sodium chromate damaged 7, 17, 20 and 31 percent of metaphases, respectively, and induced 7, 16, 24 and 40 total aberrations, respectively. At the highest dose cell cycle arrest occurred and no metaphases were seen. In right whale cells 1, 2.5, 5 and 10 uM sodium chromate damaged 7, 15, 23 and 33 percent of metaphases, respectively, and induced 8, 18, 30 and 48 total aberrations, respectively (Figure 3A and 3B). For lead chromate, 0.5, 1, 3 and 5 ug/cm^2 lead chromate damaged 6, 12, 15 and 27 percent of metaphases and induced 7, 13, 24 and 28 total aberrations, respectively in sperm whale cells. 0.5, 1, 2, 4 and 5 ug/cm^2 lead chromate damaged 16, 19, 23, 32 and 26 percent of metaphases and induced 17, 22, 28, 40 and 30 total aberrations, respectively (Figure 3C and 3D).

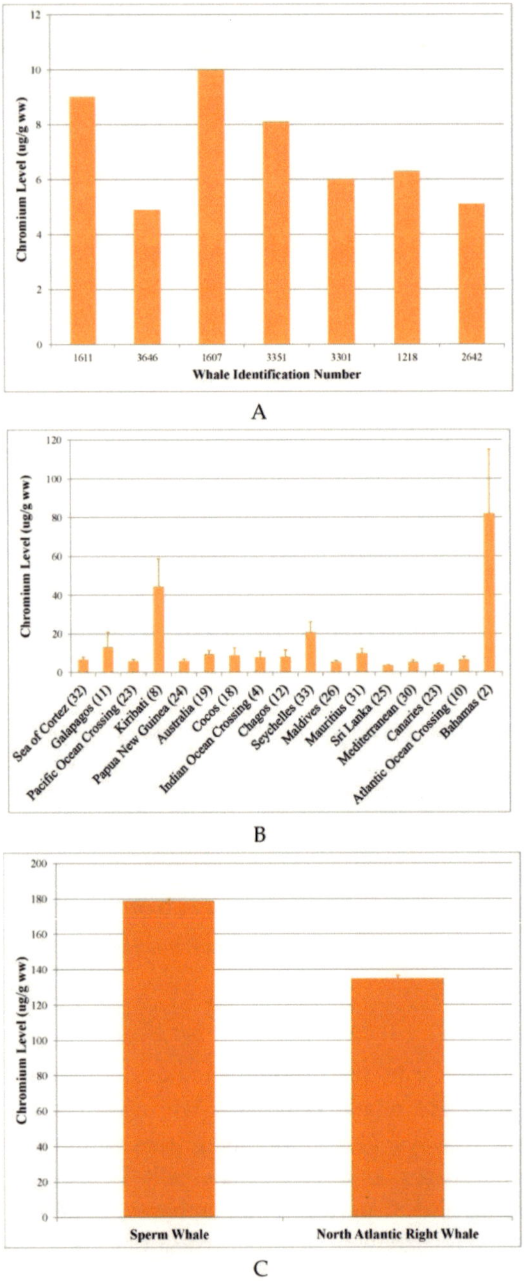

Figure 1. Chromium Levels in Whale Skin. This figure shows levels of Cr found in whale skin biopsies. A) This panel shows individual levels of Cr detected in each of seven North Atlantic right whales; B) This panel shows the mean levels of Cr in sperm whales grouped by region +/- the standard error. Number in parentheses on the x-axis indicate the number of whales sampled in that region; C) This panel shows the average level of Cr found in North Atlantic right whale and sperm whale skin. The North Atlantic right whale average is based on 7 animals biopsied in the Bay of Fundy. The sperm whale average is based on 331 animals biopsied worldwide.

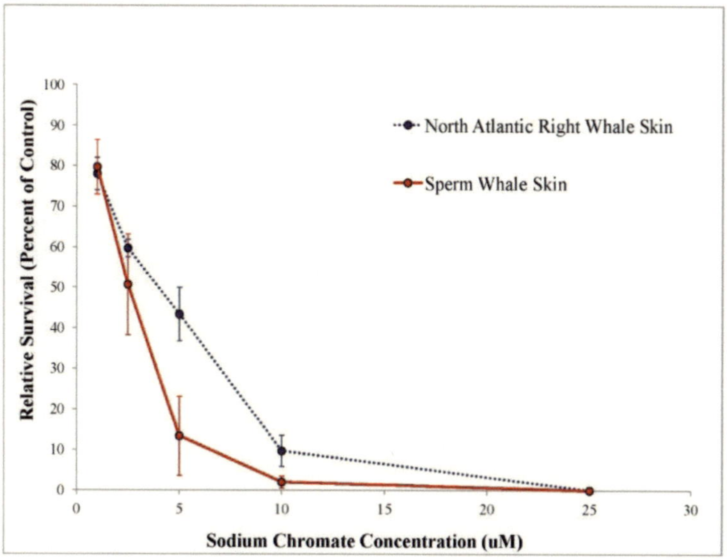

A

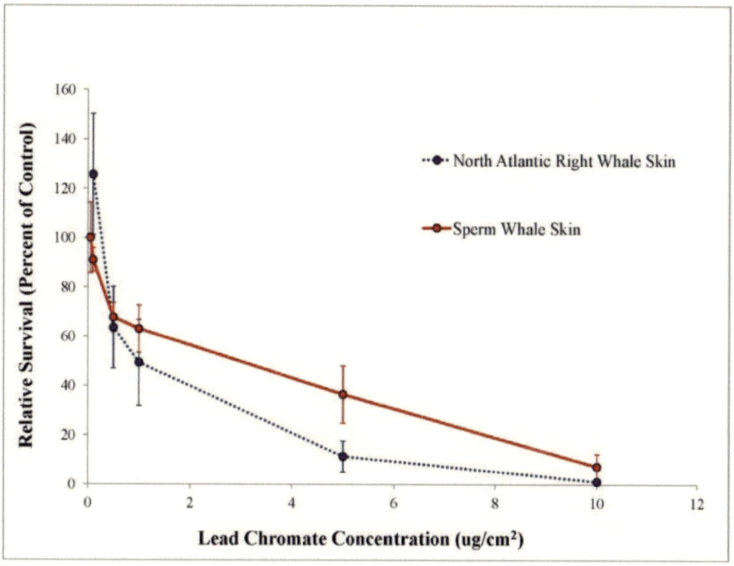

B

Figure 2. Cytotoxicity of Chromium in North Atlantic Right Whale and Sperm Whale Skin Cells. This figure shows the relative survival of cells treated for 24 h with chromium. There is a concentration-dependent decrease in the number of surviving cells. Error bars represent the standard error of the mean from three independent experiments. A) Cells treated with sodium chromate; B) Cells treated with lead chromate.

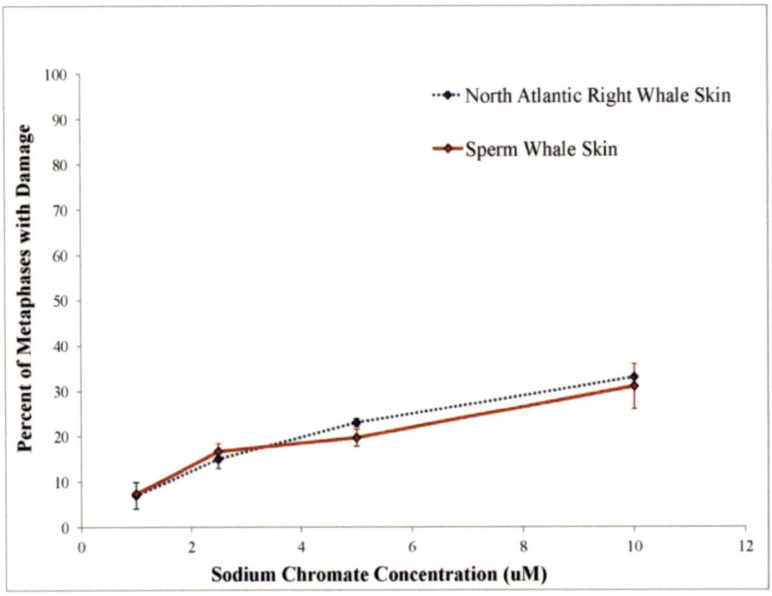
North Atlantic Right Whale Skin
Sperm Whale Skin
Percent of Metaphases with Damage
Sodium Chromate Concentration (uM)
0
10
20
30
40
50
60
70
80
90
100
0
2
4
6
8
10
12

A

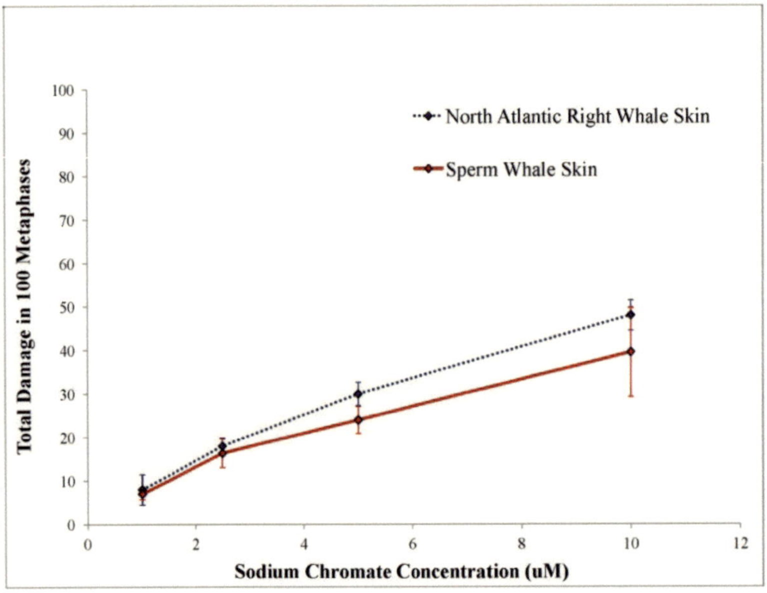
North Atlantic Right Whale Skin
Sperm Whale Skin
Total Damage in 100 Metaphases
Sodium Chromate Concentration (uM)
0
10
20
30
40
50
60
70
80
90
100
0
2
4
6
8
10
12

B

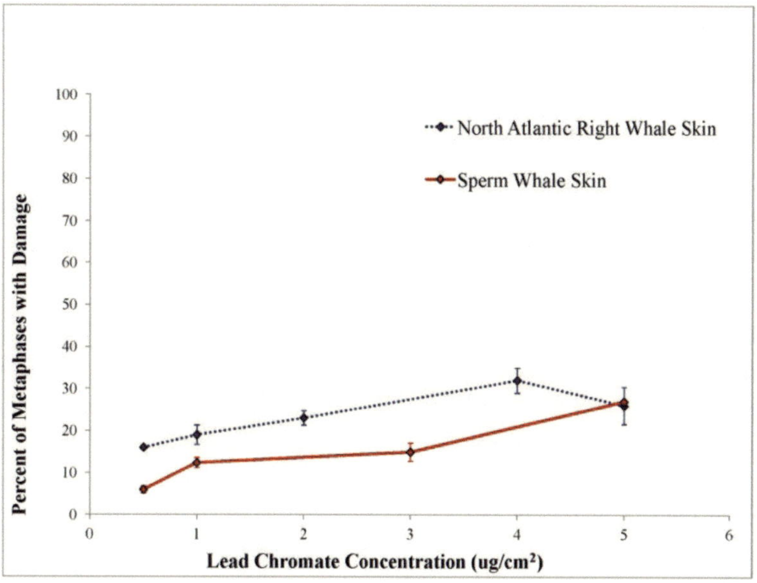

C

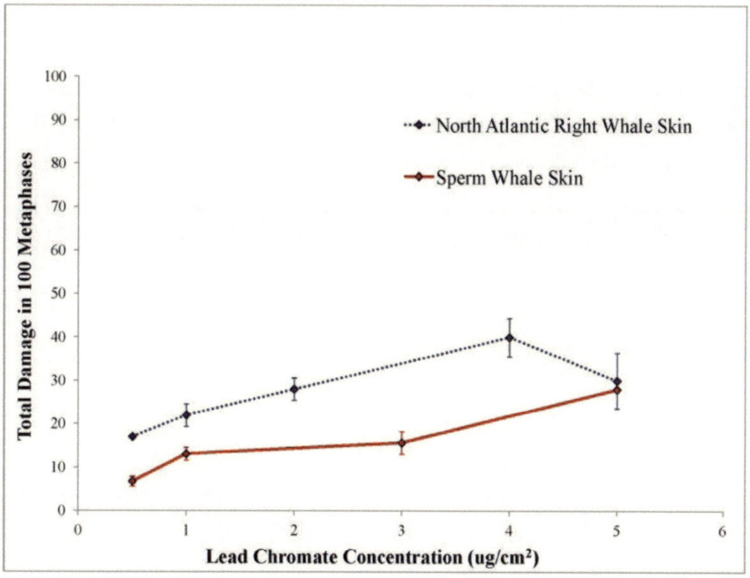

D

Figure 3. Genotoxicity of Chromium in North Atlantic Right Whale and Sperm Whale Skin Cells. This figure shows the genotoxicity of a 24 h chromium treatment in whale cells. There is a concentration-dependent increase in the amount of damage. Error bars represent the standard error of the mean from three independent experiments. A) Percent of metaphases with damage in cells treated with sodium chromate; B) Total aberrations in 100 metaphases in cells treated with sodium chromate; C) Percent of metaphases with damage in cells treated with lead chromate; D) Total aberrations in 100 metaphases in cells treated with lead chromate.

3.3. Toxic context

The next challenge is to compare the tissue culture doses to the levels in the whales. Comparisons can be made to gain toxicological context; however, one must always bear in mind that cells grown on a dish are in a different environmental context than cells in a tissue and so the comparisons will not be precise. To contextualize the chromium toxicity data, we have converted our treatment concentrations to parts per million and converted the levels observed in the biopsies to molarity. This way we can determine if the levels we are using in the cell cultures are environmentally relevant concentrations.

Our sodium chromate treatments convert to a range of 0.052-1.3 ppm and our lead chromate treatments convert to a range of 0.34-6.8 ppm (Table 1). Considering that average tissue levels for sperm whales and right whales were 9.3 and 7.0 ppm, respectively, our treatment concentrations are well below the average levels measured in whale skin (Tables 1 and 2). In fact, if we were to treat the cell cultures with the lowest sperm whale regional average measured in Sri Lankan waters (63.8 uM) we would kill all of the cells because our highest treatment concentration of 25 uM was highly cytotoxic (Figure 2) as measured by our cytotoxicity assay. Even our lowest detectable level in the sperm whale, 0.9 ug/g, is equivalent to treating cells with 6.38 uM and would induce cytotoxicity in approximately 65% of North Atlantic right whale cells and 90% of sperm whale cells. This level of treatment would also induce DNA damage in approximately 20% of both sperm whale and right whale metaphases. Given that our experimental doses are so much lower than the levels found in whales; this outcome raises concern about the impact of chromium pollution on whales regardless of the difference in environmental context between cells in a dish and cells in a tissue.

Sodium Chromate (uM)	Total Chromium(ppm)
1	0.05
2	0.10
2.5	0.13
3	0.16
5	0.26
10	0.52
25	1.3
Lead Chromate (ug/cm^2)	**Total Chromium (ppm)**
0.5	0.34
1	0.68
2	1.4
3	2.0
4	2.7
5	3.4
10	6.8

Table 1. Chromate Treatment Conversions.

Right Whales (individual levels)		
Whale ID Number	**Cr Tissue Level (ppm)**	**Total Cr (uM)**
3646	4.9	94
2642	5.1	98
3301	6	115
1218	6.3	121
3351	8.1	156
1611	9	173
1607	10	192
Mean all right whales	7.1	135
Sperm Whales (average by region)		
Location	**Cr Tissue Level (ppm)**	**Total Cr (uM)**
Sri Lanka	3.32	63
Canaries	3.69	70
Maldives	5.18	99
Mediterranean	5.21	100
Pacific Crossing	5.55	106
Papua New Guinea	5.72	109
Atlantic Ocean Crossing	6.28	120
Sea of Cortez	6.51	125
Indian Ocean Crossing	7.65	147
Chagos	7.95	152
Cocos	8.63	166
Australia	9.19	176
Mauritius	9.59	184
Galapagos	12.9	248
Seychelles	20.6	396
Kiribati	44.3	852
Bahamas	81.9	1575
Mean all sperm whales	9.3	179

Table 2. Tissue Level Conversions

4. Conclusions

Ocean pollution is emerging as a global priority. No longer can the world's oceans be considered an easy collective dumping site because even the most remote areas of the ocean are accumulating high levels of waste and pollutants. We now understand that any kind of pollution eventually ends up in the ocean. It may be carried by the wind, freshwater rivers and/or streams, coastal erosion, or watercraft of any sort. We have even found the most remote places on earth to be impacted. However, the true extent and the impacts of our wastes are poorly understood. One reason for this lack of understanding is due to the difficulties the ocean environment poses. Many places are geographically difficult to access; these areas require a large financial, technical, personnel and personal burden to study. In the rare cases that these obstacles are overcome, there remain the additional challenges of contending with environmental challenges (e.g. weather) and the simple fact that marine life is difficult to see, and can only be observed for a short period of time. Finally, the vast majority of marine animals cannot be used in laboratory experiments, and we do not currently have a useful model species to use in their place. Yet, despite these difficulties there are ways to study the impacts of pollution on marine life without sacrificing a large number of organisms and threatening their populations.

Thus, marine mammals provide the best model for studying marine pollution. They are found near-shore (e.g. sea otters, seals, sea lions, sirenians, and some cetaceans) and off-shore (e.g. most cetaceans, some seals, and some sea lions), and their diets range the entirety of the food chain from plankton and krill to giant squid and other marine mammals. A handful of marine mammals have cosmopolitan or near-cosmopolitan distribution, which enables for a more controlled comparison of marine pollution in different areas; these include sperm whales, bottlenose dolphins, orcas, and humpback whales. All marine mammals attract ecotourism, thus there are many watercraft that seek them out for observation. Along with these watercraft come the exhaust from burning fossil fuels, noise pollution that can interfere with the animals' communication and any trash that may be accidently lost overboard.

The health of all marine mammals depends on clean air and water; they all live in the water exclusively or near-exclusively, and since they are mammals they need air to breathe. Not all are directly affected by pollution in the sediment, but may still be indirectly affected through their prey. Marine mammal prey is highly varied from species to species. Studying a variety of them, we can determine if the effect of pollution is specific to animals that are higher on the food chain (likely caused by biomagnification), or if the effect is conserved throughout the food chain (likely caused by bioaccumulation).

Furthermore, marine mammals are suitable organisms to study because they capture the attention and support of the mass media and general public. These animals are one of the major bases for marine ecotourism because of their charismatic behaviors and their similarities with humans; they breathe air, are warm-blooded, nurse their young, and communicate with songs and chatter. Their similarities also enable them to be studied as a model species for humans [8]. The ocean is a finite resource, and has a finite capacity for pollution intake. Marine mammals can serve as a model species for humans to demonstrate how we will be affected by marine pollution in the future if it is not monitored and regulated better. They spend either all or nearly all their lives in the ocean, and they will exhibit impacts by marine pollution sooner than humans will.

Skin biopsies can provide a wealth of information about marine mammal pollution. Here we have shown how chromium, a little studied chemical in the marine environment, has contaminated whales in even the remotest regions. Using the levels of chromium we reported in the biopsies we have determined that the treatment concentrations that induced DNA damage in our cell culture toxicological analysis were orders of magnitude lower. To add further supporting context, human workers exposed occupationally to Cr(VI) had levels of chromium in lung tissue of 20.4 ug/g [9] while levels of skin were reported to be 0.05 ug/g [10]. These workers died of lung cancer. Considering that the average skin levels found in sperm whales and North Atlantic right whales of 9.3 ug/g and 7 ug/g, these levels appear to be extremely high compared to occupationally exposed humans and are of concern.

By measuring levels of metals and other pollutants we can inform our in vitro toxicological experiments by using more environmentally relevant treatment concentrations. In addition, we have used the biopsies to develop cell lines to test potential environmental pollutants like chromium in a species specific model rather than use human or mouse models to inform on pollutants specifically being exposed to a marine mammal population. These results demonstrate that environmental pollutants found in the ocean are accumulating in the whales and are likely to have important health repercussions to the animals living in the oceans.

Author details

Catherine F. Wise, John Pierce Wise, Jr., Sandra S. Wise and John Pierce Wise, Sr
Wise Laboratory of Environmental and Genetic Toxicology, Maine Center for Toxicology and Environmental Health, University of Southern Maine, Portland Maine, USA

Acknowledgement

We would like to thank Tania Li Chen, James Wise, Christy Gianios Jr., Hong Xie, Amie Holmes, Carolyne LaCerte, Shouping Huang, Julieta Martino, Fariba Shaffiey, Kaitlynn M. Levine, Marijke Grau, W. Douglas Thompson, Christopher Perkins, AbouEl-Makarim Aboueissa, Tongzhang Zheng, Scott Kraus, Yawei Zhang, Tracy Romano, Todd O'Hara, Cairong Zhu, Lucille Benedict, Gregory Buzard, Iain Kerr, Bob Wallace, Jeffrey Kunz, John Atkinson, Roger Payne and all of the Odyssey boat crew for technical assistance and professional advice.

The content is solely the responsibility of the authors and does not necessarily represent the official views of the National Institute of Environmental Health Sciences (NIEHS), the National Institutes of Health, the Army Research Office, the Department of Defense, the Environmental Protection Agency, the United States Department of Commerce or the National Oceanic and Atmospheric Administration (NOAA). It has not been formally reviewed by EPA or any other federal agency. The EPA and other federal agencies do not endorse any products or commercial services mentioned in this chapter. Work was conducted under NMFS permit #1008-1637-03 (J. Wise Sr., PI) and permit #751-1614 (Iain

Kerr, PI). This work was supported by EPA GRO Fellowship Assistance Agreement No. MA-91739301-0 (C. Wise, PI), NIEHS grants ES016893 and ES10838 (J. Wise Sr., PI), NOAA grant #NA03NMF4720478 (J. Wise Sr., PI), ARO grant # W911NF-09-1-0296 (J. Wise Sr., PI), the Maine Center for Toxicology and Environmental Health, Ocean Alliance, the Campbell Foundation, and the many individual and anonymous Ocean Alliance donors.

5. References

[1] Kraus, SD; Brown, MW. North Atlantic right whales in crisis. *Science*, 2005, 309, 561-562.

[2] Kraus, SD; Hamilton, PK. Reproductive parameters of the North Atlantic right whale. *J Cetacean Res Manag* (special issue), 2001, 2, 213-236.

[3] Costa, M. Toxicity and carcinogenicity of Cr(VI) in animal models and humans. *Crit Rev Toxicol*, 1997, 27, 431-442.

[4] Wise, Sr., J.P., Payne, R., Wise, S.S., LaCerte, C., Wise, J., Gianios, Jr., C., Thompson, W.D., Perkins, C., Zheng, T., Zhu, C., Benedict, L. and Kerr, I. A Global Assessment of Chromium Pollution using Sperm Whales (*Physeter macrocephalus*) as an Indicator Species. *Chemosphere*, 2009, 75, 1461–1467.

[5] Pettine, M., Millero, F.J. Chromium speciation in seawater: the probable role of hydrogen peroxide, *Limnol Oceanogr*, 1990, 35, 730–736.

[6] Agency for Toxic Substances and Disease Registry (ATSDR), Toxicological profile for Chromium, U.S. Department of Health and Human Services, Public Health Service, Agency for Toxic Substances and Disease Registry, Atlanta, GA, 2000.

[7] Wise Sr., J.P., Wise, S.S., Little, J.E. The cytotoxicity and genotoxicity of particulate and soluble hexavalent chromium in human lung cells. Mutat. Res. 2002, 517, 221–229.

[8] Bossart, G.D. Marine Mammals as Sentinel Species for Oceans and Human Health. Oceanography 2006, 19, 134-137.

[9] Tsuneta, Y., Ohsaki, Y. Chromium content of lungs of chromate workers with lung cancer. *Thorax*, 1980, 35,294-297.

[10] Mancuso, T.F. Chromium as an industrial carcinogen: Part II. Chromium in human tissues. *Am J Ind Med*, 1997, 2, 140-147.

Chapter 6

Cutaneous Lesions in Cetaceans: An Indicator of Ecosystem Status?

Marnel Mouton and Alfred Botha

Additional information is available at the end of the chapter

http://dx.doi.org/10.5772/54432

1. Introduction

For countless generations, civilization had a relative insignificant impact on the marine environment, particularly on marine mammal species. However, with the dawning of the industrial revolution, a dramatic increase in the utilization of marine species (including whales) as a food source and other industrial purposes resulted in a radical reduction in the numbers of some species during the subsequent years, leading to a notable decrease in marine biodiversity during the 20th century. Fortunately, this exploitation was accompanied by increased awareness and campaigns by environmental and ethical lobby groups, resulting in more controlled and managed whaling activities, which in turn lead to markedly improved whale numbers [1]. However, the tide shifted once again with the dawning of the new millennium, characterized by ever growing industrial development and rapid human population growth. As a result, natural resources came under increased pressure and in particular, development and growth led to the creation of massive amounts of waste and pollutants. Consequently, high levels of marine pollution, especially near urban regions, became a serious threat to the health and well-being of marine mammals, including cetaceans.

The first reports of skin disease in cetaceans date back to the 1950's [2]. However, over the last 60 years the frequency of these reports steadily increased. The question, therefore, arose as to whether scientists are just more aware of this phenomenon and consequently report more cases, or whether the occurrence of these lesions is indeed on the increase. Many natural factors such as ecto-parasites, water temperature and salinity are role players in these diseases. However, anthropogenic impact can no longer be ignored, since these influences can even affect the natural factors by escalating their effect. This brings one to another imperative question; whether these factors are linked, or whether they are merely the result of a coincidence. The aim of this review is therefore, to take a closer look at the skin of cetaceans, the occurrence of skin lesions among these mammals, the microbes that seem to be the causative agents, as well as contributing factors such as anthropogenic activities.

2. Cetacean skin

2.1. The Barrier: First line of defence

The first study on the anatomy of whale skin was conducted early in the 20th century by Japha [3]. He noted the unusual thickness of the epidermis when compared to other mammals (15 to 20 times thicker as in humans). Later, Parry [4] distinguished between the layers of epidermis, dermis and hypodermis. He differentiated between the epidermal layers and recognised the stratum corneum and the stratum germinativum (Figure 1), consisting of the more superficial prickle cells and the deep cylindrical cells. The hypodermis was found to be thick and fatty, merging into the dermis that consists of white fibres. The dermis stretches into the epidermis by means of 'dermal papillary ridges', and therefore interdigitate with the epidermal papillae (Figure 1) [4]. Parry also studied the vascular system and conductivity of blubber, which affect temperature regulation in these mammals. A few decades later, Sokolov [5] summarized cetacean skin anatomy by stating that the skin of these animals is relatively smooth, and with a ratio of 0.3 - 1.5 % in relation to body length, may reach greater 'absolute thickness' compared to other mammals. He noted that although the stratum corneum is very thick in cetaceans, layers of strata granulosum and lucidum are absent, with the upper layers of the epidermis not fully cornified. Sokolov concluded that sebaceous and sweat glands, as well as pelage, are absent in the skin of cetaceans. Using bottlenose dolphins (*Tursiops truncatus*) as model, it was found that the epidermal layer of cetaceans has a large capacity for cell population and a long turn over time, accompanied by rapid sloughing [6]. These characteristics account for the unusual thickness, as well as the smooth surface of the skin, thereby enhancing the barrier properties and ability to limit attachment by microbes [6, 7].

The stratum corneum of cetaceans (Figure 1) is often referred to as the parakeratotic layer and is composed of moderately flattened cells, characterized by retained elongated nuclei and prominent organelles (including mitochondria), representing a form of parakeratosis [7, 8]. The latter process was attributed to a type of cornification, associated with evolutionary hair follicle loss. The phospholipid-rich cornified layer presumably also aids in waterproofing the skin of these mammals [8]. In addition to these general features of cetacean skin, several unique ultra-structural characteristics were reported for the stratum corneum of the southern right whale (*Eubalaena australis*) [9]. These include lipid droplets occurring in close association with the nucleus, as well as an abundance of intra-nuclear inclusions similar to small fragments of cytoplasmic keratin.

Keratins are scleroproteins responsible for mechanical support in epithelial cells [10]. These macromolecules are mechanically hard, chemically unreactive, insoluble, fibrous, and very tough as a result of the numerous disulfide cross linkages [11]. In terrestrial mammals, keratins are produced by the so-called keratinocytes, found in the stratum basale or germinativum. The primary function of these keratinocytes, constituting 95% of the epidermis, is to provide a barrier against adverse environmental conditions, such as heat, radiation, water loss and penetration by pathogens. In cetaceans, this barrier is provided by lipokeratinocytes, responsible for the production of both keratin and lipid droplets [12, 13]. These lipids enhance the capability of the lipokeratinocytes to act as physical barrier within a hypertonic environment, and contribute to the unique buoyancy, streamlining, insulation and caloric characteristics of cetacean skin [12]. This physical barrier represents the first line of defence against the environment and prospective invaders.

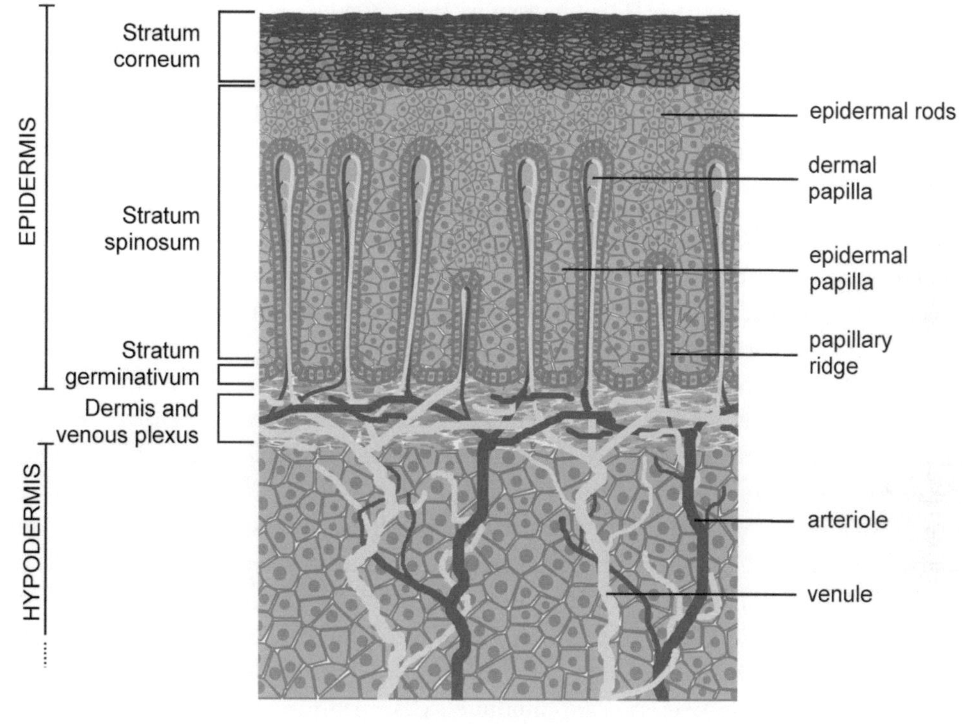

Figure 1. A cross section of cetacean skin, showing the general anatomy of the epidermal, dermal and hypodermal layers (Illustrated according to previous findings [4, 30]; as well as own unpublished data)

2.2. The Barrier: Second and third lines of defence

The primary role of skin is to provide a physical barrier against the environment and this presents the first line of defence. However, crucial immune components form integral elements of the skin, providing the subsequent layers of defence with increasing specificity. Potential microbial invaders will be confronted by the physical, innate (non-specific), chemical and granulocyte barrier (adaptive or specific immune system) that evolved over time to eradicate these invaders, digest invading cells into smaller antigens, and finally to programme lymphocytes in order to provide long term protection against the particular microbe [13].

A microbe that manages to penetrate the subcutaneous layers of the skin will be met by the second line of defence; the innate or non-specific immune system. This defence mechanism involves the production pro-inflammatory substances, such as chemokines by the lipokeratinocytes, resulting in the migration of immune cells such as leucocytes to the site of infection. Leucocytes include among others, the phagocytes (macrophages, neutrophils and dendritic cells). Upon detection of the foreign proteins originating from potential bacterial, fungal or parasitic invaders, Langerhans cells will phagocytise these antigens and migrate to adjacent lymph nodes. In the lymph nodes, these cells will develop into mature dendritic cells which will process the antigen (into smaller fragments), to activate the adaptive immune system (lymphocytes) [14]. On top of this chemical defence mechanism, non-specific antimicrobial substances such as lysozyme and the peptide β-defensin were found in cetacean integument [15]. It was found that lysozyme occurs between the layers of the stratum corneum, within cells of the stratum spinosum, dermis and endothelial cells of the dermal blood vessels. Also, β-defensin was found to be concentrated in the upper five or six layers of the stratum corneum, as well as within the cells of the upper stratum spinosum.

In addition to the above-mentioned non-specific defence mechanisms, intra-epidermal lymphocytes indicate the presence of the adaptive immune system in the epidermis of mammals. These specialized T cells have powerful cytolytic and immuno-regulatory effects on antigens and will confine antigens that overcame the first and second lines of skin defence [14]. Cells associated with this specific or adaptive immune response were also detected in cetacean skin studies [13]. Zabke and Romano [13] reported that their study on dolphin skin revealed the presence of MHC II (+) antigen cells, predominantly situated in the dermal papillae, along the epidermal-dermal border. These cells were found to have a dendritic-like morphology and form patterns, similar to those of Langerhans cells. The latter are known primary antigen presenting cells in the integument of terrestrial mammals, and thus the authors concluded that they were most likely Langerhans cells and not macrophages or dendritic cells. The latter two types were found deeper into the dermis. Zabke and Romano [13] further suggested that pathogen invasion resulting from a wound may lead to an inflammatory response, causing immune cells (neutrophils, macrophages and/or lymphocytes) to migrate from the dermal papillae to the site of infection. However, the authors indicated that inflammation is usually absent in these animals, because this barrier is normally sufficient against small injuries sustained via interactions with other dolphins. The authors further noted that wound healing in dolphins is not accompanied by scab formation. In these mammals hydrophobic changes within in the stratum spinosum, causes rapid sloughing and replacement with cells of the stratum germinativum undergoing mitosis.

From the above it is evident that cetacean skin is an effective physicochemical barrier. To overcome such a challenging obstacle would require a failure of the barrier itself, or creative strategies and unique properties in the prospective invader.

Immune response to cutaneous fungal infections. The type of fungal invader, whether a unicellular yeast or hyphal fungus, as well as anatomical site of infection, will determine the immune response of the host [16,17]. Yeast cells are usually phagocytosed, whereas the larger size of hyphae prevent them from being ingested. Over time, pathogenic fungi evolved and developed different strategies to survive and even disseminate in mammalian tissue. The neutrophils, macrophages and monocytes are the main antifungal effector cells in the defence strategy, employing shared mechanisms [17]. Firstly, macrophages present at the site of attempted infection, will make an effort to damage or kill the fungus. The second line of defence includes effector cells, the neutrophils and monocytes, which are summoned to the infection site by inflammatory signals from cytokines, chemokines and complement components. These effector cells damage or kill the fungal invader using strategies such as producing reactive oxygen intermediates and antimicrobial peptides [17].

There are four main groups of fungal infections [16]. The first, superficial mycoses, does not provoke any immune response from the host since the fungus would only grow on compounds associated with the skin. The remaining three categories provoke immune responses and include cutaneous, subcutaneous and deep mycoses. In cases of local trauma, subcutaneous mycoses may develop with the subcutis as the primary site of infection. In such cases, leucocytes and eosinophils will respond, leading to the formation of cysts or granulomas. Deep mycoses usually occurs in immuno-suppressed mammals, with entry through the lungs, paranasal sinuses, digestive system, or mucous membranes. Total fungal dissemination is usually a sign of severe immune failure.

3. Microbes – The enemy?

A limited number of organisms have the ability to degrade and utilize keratin, the key structural component of mammalian skin [18]. These include a few insect species, as well as a number of bacteria and fungi. Higher vertebrates are also not known to digest keratin [11]. Microbes degrade keratin by the secretion of extracellular proteolytic enzymes, known as keratinases; members of the serine proteinase group of enzymes [19, 20]. These enzymes are robust with a wide temperature and pH activity range, and have the ability to hydrolyze both natural and denatured keratin [21]. Keratinases from the fungus *Microsporum gypseum* were found to cleave the disulfide bridges in the keratin (sulfitolysis), which were followed by a further attack on the keratin structure by extracellular proteases [18]. Tsuboi and co-workers [22] found a keratinolytic proteinase in another fungus, i.e. *Trichophyton mentagrophytes,* with an optimal pH of 4.5 for keratin and 3.9 for haemoglobin. They showed that this fungus could potentially invade healthy skin (with a weakly acidic pH), by breaking down the keratin and thereby making it possible for the organism to invade the stratum corneum. This ability to degrade keratin, and other molecules associated with skin, can be regarded as a putative microbial virulence factor [20].

Keratinophilic (keratin loving) fungi (Figure 2) represent the largest group of organisms with the ability to degrade and utilize keratin as a source of carbon and nitrogen [11, 16]. These fungi commonly occur in soil and sewage sludge, which contain high concentrations of keratin remnants with specific physiochemical properties and associated microbial populations [23]. Releasing sewage sludge into the environment, or using it for fertilizing purposes, can therefore lead to spreading of potentially pathogenic fungi into new ecosystems. Plants, humans and animals are subsequently exposed to a variety of potentially infectious microbes. Fungal genera known to harbour keratinophilic species include: *Acremonium, Alternaria. Aspergillus, Candida, Chaetomium, Chrysosporium, Cladosporium, Curvularia, Fusarium, Geotrichum, Gliocladium, Gymnoascus, Microsporum, Monoascus, Mucor, Paecilomyces, Penicillium, Scopulariopsis, Sporothrix, Trichoderma, Trichophyton,* as well as *Verticillium* [16, 23-25].

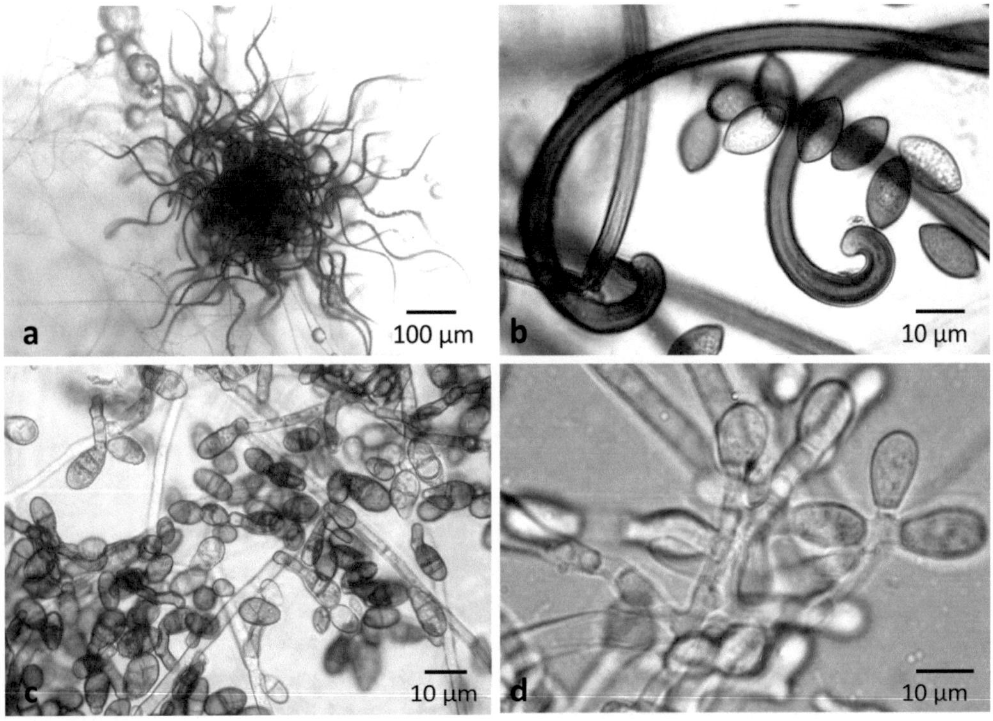

Figure 2. Light micrographs of keratinophilic fungi. a. Fruit body of *Chaetomium murorum* with long ascomatal setae; b. ascomatal setae and ascospores of *Chaetomium murorum*; c. conidia of *Alternaria alternata*; d. Fertile hyphae of *Chrysosporium keratinophilum* bearing conidia

Some keratinophilic fungi (Figure 2) are known to be pathogenic, and often the cause of cutaneous skin infections. These include the so-called dermatophytes, belonging to the genera *Epidermophyton, Microsporum* and *Trichophyton,* as well as non-dermatophytes such as *Aspergillus, Candida, Fusarium* and *Scopulariopsis* spp. Moreover, De Hoog et al. [16] remarked that more cases of fungal infections seem to be caused by fungal spp. formerly known as being saprobic, and appear to be associated with the increasing numbers of immuno-compromised patients. These fungi are mostly opportunists that cause infection when the immune system of the host is breached [16, 17]. Interestingly, of more than 100 000 known fungal species, only about 100 have been reported regularly, as infectious agents of animals, as well as humans [16].

Certain species of the bacterial genus *Bacillus* also have the ability to produce keratinases. *Bacillus cereus, Bacillus licheniformis* and *Bacillus subtilis* have been identified and studied in this regard for commercial purposes, specifically for the biodegradation of feathers in the poultry industry [21, 26]. Other bacterial genera with keratinolytic ability include: *Lysobacter, Nesternokia, Kocurica, Microbacterium, Vibrio, Xanthomonas, Stenotrophomonas* and *Chryseobacterium* [24].

4. Case studies of skin lesions in cetaceans – The signs?

Skin lesions in cetaceans have been reported since the 1950's [2]. Some examples of the steady stream of reports in later years are presented below.

Skin lesions containing *Staphylococcus* were reported for the first time in cetaceans in 1988 [27]. In that study, two isolates of *Staphylococcus* were obtained from purulent tissue that occurred in two captive dolphins with multiple, suppurating lesions, and subsequently described as a new species, *Staphylococcus delphini*.

A study conducted over a four year period found that a range of microbes were associated with lesional and non-lesional skin in a group of bowhead whales (*Balaena mysticetus*), characterized by dozens to hundreds of roughened areas on their skin surfaces [28]. The majority of microbes, isolated during the study, were associated with the lesional skin: 56% of Gram positive bacteria, 75% of Gram negative bacteria and 64% of the yeasts. Also, the lesional skin was characterized by the presence of *Corynebacterium* spp., *Acinetobacter* spp., as well as representatives of *Moraxella*. *Candida* spp. were the dominant yeast species, followed by representatives of *Cryptococcus* and *Rhodotorula.* Subsequent tests on the isolates showed the production of enzymes able to cause necrosis by microbes originating from both lesional and non-lesional skin. These whales occurred in regions with increased industrial activities (gas and oil exploration) in the Beaufort Sea; the authors speculated that the roughened skin areas might have been associated with the adherence of spilled oil [28].

A study examining skin diseases among wild cetaceans from British waters found 69% of individuals to be affected [29]. The authors reported wounds and other traumatic injuries, as well as lesions caused by pox and herpes viruses, as well as bacteria, ectoparasites and non-specific ulcers. Concerns that the lesions were associated with pollution were raised but not confirmed [29].

Henk and Mullan [30] examined 23 bowhead whales and reported shallow lacerations, circular depressions and epidermal sloughing on these whales' skin. The authors also found abundant bacteria and diatoms associated with these lesions, and even higher numbers where the stratum spinosum was exposed. The bacterial isolates were found to include cocci, bacilli and filamentous spp., with increasing numbers associated with higher levels of necrotic decay. Protozoa and fungi were also observed and also increased in incidence with more disturbed epidermal surfaces. The authors also identified several erosive enzymes from these microbes and suggested an association between the whales' skin and spilled oil.

A pygmy sperm whale (*Kogia breviceps*) and an Atlantic white-sided dolphin (*Lagenorhynchus acutus*) were reported with mycotic dermatitis, in the form of raised, firm, erythemous, cutaneous nodules on parts of their bodies [31]. The dermatitis cases were the result of infections caused by a species of *Fusarium*, and in both cases, the disease was preluded by stress factors, such as stranding, which presumably induced immuno-suppression.

Resident bottlenose dolphins, from the Sado estuary in Portugal, were examined and 85% of the community showed signs of skin disorders [32]. The authors compared these results to observations from other areas, and came to the conclusion that habitat degradation played a significant role in these disorders since eutrophication seemed to a serious problem in this estuary. Consequently, the authors attributed these conditions to apparent depressed immune systems caused by stress, habitat degradation and pollution.

Ten coastal populations of bottlenose dolphins served in a photographic study that compared levels of epidermal disease among populations exposed to a wide range of natural and anthropogenic conditions [33]. It was found that epidermal lesions were common in all populations. However, the severity and prevalence of the different classes of lesions, varied among the populations. Those occurring in areas with lower water salinity and temperature had a higher lesion incidence and severity. On the other hand, no direct correlation was found between the lesion characteristics and toxicology data. The authors concluded that the oceanographic variables might influence the epidermal integrity of the skin or cause physiological stress, thereby rendering these mammals more prone to natural infections or impact by anthropogenic activities.

In 2001, Mikaelian and co-workers [34] reported a case of six beluga whales (*Delphinapterus leucas*) with slightly depressed, greyish round lesions, found dead on the shores of the St. Lawrence estuary. Histology of the lesions revealed *Dermatophilus*-like actinomycetes that had invaded the stratum corneum of the epidermis [34]. Even the stratum spinosum of these animals were characterized by marked spongiosis and vacuolar degeneration. These whales were furthermore free from ectoparasites, thereby eliminating the possibility of this predisposing factor. Immunodeficiency seemed to be the most likely cause of these *Dermatophilus*-like infections, since four of the six individuals were characterized by chronic debilitating diseases. Martineau [35] followed this up by reporting on an extensive survey (1983-1999) on the beluga whales from this estuary. They found cancer rates in these animals to be much higher than in any other population of cetaceans, similar to that of other mammals (including humans) from the same area. They believed that environmental contaminants, such as polycyclic hydrocarbons (PAH) were involved in the aetiology of these conditions.

A visual health assessment study in North Atlantic right whales (*Eubalaena glacialis*) revealed a variety of skin conditions, as well as the calving intervals in these mammals increasing from 3.67 to 5 years, thereby significantly impacting on the population growth [36]. The authors speculated that possible causes might include environmental contaminants, marine bio-toxins, nutritional stress, genetic influences, as well as infectious diseases.

Hamilton and Marx [37] presented results of a study conducted from 1980 until 2002 on skin lesions among North Atlantic right whales. The authors documented white, blister, swath and circular lesions. White lesions appeared to represent episodic events, the incidence of blister lesions were more constant, whereas swath lesions were often associated with fatal conditions. They reckoned that skin lesions are indicators of compromised health, possibly caused by deteriorating habitat quality in coastal areas.

Between the years 2006 and 2009, Van Bressem and co-workers published extensively on many cases of skin lesions in a range of cetaceans. In one such study [38], the authors examined 'tattoo' skin lesions caused by poxvirus infections in four species of small cetaceans (*Lagenorhynchus obscurus, Delphinus capensis, T. truncatus* and *Phocoena spinipinnis*) near Peru. They reported a possible increase in the disease since 1990 in two of the species (*L. obscurus* and *P. spinipinnis*). They also found that male *P. spinipinnis* individuals were two times more infected than their females [38].

A population of long-beaked common dolphins (*D. capensis*) was studied between 1985 and 2000 in the Southeast Pacific near Peru, characterised by extensive fisheries activities [39]. The authors reported the presence of a variety of cutaneous lesions, abnormalities and scars, on between 1.8% and 48.2% of these dolphins. Tattoo lesions, punctiform and round marks, dark circle lesions, coronet marks and abnormal pigmentation were described, among many other abnormalities in this population. These lesions were attributed to pox- and herpes viruses, as well as other unknown viruses and parasites.

An extensive survey conducted from 1984 until 2007 and dealing with previously unreported cases of skin and skeletal diseases in cetaceans from Ecuador, Colombia, Peru, Chile, Argentina, Uruguay, Brazil and Venezuela, was presented as an overview in 2007 [40]. The authors reported tattoo skin disease, lobomycosis-like disease and other cutaneous infections with unknown aetiology in 590 cases, out of a total of 7635 specimens, including 12 different odontocete spp. that were examined. It was suggested that anthropogenic factors, including aquaculture, fish factories, untreated waste water, ballast water and chemical pollution, play a major role in the degradation of the habitats of these cetaceans, thereby contributing to the poor health status of many individuals in these populations.

Chronic mycotic disease of the skin and subdermal tissues in Indo-Pacific bottlenose dolphins (*Tursiops aduncus*), caused by the fungus *Lacasia loboi,* was reported in 2009 [41]. These dolphins lived in the tropical lagoon of Mayotte, situated in the Indian Ocean between Mozambique and Madagascar, and were characterized by numerous raised, greyish nodules on the head, flanks, dorsal fin, belly, back and tail. In some individuals, the lesions appeared to be quite severe and lesions resembling other unknown fungal infections were also observed. Habitat degradation, especially along the coastal areas where rapid urbanization, agricultural activities and untreated waste water are evident, was mentioned as a contributing factor to the aetiology of the disease.

Inshore and offshore surveys conducted, from 1997 to 2007, on bottlenose dolphins in the larger Santa Monica Bay, California, revealed a very high incidence and extent of skin lesions [42]. Causative agents found included bacteria, viruses, fungi, vitamin deficiencies, diatom growth and parasites. However, anthropogenic activities were thought to be a major contributing factor and seemed to be linked to especially viral outbreaks. Moreover, this area (the Southern California Bight) is known for high concentrations and volumes of pollutants entering the coastal and offshore environment, as well as contaminated sediments that cover a 44 km^2 on the ocean floor, containing dichloro-diphenyl-trichloro-ethane (DDT) and polychlorinated biphenyls (PCBs).

A study conducted during 2006 and 2007 revealed notable differences in skin diseases between two communities of Guiana dolphins (*Sotalia guianensis*) from Brazil [43].The authors found that the community, living in the chemically and biologically polluted Paranaguá estuary, was characterized by the occurrence of lobomycosis-like and nodular skin diseases. This estuary is known for its high levels of chlorinated hydrocarbons, as well as regular oil spills. In contrast, another community of Guiana dolphins, living in the less polluted Cananéia estuary, was free from these skin diseases and had relatively low tissue levels of organochlorines. The authors consequently proposed that lobomycosis-like and nodular skin diseases might act as indicators of environmental burden.

A southern right whale neonate that beached along the Southern coast of South Africa and suffering from extensive skin lesions were found to be infected with a number of cutaneous fungi including *Chaetomium globosum, Chaetomium murorum* and *Penicillium coprophilum* [44]. During the same period, another southern right whale neonate was found beached and suffering from a yeast infection caused by *Candida zeylanoides* [45]. In both cases the authors speculated about anthropogenic factors contributing to the condition of the animals.

Maldini and co-workers [46] reported on bottlenose dolphins living in Monterey Bay, California. Their research, conducted between 2006 and 2008, showed that approximately 90% of these cetaceans were characterized by skin lesions. They identified five skin conditions, with pox-like lesions being the most frequently found. The authors suspected that contaminants such as persistent organic pollutants (POPs) and heavy metals were contributing factors that weakened the immune systems of the dolphins, thereby rendering them more susceptible to viral infections.

A recent study on North Pacific humpback whales (*Megaptera novaeangliae*) examined bacterial species associated with the skin of these mammals [47]. It was found that healthy individuals were characterized by similar microbial communities, as opposed to health-compromised individuals that harboured different populations. Also, the microbial populations on the skin of these animals were found to be different to that of the seawater, which led to the conclusion that the skin-associated bacteria were adapted to live on the epithelium and its constituents. The study also reported that the bacterial phylum found most on healthy individuals is Bacteroidetes, in contrast to the health-compromised individuals which harboured Gammaproteobacteria as the dominant group.

Cetaceans are also exposed to UV radiation and often exhibit lesions similar to skin cancer in humans [48]. Interestingly, cetacean species with darker pigmentation have less UVR induced skin damage. Also, latitude affects the incidence of this phenomenon, since UVR dosage at lower latitudes, are 5 times higher than at mid-latitudes.

A recent study by Fury and Reif [49] reported poxvirus lesions in two estuarine populations of bottlenose dolphins from Australia. This was the first report of poxvirus-like lesions in Australian cetaceans. Their results suggested that these infections were accompanied by climatic events, such as flooding in this case, which lead to lower water salinity and higher occurrence of dolphin pox. They concluded that diseases such as dolphin pox, might act as indicators of environmental stress.

It seems evident from the above mentioned studies that the occurrence and high prevalence of skin lesions in many cetacean populations are linked to environmental factors, including water salinity and temperature, as well as pollution and eutrophication. These skin lesions may be caused by a wide diversity of microbes that will be discussed in the next section.

5. Microbes causing skin lesions in cetaceans

5.1. Virus infections of cetacean skin

5.1.1. Dolphin pox

Poxviridae represent the largest family of viruses known to cause diseases in marine and terrestrial mammals [50]. Among cetaceans, the odontocetes seem to be more affected than the mysticetes [49]. Species of cetaceans reported to be affected by pox viruses, include Atlantic bottlenose dolphins [7, 50], bottlenose dolphins from Australia [49], Atlantic white-sided dolphins [7, 50], common dolphin (*Delphinus delphis*), dusky dolphins, white-beaked dolphins (*Lagenorhynchus albirostris*) [29, 50], striped dolphins (*Stenella coeruleoalba*) [29, 50], Hector's dolphins (*Cephalorhynchus hectori*) [7, 29, 50], long finned pilot whales (*Globocephala melaena*) [29, 50], as well as a few spp. of porpoises [50].

Dolphin pox has been associated with a variety of lesions, referred to as 'targets', watered-silk', 'ring', 'pinhole', 'circle' and 'tattoo'-like [7]. These lesions emerge as single or overlapping circular grey spots. Later, these ring lesions may develop into black punctiform stippled patterns. Histological studies revealed a thickened stratum corneum with ballooning degeneration, and eosinophilic inclusions containing virus particles, inside the cytoplasm of stratum intermedium cells. Studies on this disease pointed to one consistent feature; its relationship with compromised environmental conditions and consequent general health of affected individuals [7, 51].

5.1.2. Herpes virus

Herpes virus had been reported as the causative agent of focal dermatitis in captive, as well as free-swimming beluga whales [51, 52]. Lesions caused by this virus appeared as multiple grey, raised, pale grey regions on the skin, which eventually ulcerated and healed very slowly. At the time of infection, these whales appeared to be in poor health and under stress. Histological analyses showed epithelial cells that underwent intercellular oedema, necrosis and the development of microvescicles. Prominent eosinophilic, intra-nuclear, inclusion bodies were evident in infected epithelial cells [51].

5.1.3. Papillomavirus in cetaceans

In cetaceans, papillomas have been reported on the skin, as well as the tongue, penis, pharynx and first gastric compartment [51].

5.1.4. Calicivirus induced vesicular disease

Smith and co-workers [53] reported an Atlantic bottlenose dolphin with vesicular skin disease caused by calicivirus. This disease was apparently transferred to another dolphin, but via a sea lion. The lesions eventually eroded and left shallow ulcers. Interestingly, serological studies on mysticete spp. from the North Pacific, showed the presence of neutralizing bodies to several marine vesiviruses, including calicivirus [54].

5.2. Bacterial infections of cetacean skin

A variety of bacteria have been reported from skin lesions in cetaceans. These include species of *Brucella, Corynebacterium, Dermatophilus, Escherichia, Erysipelothrix, Klebsiella, Mycobacterium, Pseudomonas, Staphylococcus, Streptococcus* and *Vibrio* (Table 1) [27, 47, 55]. Bacteria also cause death of cetaceans in many cases [55]. However, it must be noted that these infections are usually of a secondary nature and occur as a result of other stresses, such as parasites and immuno-suppression following exposure to toxins.

Bacterial genus	Bacterial species occurring on cetaceans
Brucella	*Brucella cetacea* had been reported from sub-blubber abscesses in several species of cetaceans [56]. Also, *Brucella ceti* had been isolated from subcutaneous and skin lesions in cetaceans [57].
Corynebacterium	A species of *Corynebacterium* had been isolated from skin lesions in an Atlantic bottlenose dolphin [55].
Dermatophilus	Lesions caused by *Dermatophilus*-like actinomycetes were observed in six deceased beluga whales that were found in the St. Lawrence estuary [34]. These lesions presented as slightly depressed, round and grey areas on the skin.
Enterobacter	*Enterobacter agglomerans* had been reported from skin lesions on an Atlantic bottlenose dolphin from Florida, USA [55].
Escherichia coli	*E.coli* isolates were obtained from skin lesions on Atlantic bottlenose dolphins from Florida and New York, USA [55].
Erysipelothrix	*Erysipelothrix rhusiopathiae* had been reported as the cause of skin disease in dolphins, characterized by dermal infarction causing dermal sloughing [51]. This disease is usually subacute and chronic in cetaceans or acute septicaemia [56]. As a small, pleomorphic, Gram positive rod, this bacterium is a commonly found in the mucous of fish, suggesting that the dolphins become infected by ingesting these fish [51, 56, 58, 59]. Injuries caused by the teeth of other cetaceans, is another possible route of infection [56].
Klebsiella	*Klebsiella oxytoca* was isolated from skin lesions of an Atlantic bottlenose dolphin, while another *Klebsiella* sp. was obtained from a goosebeak whale (*Ziphius cavirostris*), both from Florida (USA) [55].

Bacterial genus	Bacterial species occurring on cetaceans
Mycobacterium	Species of *Mycobacterium* have been associated with infections in cetaceans, including bottlenose dolphins, belugas and pseudorca's. In the infected animals, non-healing, chronic cutaneous or subcutaneous lesions were present and associated with other symptoms, such as granulomas in various organs and lymph nodes, as well as pulmonary infections [56].
Pseudomonas	This genus is known for Gram negative, motile, slender bacillus bacterium cells, and generally occurs in water. These bacteria often colonize wounds, which can lead to septicaemia [14]. *Pseudomonas aeruginosa* had been reported to form large cutaneous ulcers in Atlantic bottlenose dolphins, penetrating deep into the tissue and consequently leading to serious conditions in the affected animals. Septicaemia develops when the bacteria proliferate into the walls of the blood vessels [56]. A *Pseudomonas* sp. and *Pseudomonas putrefaciens,* had also been isolated from skin lesions in Atlantic bottlenose dolphins by Buck and co-workers [55].
Staphylococcus	A new species in this genus, *Staphylococcus delphini,* was described by Varaldo and co-workers [27]. Moreover, this represented the first report of an association between this genus and cetaceans. *S. delphini* isolates were obtained from suppurative skin lesions in two captive dolphins which recovered rapidly after antibiotic treatment [27].
Streptococcus	*Streptococcus* bacteria are Gram positive diplococci and common residents of cetacean's skin and upper respiratory tract. Cutaneous infections caused by these species, are therefore usually opportunistic and associated with animals under stress of some sort [51]. Species from this genus have been isolated from cetaceans with skin lesions [55], septicaemia, metritis and pneumonia. Amazon River dolphins have been reported with a specific dermatological condition, commonly known as 'golf ball disease', caused by *Streptococcus iniae,* and characterized by the presence of slow-growing, nodular, subcutaneous abscesses [56].
Vibrio	Buck and co-workers [55] found that *Vibrio* spp. were the most commonly isolated bacteria from stranded cetaceans. The two species specifically associated with skin lesions, were *Vibrio alginolyticus* and *Vibrio parahaemolyticus* originating from Atlantic bottlenose dolphins (*T. truncatus*) from Florida and New Jersy (USA). Dhermain and co-workers [58] also implicated bacteria from this genus, in cases of cetacean septicaemias.

Table 1. Bacterial species reported to cause skin lesions in cetaceans

5.3. Fungal infections of cetacean skin

5.3.1. *Candida*

Species of *Candida* are commonly associated with the mucous membranes of animals in limited numbers, and occur predominantly in the regions of the blowhole, oesophagus, vagina and anal area in cetaceans [59-61]. *Candida* spp. reported from cetaceans include *Candida albicans, Candida glabrata, Candida krusei, Candida tropicalis, Candida parapsilosis, Candida guiliermondii, Candida lambica* and *Candida ciferrii* [59, 60]. *Candida zeylanoides* was also found to be associated with the skin of a southern right whale neonate from South Africa [45]. Infections caused by *Candida* spp. mostly affect captive cetaceans and are usually associated with immuno-suppressed individuals, where the *Candida* infection may proliferate and cause severe local infection of the skin or mucosal membranes. In captive cetaceans, infections are usually observed after long-term antibiotic therapy [30, 61], but had also been reported following corticosteroid treatment of the animals, as well as after overtreatment of tank water [59]. These lesions usually occur as whitish, creamy plaques on the skin or mucosal surfaces. Histological examinations usually show colonies of pseudohyphae, septate hyphae and blastospores [16, 60].

In cutaneous *Candida* infections, the skin or mucosal membranes, may suffer acanthosis (hyperplasia and thickening of the stratum spinosum) with pseudoepitheliomatous hyperplasia, with the fungus growing in the epithelial tissue. These cutaneous *Candida* infections had been reported in a number of cases pertaining to especially captive cetaceans and varied from ulcerative dermatitis, to inflammation without ulcers and healed ulcers. However, significantly more cases of visceral lesions and systemic candidiasis had been reported in these mammals [60].

5.3.2. *Fusarium*

Fusarium spp. are well known saprophytes in soil, as well as the cause of a range of plant diseases. However, they also often cause hyalohyphomycosis after traumatic inoculation in humans and seem to be an emerging pathogen in immuno-compromised patients [16]. In cetaceans, mycotic dermatitis caused by a *Fusarium* sp., was reported in a pygmy sperm whale and an Atlantic white-sided dolphin. These cetaceans were characterized by elevated, firm, erythematous, cutaneous tubercles mostly found on the heads, trunks and caudal portions of the cetaceans' bodies [31].

5.3.3. *Lacazia loboi*

The yeast-like, dimorphic fungus, *Lacazia loboi* (formerly known as *Laboa loboi*) causes invasive cutaneous lesions in dolphins and humans [58], known as lobomycosis, lacaziosis or keloidal blastomycosis [41, 59, 62]. The first case of keloidal blastomycosis was described in 1931 in a human patient from the Amazon valley [63]. This disease was only known to affect humans until 1971, when Migaki and co-workers described a case of lobomycosis in an Atlantic bottlenose dolphin [64]. In 1973, De Vries and Laarman [63] described another case of this disease in a Guinana dolphin. Interestingly, no other cetaceans other than dolphins have ever been reported with this disease [59, 60]. It is generally accepted that the disease is the consequence of injuries sustained by the animal [61], and can therefore be transmitted to humans during necropsies [58]. It presents itself as white, elevated, crusty, nodular lesions on the animal's body [59-61], although mainly on the head, flippers, abdomen, fins, back tail stocks and flukes [60]. On cellular level, the disease presents with superficial granulomatous dermatitis, associated with macrophages and multinucleated giant cells containing a variety of round yeast cells [61]. More cases of this disease affecting cetaceans have been added to the list and recently some authors suggested that the incidence of lobomycosis might represent opportunistic infections in immuno-compromised hosts [62, 65]. The bio-accumulation of environmental contaminants in the affected dolphins was thought to possibly contribute to susceptibility to this disease [62].

5.3.4. *Dermatophytes*

Dermatophytes are fungi that grow on the outermost layers of the skin of animals, including muco-cutaneous membranes, genitalia, external ears, as well as dead skin or hair. Ringworm presents one type of dermathophyte and includes the genera *Tinea, Trichophyton, Epidermophyton* and *Microsporum* [16]. Infections caused by dermatophytes, seem to be rare in marine mammals, and therefore also in cetaceans. Limited reports include discrete nodules on the back of a captive Atlantic bottlenose dolphin caused by a sp. of *Trichophyton* [60].

6. Marine pollution and the impact of industrialization on cetaceans – A possible cause?

Rapid population growth and industrialization have characterized urban development. However, the rate of population growth is far higher than the rate of waste and wastewater infrastructure planning and development [1, 66]. This phenomenon has led to ever increasing pressure on natural resources and the creation of massive volumes of waste and waste water [67]. Marine pollution had been defined as deleterious effects resulting, either directly or indirectly, from the introduction of substances or energy by humans to the marine environment [66]. High levels of marine pollution especially along urban areas, pose a serious threat to the health of humans, as well as marine mammals. Although numbers of cetaceans have markedly improved over the past few years as a result of conservation efforts, the quality of the habitat in and near bay areas are critical, because these areas are used by these marine mammals for feeding, mating and calving [1]. Waste entering the marine environment had been categorized into the following groups of activities [68]:

- Waste from land-based sources such as sewage and industrial effluent discharges, storm-water run-off, agricultural and mining return flows, as well as seepage from contaminated ground water. These account for the highest percentage of waste to enter the sea environment.
- Atmosphere pollutants, including persistent organic compounds from vehicle exhausts and industries.

- Waste from ships, oil spills and the discharge of ballast water, as well as other waste.
- Dumping at sea, e.g. dredge spoil.
- Offshore exploration, e.g. oil exploration platforms.

6.1. Human waste into the marine ecosystem

There are 33 megacities in the world, and 21 of them are situated on coastlines. If one considers that at least 50% of the world's population lives in urban areas, where wastewater management systems are often outdated or completely inadequate, the sensitive urban coastal ecosystems are under serious threat [66]. It is estimated that over 90% of wastewater in developing countries is discharged, without any form of treatment into the marine environment, as well as into rivers and lakes. In more developed countries, at least primary treatment is required before discharging wastewater into the environment. Consequently, millions of litres of raw waste water or digested sludge are released into the marine environment. This practice often represents an alternative to tertiary treatment, as well as the most inexpensive means of disposal. The effluents released in ocean outfalls contain faecal material from domestic waste water and industrial discharges with persistent toxic substances and heavy metals. Harmful impacts include risk to public health, beaches, marine animals, as well as contamination of shellfish. Moreover, this release of high volumes of organic matter into the marine environment or contributory rivers also results in eutrophication, which leads to so-called dead zones in the seas and oceans. Approximately 245 000 km^2 of marine environment is thought to be affected by these dead zones [66, 69].

Bitton's [69] book summarized several global surveys of enteric pathogens in contaminated sea water. A number of authors reported enteric viruses of the coast of countries such as Brazil, France, Israel, Italy, Spain and the USA. The virus types found included polioviruses, coxsackie A and B viruses, echo-, adeno- and rotaviruses. These were also detected in sediments in close proximity of sewage outfalls, and found to prevail for extended periods of time. Also, pathogenic bacteria such as *Salmonella* and *Vibrio cholerae* were reported in coastal waters near ocean outfalls. Several illnesses among the human population have been reported as a result of exposure to bacteria from sewage contaminated water, including spp. of *Aeromonas, Leptospira, Mycobacterium, Legionella, Pseudomonas, Vibrio* and *Staphylococcus.* Adenoviruses were also associated with these illnesses, as well as the protozoa, *Naegleria fowleri* and *Acanthamoeba.* These microbes were found in cases of wound infections, skin and subcutaneous lesions, dermatitis, subcutaneous abscesses, septicaemia, conjunctivitis, pharyngitis, meningoencephalitis, Legionnaire's disease and ear infections. Outbreaks of gastroenteritis, on the other hand, were attributed to *Giardia* and *Cryptosporidium,* as well as *E. coli, Shigella* and gastroenteritis of unknown origin.

Some autochthonous microbes have also been documented as agents of disease among humans. *Pseudomonas aeruginosa,* for example, had been reported in immuno-compromised patients, and was also detected in high numbers in faecal polluted recreational waters [69]. Utilizing such recreational areas, was found not only to be associated with enteric diseases, but also with upper respiratory infections. Cases of pneumonia have been attributed to

Staphylococcus aureus, Pseudomonas putrefaciens, Aeromonas hydrophila and *Legionellae pneumophila.* Cases of skin infections were found to be caused by opportunistic spp. of *Aeromonas, Mycobacterium, Staphylococcus* and *Vibrio,* after swimming in contaminated recreational waters.

Sewage pollution may also influence the diversity of potential fungal pathogens that are known to infect humans and animals. Keratinophilic microbes occur in high numbers in sewage sludge since it contains high concentrations of keratin remnants [16]. These fungi, including the so-called dermatophytes are known as causative agents of a wide variety of cutaneous and subcutaneous mycoses. Awad and Kraume [23] found the following fungal species, retained in the following genera, in aerobic and anoxic sludge from wastewater plants in Berlin, Germany: *Chrysosporium, Microsporum, Trichophyton, Acremonium, Alternaria, Aspergillus, Candida, Chaetomium, Cladosporium, Fusarium, Geotrichum, Gliocladium, Gymnoascus, Mucor, Paecilomyces, Penicillium, Scopulariopsis, Sporothrix, Trichoderma* and *Verticillium.* It has been stated that the release of these fungi into the environment via activated sludge, presents an indisputable health risk [11, 23].

6.2. Heavy metals

Heavy metals are metallic chemical elements with a relatively high density and are also known as 'toxic metals'. This group includes elements such as copper (Cu), zinc (Zn), lead (Pb), mercury (Hg), nickel (Ni), cobalt (Co), arsenic (As), thallium (Tl) and chromium (Cr). These metals are common components of the earth's crust and occur naturally in ecosystems at different concentrations. However, the bio-available concentrations of these elements have increased significantly over time since the dawn of the Industrial Revolution. The resulting bio-accumulation of these metals has reached critical levels in many ecosystems. These elements enter the water supply through industrial and consumer waste, or acidic rain that in turn causes chemical reactions in soil releasing the metals into streams, rivers and ground water. Additional sources include waste from chemical, electro-plating, tanning, smelting and especially the mining industry [70]. Previously, the lead added to motor fuels as an anti-knock agent was released on large scale into the atmosphere in exhaust fumes with a substantial proportion settling on road surfaces. Run-off water from these surfaces therefore contained high levels of lead. Urbanized and heavy industrialized areas are consequently the foci of heavy metal pollution [71].

Living organisms require small amounts of certain heavy metals or trace elements for normal metabolic processes, in the form of co-factors in enzymes. Some heavy metals such as Cd, As, Pb and Hg pose the biggest threat to the health of humans and animals, and daily intake of these metals are toxic and often fatal. Long-term exposure of humans to As in drinking water for example, had been reported to cause increased risk to skin and other types of cancer, as well as other skin lesions including hyperkeratosis and pigmentation anomalies. Similarly, Cd exposure is associated with kidney damage and increased incidence of bone fractures [70]. Marine mammals often occupy the top levels of marine food chains, and therefore, heavy metals accumulate in these animals [72]. In water, heavy metals may be found in solution in the water column, as colloids, as suspended particles or absorbed to particulate matter [71]. The latter form often ends up as a constituent of the sediment and at least a portion is released again into the water column. It is also important to realize and take into account that not all sources of heavy metals are anthropogenic. The Mediterranean Sea for example is known for its natural high levels of Hg, and likewise the Arctic Sea for higher cadmium (Cd) concentrations [72]. Furthermore, different organs tend to accumulate different heavy metals, and are therefore metal specific. Heavy metals enter cetaceans through their lungs, skin (absorption), from the mother during gestation, through milk during nursing, as well as by ingestion of sea water and food. Also, mysticetes seem to be less affected by heavy metal accumulation as a result of their position in the food chain, when compared to odontocetes that occupy the top level. Interestingly, heavy metal levels are higher in older individuals and non-breeding females. All these factors should be taken into account in studies on the effects of these metals on marine mammals.

A recent multi-factorial study [73] on mass stranded sperm whales (*Physeter macrocephalus*) revealed relatively high levels of environmental pollutants, including organic Hg (MeHg), Se, Cd, with a Hg:Se ratio of 1:1. Also, opportunistic bacteria cultured from selected organs included representatives of *Vibrio, Aeromonas hydrophila* and *Enterococcus*. The authors concluded that these whales were presumably starved causing the mobilization of lipophilic contaminants which accumulated in the adipose tissue. These chemical compounds entered the blood circulation causing immuno- and neurotoxic effects which led to impaired orientation and space perceptions in these sperm whales [73].

6.3. POPs

It is a generally accepted fact that that persistent organic pollutants (POPs) cause harm to human and animal tissue, earning them a reputation of 'the most widespread and toxic group of pollutants', as well as 'poisons without passports' [74-78]. Humans and animals worldwide were found to be carriers of POPs; from trace up to harmful amounts in their bodies [79-81]. Persistent organic pollutants include the chlorinated pesticides aldrin, dieldrin, dichlorodiphenyltrichloroethane (DDT), endrin, heptachlor, hexachlorobenzene, mirex and toxaphene, as well as industrial chemicals like PCBs, and the unwanted waste by-products polychlorinated dibenzo-p-dioxin and dibenzofurans (PCDD/F), and brominated flame retardants [75]. These chemicals have common, but distinct chemical and physical characteristics including:

- POPs 'persist' in a certain environment resisting physical, chemical and/or biological processes of break-down;
- POPs are transported via air/water currents over long distances because they are semi-volatile causing them to evaporate slowly and enter the atmosphere. They then return to earth in rain and snow in colder areas leading to accumulation in regions such as the Arctic which is thousands of kilometres away from their source;

- POPs have adverse effects in human and animal tissue even at very low concentrations. Some of these compounds disrupt normal biological functions, including hormone and other chemical messengers leading to metabolic conditions [79];
- POPs have a low water and high lipid solubility, resulting in bio-accumulation in the fatty tissue of living organisms and cannot be excreted readily. Also, bio-magnification occurs causing an accumulation effect by factors of many thousands or even millions, as these compounds move up food chains [74, 77, 80, 82].

These chemical compounds are primarily products and by-products of industrial processes, synthetic chemical manufacturing and waste incineration ([80, 81]. Their existence dates back to the industrial boom after World War II, but are currently ubiquitous, and are found in food as well as soil, the atmosphere and various water bodies. In 2001, the United Nations Environment Program (UNEP) completed global negotiations with the signing of the so-called Stockholm convention on banning certain POPs, collectively known as the 'dirty dozen'. During this assessment, certain criteria for the identification and listing of chemicals under the convention were identified. New POPs that were identified during the survey include butylated tin, methylated mercury and polyaromatic hydrocarbon (PAH), as well as other less studied compounds such as chlorinated paraffin's, brominated diphenyl ethers and other flame retardants.

The big concern caused by organo-chlorines is because of growing evidence that these compounds act as endocrine disruptors [79, 83]. The U.S. EPA [83] report defined endocrine disruptors as 'exogenous agents that interfere with the synthesis, secretion, transport, binding, action or elimination of natural hormones in the body that are responsible for the maintenance of homeostasis, reproduction, development and/or behaviour'. The science of endocrine disruption is still very new (only about 2 decades) and long term studies with effect results are therefore still preliminary [80]. The hormone disrupting effects differ according to the exposure situation and depend on the relative occurrence of the active congeners, specifically on the trophic level from which the food originates. For example, contaminants in the marine food chain of the Arctic Inuit population follow a very long passage. The result is that the higher, slowly metabolized, higher chlorinated PCBs will dominate over the lower chlorinated and more readily metabolized congeners. This situation will therefore create a specific effect on the hormonal balance, which will be different from other populations exposed more directly to the sources of these contaminants [76]. Toxic contaminants can also act as causal or aggravating factors in the development of a range of metabolic disorders. Several studies reported a correlation between metabolic disorders such as diabetes and cardiovascular disease, and lipid adjusted serum levels of substances like PCBs and dioxins. At present, 80% of the adult Greenlandic human population has PCB serum values in excess of 10 μg/l; a concentration shown to cause increased incidences of diabetes. Processes related to the development of metabolic disorders in which toxic contaminants may play an aggravating role include: (1) Pro-inflammatory effect through the formation of pro-inflammatory cytokines, oxidative stress and/or the formation of reactive oxygen species, (2) modulation of fatty acid metabolism, (3) influence on nuclear receptors, (4) effects on steroidogenesis, and (5) influence on uric acid levels [76].

In cetaceans, POP pollution has been documented to cause a variety of species-specific and congener-specific toxic and physiological effects. These include the formation of cancers, reproductive and endocrine impairment, skeletal anomalies, immune-suppression, as well as organ-specific disorders [1, 79, 84]. Also, the POP effects are often not seen in the exposed generation but rather in the second or third generation offspring [83, 85]. Since these compounds are lipophilic, they accumulate in the blubber of cetaceans and other marine mammals [84]. In cetaceans, PCBs are a recognized immuno-suppressant and many researchers believe that high levels of these and other POPs reduce resistance of these animals to disease due to their poor ability to metabolize these compounds [1, 58, 79, 84]. If contamination levels are high enough, it is possible for marine pollution to cause outright deaths of cetaceans. One such case was in the St. Lawrence Estuary in Canada where a marine reserve was established for a resident population of beluga whales. At the time of the article, about one beluga corpse was being washed up every week. These whales had signs of depressed immune systems, complications with digestive systems and carcinogenic tumours. Clinical testing revealed levels of contamination so high that the corpses had to be treated as toxic waste under the Canadian legislation [86].

6.4. Plastics

Plastics are widely used globally for packaging and storage, because they are relatively inexpensive, light weight, convenient and do not break easily. However, they also create widespread environmental concerns, because they are manufactured from petroleum, which is a non-renewable and usually imported resource [87]. Moreover, plastics degrade very slowly, resulting in alarming volumes ending up in landfills and the marine environment. Plastics also present a health hazard. Cooking and storing food in plastic containers cause migration of chemicals, such as Bisphenol A (BPA), into food and beverages. Types of plastics that have been shown to leach these substances are polycarbonate, polyvinyl chloride (PVC) and styrene. Interestingly, the leaching effect increases with heating, freezing and contact with oily or fatty food. A number of studies revealed evidence that human and wildlife populations are exposed to levels of BPA high enough to cause harmful developmental and reproductive effects in a number of species and laboratory animal studies [88]. Canada's government ruled in April 2008 that BPA is harmful to infants and toddlers and announced plans to ban certain products. Some states in the USA are also considering bills to restrict or ban BPA from children's products (Reuters, 14 September 2008). However, after considering extensive research, the European Commission concluded (2008) that products containing BPA were safe for consumers, as long as the products were used as indicated by the manufacturer. In Canada, a ban on the importation, sale and advertising of polycarbonate baby bottles was enforced, together with implementation of efforts to reduce levels of BPA in infant formula, to the lowest achievable concentrations. The use of water bottles, sport bottles, sport equipment, etc. was considered to be safe and exposure from these, regarded to be very low. However, BPA was listed as 'CEPA toxic' in Canada in October 2010, to allow Environment Canada to establish water quality standard to restrict BPA levels in effluent discharges to the environment [89]. BPA is a common component of wastewater entering the oceans [90-92].

7. Conclusions

Under natural conditions and in a pristine environment, cetacean skin usually acts as an effective barrier against the environment, as well as against potential pathogenic microbes. The relatively thick, keratin-rich skin provides a physical barrier against injury and penetration by pathogens, while immune cells of both the innate and adaptive immune system occur in the skin. Despite this seemingly formidable barrier numerous different skin lesions were reported on cetaceans all over the globe during past decades. Many microbial species, including known pathogens, were found to be associated with the lesions. Viruses, bacteria and fungi were frequently encountered in the lesions, which seems to be more prevalent among immuno-compromised cetacean populations, and those subjected to pollution.

Typical toxic pollutants that enter the coastal waters via urban effluents all over the globe are heavy metals, POPs and plastics. The detrimental effects of these toxins on mammalian physiology are well known and include neuro-toxic effects, endocrine disruption, harmful effects on the reproductive system, as well as an impaired immune system. The latter would render mammals, including cetaceans, more susceptible to microbial infections. Moreover, taking into account that the large volumes of sewage effluents entering the oceans contain opportunistic pathogens, such as keratinophilic fungi, the development of skin lesions on a cetacean with an already compromised immune system as a result of toxic pollutants, seems inevitable.

Numerous case studies were reported where skin lesions were found on cetaceans from polluted waters. However, a clear connection between pollution levels and lesion incidence based on sound statistical analyses was not established. Nevertheless, surveys conducted on similar dolphin populations subjected to different levels of pollution, indicated that skin lesions among these animals were more prevalent in populations subjected to a polluted environment. Considering these results, together with known physiological effects of toxic pollutants, skin lesions among cetaceans may be indicative of an ecosystem under severe pressure as a result of anthropogenic activities. Since pollution levels are increasing in all the oceans of the world, it is imperative that more correlations between cetacean skin lesions and pollution levels be expediently studied on ecosystem, organismal, cellular and sub-cellular levels. The findings of such studies can be used by decision makers to manage anthropogenic activities, in such a manner that pollution to the marine environment is reduced for a sustainable planet.

Author details

Marnel Mouton* and Alfred Botha
Department of Microbiology, University of Stellenbosch, Stellenbosch, South Africa

* Corresponding Author

Acknowledgement

Dr. Nicolene Botha for the preparation of the drawing and assistance with artwork.

8. References

[1] Raaymakers S. Marine Pollution & Cetaceans – Implications for Management. Encounters with Whales '93 Workshop Series No 20, Proceedings of a conference to further explore the management issues relating to human/whale interactions, 6-10 September 1993, Great Barrier Reef Marine Park Authority, Townsville, Australia; 1993. p82-87.

[2] Simpson CF, Wood FG, Young F. Cutaneous lesions on a porpoise with Erysipelas. Journal of the American Veterinary Medical Association 1958; 133 558-560.

[3] Japha A. Über den Bau der Haut des Seihwales *Balaenoptera borealis* (Lesson). Zoologischer Anzeiger 1905; 29 442-445.

[4] Parry DA. The structure of whale blubber, and a discussion of its thermal properties. Quarterly Journal of Microscopic Science 1949; S3-90 13-25.

[5] Sokolov VE. Comparative morphology of skin of different orders: Ordo Cetacea, In: Mammal Skin. University of California Press Ltd.; 1982. p284-324.

[6] Hicks BD, St. Aubin DJ, Gerach JR, Brown WR. Epidermal growth in the bottlenose dolphin, *Tursiops truncatus.* The Journal of Investigative Dermatology 1985; 85 60-63.

[7] Geraci JR, Hicks BD, Aubin DJ. Dolphin Pox: A skin disease of cetaceans. Canadian Journal of Comparitive Medicine 1979; 43 399-404.

[8] Spearman RIC. The epidermal stratum corneum of the whale. Journal of Anatomy 1972; 113(3) 373-381.

[9] Pfeiffer CJ, Rowntree VJ. Epidermal ultrastructure of the southern right whale calf (*Eubabaena australis*). Journal of Submicroscopic Cytology and Pathology 1996; 28(2) 277-286.

[10] Schweizer J , Bowden PE, Coulombe PA, Langbein L, Lane EB, Magin TM, Maltais L, Omary MB, Parry DAD, Rogers MA, Wright MW. New consensus nomenclature for mammalian keratins. Journal of Cell Biology 2006; 174(2) 169-174.

[11] Sharma R, Rajak RC. Keratinophilic fungi: Nature's keratin degrading machines! Resonance September 2003; 28-40.

[12] Menon GK, Grayson S, Brown BE, Elias PM. Lipokeratinocytes of the epidermis of a cetacean (*Phocena phocena*). Cell and Tissue Research 1986; 244 385-394.

[13] Zabka TS, Romano TA. Distribution of MHC II cells in skin of Atlantic bottlenose dolphins (*Tursiops truncatus*): An initial investigation of dolphin dendritic cells. The Anatomical Record Part A 2003; 273A 636-647.

[14] Willey JM, Sherwood LM, Woolverton CJ. Precott's Microbiology. 8th Edition. McGrawHill; 2011.

[15] Meyer W, Seegers U. A preliminary approach to epidermal antimicrobial defense in the Dephinidae. Marine Biology 2004; 144 841-844.

[16] De Hoog GS, Guarro J, Gené J, Figueras MJ. Atlas of Clinical Fungi, Second Edition, Centraalbureau voor Schimmelcultures ; 2000.

[17] Shoham S, Levitz SM. The immune response to fungal infections. British Journal of Haematology 2005; 129 569-582.

[18] Kunert J. Biochemical mechanism of keratin degradation by the actinomycete *Streptomyces fradiae* and the fungus *Microsporum gypseum*: A comparison. Journal of Basic Microbiology 1989; 29(9) 597-604.

[19] Gradišar H, Friedrich J, Križaj I, Jerala R. Similarities and specifities of fungal keratinolytic proteases: Comparison of keratinases of *Paecilomyces marquandii* and *Doratomyces microsporus* to some known proteases. Applied and Environmental Microbiology 2005; 71(7) 3420-3426.

[20] Marchisio VF. Keratinophilic fungi: Their role in nature and degradation of keratinic substrates. Revista Iberoamericana de Micologia, Apdo 699, E-48080, Bilbao, Spain; 2000. p86-92.

[21] Poovendran P, Kalaigandhi V, Kanan VK, Jamuna rani E, Poongunran E. A study of feather keratin degradation by *Bacillus licheniformis* and quantification of keratinase enzyme produced. Journal of Microbiology and Biotechnology Research 2011; 1(3) 120-126.

[22] Tsuboi R, Ko I-J, Takamori K, Ogawa H. Isolation of keratinolytic proteinase from *Trichophyton mentagrophytes* with enzymatic activity at acidic pH. Infection and Immunity 1989; 57(11) 3479-3483.

[23] Awad MF, Kraume M. Keratinophilic fungi in activated sludge of wastewater treatment plants with MBR in Berlin, Germany. Mycology 2011; 2(4) 276-282.

[24] Gupta R, Ramnani P. Microbial keratinases and their prospective applications: An overview. Applied Microbiology and Biotechnology 2006; 70 21-33.

[25] Kim J-D. Keratinolytic activity of five *Aspergillus* species isolated from poultry farming soil in Korea. Mycobiology 2003; 31(3) 157-161.

[26] Mazotto AM, De Melo ACN, Macrae A, Rosado AS, Peixoto R, Cedrola SML, Couri S, Zingali RB, Villa ALV, Rabinovitch L, Chaves JQ, Vermehlo AB. Biodegradation of feather waste by extracellular keratinases and gelatinases from *Bacillus* spp. World Journal of Microbioly and Biotechnology 2011; 27 1355-1365.

[27] Varaldo PE, Kilpper-Balz R, Biavasco F, Satta G, Schleifer KH. *Staphylococcus delphini* sp. nov., a coagulase-positive species isolated from dolphins. International Journal of Systematic Bacteriology 1988; 38(4) 436-439.

[28] Shotts EB Jr., Albert TF, Wooley RE, Brown J. Microflora associated with the skin of the bowhead whale (*Balaena mysticetus*). Journal of Wildlife Diseases 1990; 26(3) 351-359.

[29] Baker JR. Skin disease in wild cetaceans from British waters. Aquatic animals 1992; 18(1) 27-32.

[30] Henk WG, Mullan DL. Common epidermal lesions of the bowhead whale, *Balaena myticetus*. Scanning Microscopy 1996; 10(3) 905-916.

[31] Frasca S Jr, Dunn JL, Cooke JC, Buck JD. Mycotic dermatitis in an Atlantic white-sided dolphin, a pygmy sperm whale, and two harbor seals. Journal of the American Veterinary Medical Association 1996; 208(5) 727-729.

[32] Harzen S, Brunnick BJ. Skin disorders in bottlenose dolphins (*Tursiops truncatus*), resident in the Sado estuary, Portugal. Aquatic Mammals 1997; 23(1) 59-68.

[33] Wilson B, Arnold H, Bearzi G, Fortuna CM, Gaspar R, Ingram S, Liret C, Pribanić S, Read AJ, Ridoux V, Schneider K, Urian KW, Wells RS, Wood C, Thompson PM, Hammond PS. Epidermal diseases in bottlenose dolphins: impacts of natural and anthropogenic factors. Proceedings of the Royal Society of London B 1999; 266 1077-1083.

[34] Mikaelian I, Lapointe J-M, Labelle P, Higgins R, Paradis M, Martineau D. Case report. *Dermatophilus*-like infection in beluga whales, *Delphinapterus leucas,* from the St. Lawrence estuary. Veterinary Dermatology 2001; 12 59-62.

[35] Martineau D, Lemberger K, Dallaire A, Labelle P, Lipscomb TP, Michel P, Mikaelian I. Cancer in wildlife, a case study: Beluga from St. Lawrence estuary, Québec, Canada. Environmental Health Perspectives 2002; 110(3) 285-292.

[36] Pettis HM, Rolland RM, Hamilton PK, Brault S, Knowlton AR, Kraus SD. Visual health assessment of North Atlantic right whales (*Eubalaena glacialis*) using photographs. Canadian Journal of Zoology 2004; 82 8-19.

[37] Hamilton PK, Marx MK. Skin lesions on North Atlantic right whales: categories, prevalence and change in occurrence in the 1990s. Diseases of Aquatic Animals 2005; 68 71-82.

[38] Van Bressem M-F, Van Waerebeek K. Epidemiology of poxvirus in small cetaceans from the eastern south Pacific. Marine Mammal Science 2006a; 12 371-382.

[39] Van Bressem M-F, Van Waerebeek K, Montes D, Kennedy S, Reyes JC, Garcia-Godos A, Onton-Silva K, Alfaro-Shigueto J. Diseases, lesions and malformations in the long-beaked common dolphin *Delphinus capensis* from the Southeast Pacific. Diseases of Aquatic Organisms 2006b; 68 149-165.

[40] Van Bressem M-F, Van Waerebeek K, Reyes JC, Felix F, Echegaray M, Siciliano S, Di Beneditto AP, Flach L, Viddi F, Avila IC, Herrera JC, Toron IC, Bolaňos-Jimenez J, Moreno IB, Ott PH, Sanino GP, Castineira E, Montes D, Crespo E, Flores PA, Haase B, Mendonca De Souza SMF, Laeta M, Fragoso AB. A preliminary overview of skin and skeletal diseases and traumata in small cetaceans from South American waters. Latin American Journal of Aquatic Mammals 2007; 6(1) 7-42.

[41] Kiszka J, Van Bressem M-F, Pusineri C. Lobomycosis-like disease and other skin conditions in Indo-Pacific bottlenose dolphins *Tursiops aduncus* from the Indian Ocean. Diseases of Aquatic Organisms 2009; 84 151-157.

[42] Bearzi M, Rapoport S, Chau J, Saylan C. Skin lesions and physical deformities of coastal and offshore common bottlenose dolphins (*Tursiops truncatus*) in Santa Monica Bay and adjacent areas, California. Ambio 2009; 38(2) 66-71.

[43] Van Bressem M-F, De Oliveira Santos MC, De Faria Oshima JE. Skin diseases in Guiana dolphins (*Sotalia guianensis*) from the Paranaguá estuary Brazil: A possible indicator of a compromised marine environment. Marine Environmental Research 2008; 67 63-68.

[44] Reeb D, Best PB, Botha A, Cloete KJ, Thornton M, Mouton M. Fungi associated with the skin of a southern right whale (*Eubalaena australis*) from South Africa. Mycology 2010; 1(3) 155-162.

[45] Mouton M, Reeb D, Botha A, Best P. Yeast infection in a beached southern right whale (*Eubalaena australis*) neonate. Journal of Wildlife Diseases 2009; 45(3) 692-699.

[46] Maldini D, Riggin J, Cecchetti A, Cotter MP. Prevalence of epidermal conditions in California coastal bottlenose dolphins (*Tursiops truncatus*) in Monterey Bay. Ambio 2010; 39(7) 455-462.

[47] Apprill A, Mooney TA, Lyman E, Stimpert AK, Rappé MS. Humpback whales harbour a combination of specific and variable skin bacteria. Environmental Microbiology Reports 2011; 3(2) 223-232.

[48] Martinez-Levasseur LM, Gendron D, Knell RJ, O'Toole EA, Singh M, Acevedo-Whitehouse K. Acute sun damage and photoprotective responses in whales. Proceedings of the Royal Society B 2011; 278 1581-1586.

[49] Fury CA, Reif JS. Incidence of poxvirus-like lesions in two estuarine dolphin populations in Australia: Links to flood events. Science of the Total Environment 2012; 416 536-540.

[50] Bracht AJ, Brudek RL, Ewing RY, Manire CA, Burek KA, Rosa C, Beckmen KB, Maruniak JE, Romero CH. Genetic identification of novel poxviruses of cetaceans and pinnipeds. Archives of Virology 2006; 151 423-438.

[51] Moeller Jr. RB. Pathology of marine mammals with special reference to infectious diseases. In: Vos JG, Bossart GD, Fournier M, O'Shea TJ. (eds.) Toxicology of Marine Mammals. Vol. 3. , Taylor & Francis; 2003. p3-37.

[52] Martineau D, LaGacé A, Belard P, Higgins R, Amstrong D, Shugart LR. Pathology of stranded beluga whales (*Delphinapterus leucas*) from the St. Lawrence Estuary, Quebec, Canada. Journal of Comparative Pathology 1988; 98 287-311.

[53] Smith AW, Skilling DE, Ridgway S. Calicivirus-induced disease in cetaceans and probable interspecies transmission. Journal of American Veterinary Medicine Association 1983; 183(11) 1223-1225.

[54] Smith AW, Boyt PM. Calicivirus of ocean origin: A review. Journal of Zoo and Wildlife Medicine 1990; 21 3-23.

[55] Buck JD, Overstrom NA, Patton GW, Anderson HF, Gorzelany JF. Bacteria associated with stranded cetaceans from the northeast USA and southwest Florida Gulf coasts. Diseases of Aquatic Animals 1991; 10 147-152.

[56] St. Leger J. Gross morbid pathology of marine mammals. In Davis CL. (ed.) Gross Morbid Anatomy of diseases of Animals. 2007.

[57] Nymo IH, Tryland M, Godfroid J. A review of *Brucella* infection in marine mammals, with special emphasis on *Brucella pinnipedialis* in the hooded seal (*Cystophora cristata*). Veterinary Research 2011; 43 93-117.

[58] Dhermain F, Soulier L, Bompar J-M. Natural mortality factors affecting cetaceans in the Mediterranean Sea. In: Notarbartolo di Sciara G. (ed.) Cetaceans of the Mediterranean and Black Seas: state of knowledge and conservation strategies, a report to the ACCOBAMS Secretariat, Monaco; 2002. Section 15, p14.

[59] Higgins R. Bacteria and fungi of marine mammals. Canadian Veterinary Journal 2000; 41 105-116.

[60] Migaki G, Jones SR. Mycotic diseases. In: Howard EB. (ed.) Pathobiology of Marine Mammal Diseases. Boca Raton; 1983. p1-28.
[61] Moeller Jr. RB. Diseases of Marine Mammals. LTC; 1997.
[62] Murdoch EM, Reif JS, Mazzoil M, McCulloch SD, Fair PA, Bossart GD. Lobomycosis in bottlenose dolphins (*Tursiops truncatus*) from the Indian River lagoon, Florida: Estimation of prevalence, temporal trends, and spatial distribution. EcoHealth 2008; 5 289-297.
[63] De Vries GA, Laarman JJ. A case of Lobo's disease in the dolphin *Sotalia guianensis.* Aquatic Mammals 1973; 1 26-29.
[64] Migaki G, Valerio MG, Irvine B, Garner FM. Lobo's disease in an Atlantic bottle-nosed dolphin. Journal of the American Veterinary Medical Association 1971; 159 578-582.
[65] Reif JS, Fair PA, Adams J, Joseph B, Kilpatrick DS, Sanschez R. Health status of Atlantic bottlenose dolphins (*Tursiops truncatus*) from the Indian River Lagoon, FL and Charleston, SC. Journal of the American Veterinary Medical Association 2008; 233(2) 299-307.
[66] Corcoran E, Nellemann C, Baker E, Bos R, Osborn D, Savelli H. Sick Water? The central role of wastewater management in sustainable development. A Rapid Response Assessment. United Nations Environment Programme. UN-HABITAT, GRID-Arendal; 2010.
[67] De Boer MN, Baldwin R, Burton CLK, Eyre EL, Jenners M-NM, Keith SG, McCabes KA, Parsons ECM, Peddemors VM, Rosenbaum HC, Rudolph P, Simmonds MP. Cetaceans in the Indian Ocean Sanctuary: A review. A WDCS Report; 2002.
[68] Taljaard S, Monteiro PMS, Botes WAM. A structured ecosystem-scale approach to marine water quality management. Water S.A. 2006; 4535-542.
[69] Bitton G. Public health aspects of wastewater and biosolids disposal in the marine and other aquatic environments. In: Wastewater Microbiology, 2nd Edition. Wiley-Liss; 1999. p449-460.
[70] Järup L. Hazards of heavy metal contamination. British Medical Bulletin 2003; 68 167-182.
[71] Coetzee MAS. Water pollution in South Africa: its impact on wetland biota. In: Cowan GI (ed.) Wetlands of South Africa. Department of Water Affairs and Tourism; 1995. p247-262.
[72] Das K, Debacker V, Pillet S, Bouquegneau J-M. Heavy metals in marine mammals. In: Vos JG, Bossart GD, Fournier M, O'Shea TJ. (eds.) Toxicology of Marine Mammals. Taylor & Francis; 2003. p135-167.
[73] Mazzariol S, Di Guardo G, Petrella A, Marsili L, Fossi CM, Leonzio C, Zizzo N, Vizzini S, Gaspari S, Pavan G, Podestà M, Garibaldi F, Ferrante M, Copat C, Traversa D, Marcer F, Airoldi S, Frantzis A, De Bernaldo Quirós Y, Cozzi B, Fernandez A. Sometimes sperm whales (*Physeter macrocephalus*) cannot find their way to the high seas: A multidisciplinary study on a mass stranding. PLoS ONE 2011; 6(6) 1-13.
[74] AMAP Assessment. Persistent Organic Pollutants in the Arctic (Chapter 1). Arctic Monitoring and Assessment Program (AMAP). Oslo, Norway; 2002.

[75] AMAP Report. Persistent toxic substances, food security and indigenous peoples of the Russian north. Final Report (Chapter 3). Arctic Monitoring and Assessment Program (AMAP). Oslo, Norway; 2004.

[76] Jensen JC, Deutch B, Øyvind Odland J. Dietary transition and contaminants in the Arctic: emphasis on Greenland. International Association of Circumpolar Health Publishers; 2008.

[77] The International POPs Elimination Network (IPEN). Establishing the Prevalence of POPs Pesticide Residues in Water, Soil and Vegetable Samples and Creating Awareness About their Ill-effects. 2006. http://www.ipen.org.

[78] UNECE; 1994. http://www.unece.org/env/lrtap/status/94s_st.htm.

[79] Colborn T, Smolen MJ. Cetaceans and contaminants. In: Vos JG, Bossart GD, Fournier M, O'Shea TJ. (eds.) Toxicology of Marine Mammals. Taylor & Francis; 2003. p291-332.

[80] IPEP. The International POPS Elimination Project (IPEP). 2006. http://www.oztoxics.org/ipepweb/index.html.

[81] Wells M, Leonard L. DDT contamination in South Africa by Groundwork. The International POPs Elimination Project (IPEP); May 2006.

[82] Nfon E, Cousins IT, Broman D. Biomagnification of organic pollutants in benthic and pelagic marine food chains from the Baltic Sea. Science of the Total Environment 2008; 397 190-204.

[83] U.S. EPA. Exposure Factors Handbook (1997 Final Report). U.S. Environmental Protection Agency, Washington, DC, EPA/P-95/002F a-c; 1997.

[84] Borrell A, Aguilar A, Cantos G, Lockyer C, Heide-Jørgensen MP, Jensen J. Organochlorine residues in harbour porpoises from Southwest Greenland. Environmental Pollution 2004; 128(3) 381-391.

[85] Han S-L, Stone D. A case study of POPs concentrations in wildlife and people relative to effects levels. 2001. http://www.chem.unep.ch/pops/POPs_Inc/proceedings (accessed 14 February 2008).

[86] Jones P. Beluga whale deaths. Marine Pollution Bulletin, 1991; 22(1).

[87] Le HH, Carlson EM, Chua JP, Belcher SM. Bisphenol A is released from polycarbonate drinking bottles and mimics the neurotoxic actions of estrogen in developing cerebellar neurons. Toxicology Letters 2008; 176 149–156.

[88] Wetherill YB, Akingbemi BT, Kanno J, McLachlan JA, Nadal A, Sonnenschein C, Watson CS, Zoeller RT, Belcher SM. In vitro molecular mechanisms of bisphenol A action. Reproductive Toxicology 2007; 24(2) 178-198.

[89] http://www.bisphenol-a-europe.org/index.php?page+additional-egislation (accessed 15 March 2012).

[90] Gómez MJ, Martínez Bueno MJ, Lacorte S, Fernández-Alba AR, Agüera A. Pilot survey monitoring pharmaceuticals and related compounds in a sewage treatment plant located on the Mediterranean coast. Chemosphere 2007; 66 993–1002.

[91] Kang J-H, Aasi D, Katayama Y. Bisphenol A in the aquatic environment and its endocrine-disruptive effects on aquatic organisms. Critical Reviews in Toxicology 2007; 37 607–625.

[92] Stasinakisa AS, Gatidou G, Mamais D, Thomaidis NS, Lekkas TD. Occurrence and fate of endocrine disrupters in Greek sewage treatment plants. Water Research 2008; 42 1796-1804.

Chapter 7

Uruguayan Pinnipeds (*Arctocephalus australis* and *Otaria flavescens*): Evidence of Influenza Virus and *Mycobacterium pinnipedii* Infections

Juan Arbiza, Andrea Blanc, Miguel Castro-Ramos,
Helena Katz, Alberto Ponce de León and Mario Clara

Additional information is available at the end of the chapter

http://dx.doi.org/10.5772/54214

1. Introduction

1.1. Location and general characteristics of the pinniped population

1.1.1. Otariid and phocid species of Uruguay

Uruguay has 450 km of shorelines along the La Plata River and 220 km along the Atlantic Ocean (MTOP-PNUD-UNESCO, 1980). Two species of Otariids breed and reproduce on Uruguayan Atlantic islands: the South American fur seal, *Arctocephalus australis* (Zimmermann, 1783) (Fig. 1), and the South American sea lion, *Otaria flavescens* (Shaw, 1800), (Fig. 2), (Ponce de León, 2000; Ponce de León & Pin 2006; Vaz-Ferreira, 1976, 1982). Both are polygynous, gregarious and show strong sexual dimorphism (Bartholomew, 1970). South American fur seal adult males reach lengths of 1.9 m and weigh from 120 kg to 200 kg, while females can reach 1.4 m long and weigh from 40 kg to 55 kg, and newborns can be from 0.4 m to 0.5 m long and weigh from 3.5 kg to 5.5 kg (Vaz-Ferreira, 1982). Sea lion males may reach 2.8 m and weigh up to 354 kg while adult females are much smaller, reaching up to 1.9 m long and weighing as much as 150.0 kg (Ponce de León, 2000). Newborns in this species are between 0.7 m and 0.9 m long and weigh from 10.0 kg to 17.0 kg (Cappozzo et al., 1994). A third pinniped species, the southern elephant seal *Mirounga leonina* (Fig. 3), is a frequent visitor of Uruguayan islands and shorelines, although its reproductive areas are located in Argentina. Elephant seals can reach up to 5 m, 3 m or 1.3 m in length for males, females and pups respectively, and they can weigh as much as 5,000 kg, 800 kg from 40 kg to 50 kg (Reeves et al., 1992).

Figure 1. Group of South American fur seal *Arctocephalus australis* males, females and pups on Lobos Island. Photograph: A. Ponce de León.

Figure 2. South American sea lion *Otaria flavescens* reproductive groups with pups on Marco Island with pups. Photograph: A. Ponce de León.

Figure 3. Young male southern elephant seal *Mirounga leonina* on Coronilla's Islet. Photograph: A. Ponce de León.

1.2. Brief exploitation history

The exploitation of fur seals by Europeans in Uruguay is known to have begun in 1516, soon after the Spaniards explored the South Western Atlantic Ocean. During this exploration, Juan Díaz de Solís discovered the La Plata River and his crew landed on Isla de Lobos, where they killed 66 seals for their meat to be salted and consumed on their way back to Europe. The first semi-organized commercial exploitation took place in 1724, and the seal oil obtained was used for illuminating the city of Maldonado. From 1792 the Real Compañía Marítima, under direct instructions of the King of Spain, was responsible for sealing, until England invaded the territory in 1808. Shortly after, seal harvesting was carried out by private concessionaries and controlled by the local Government. From 1873 to 1900 a total of 440,000 seals were slaughtered (annual average of 16,000 pelts), whereas no records are available from 1901 to 1909. Further on, from 1910 to 1942, 72,000 South American fur seals were killed, as well as 17,000 more between 1943 and 1947. Due to the uncontrolled exploitation, populations of both seal species began to decrease. After 1950 a new management scheme started on Isla de Lobos, based on the system used for Northern fur seals (*Callorhinus ursinus*) in Pribilof Islands (Alaska), and the harvest was restricted to males. Also, private sector concessions were suspended, and the Government directly organized the harvesting program and related activities. Between 1959 and 1991 a total of 276,000 South American fur seals were removed (about 8,400 animals per year) and from 1967 to 1978, 36,400 sea lions were also slaughtered (3,000 animals per year). Products taken were crude skins, oil, meat and male genitals. Pelts were tanned and prepared in specific areas in Uruguay. Carcasses and fat were processed to obtain special oil for making soap, cosmetics and paints. In the XIX century, seal oil was used for illuminating the main streets of some cities. The meat was sometimes dried and given to the Montevideo Zoo for feeding big cats, eagles and condors. Since 1980, genitals were processed and sold for preparing medicines and aphrodisiacs (Acosta y Lara, 1884; DINARA, 2006; Pérez Fontana, 1943; Ponce de León, 2000; Vaz-Ferreira 1982; Vaz-Ferreira & Ponce de León, 1984, 1987).

Harvesting and slaughtering of Uruguayan seals stopped in 1978 for South American sea lions and in 1991 for South American fur seals. From 1992 to the present day, the conservation and preservation of pinnipeds and cetacean species are under control of the National Direction of Aquatic Resources (DINARA: Dirección Nacional de Recursos Acuáticos).

1.3. Local geographical distribution

Uruguayan South American fur seal and sea lion colonies are located on three main islands in the Atlantic Ocean: 1) Isla de Lobos and Lobos Islet, 9,260 m off Punta del Este (Department of Maldonado); 2) Torres Group Islands (Rasa Island, Encantada Island and Islet) close to Polonio's Cape (Department of Rocha) and 3) Marco Island close to Valizas (Department of Rocha). There are two more small islets close to La Coronilla (Department of Rocha), where small groups of sea lions aggregate in reproductive areas (Fig. 4). Sometimes, a few South American fur seals also appear on these islands (Ponce de León, 2000; Ponce de León & Pin, 2006; Smith, 1934; Vaz Ferreira, 1950, 1952, 1956; Vaz Ferreira & Ponce de León, 1984, 1987).

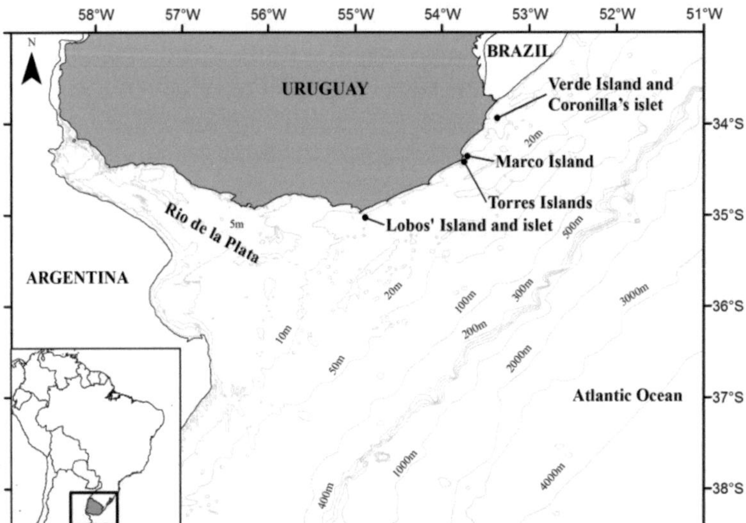

Figure 4. Location of Uruguayan South American fur seal and sea lion calving, breeding and mating islands close to the shorelines of the Departments of Maldonado and Rocha : Isla de Lobos (35º 01′ 38″ S – 54º 52′ 55″ W) and Lobos Islet; Rasa Island (34º 24′ 12″ S – 53º 46′ 10″ W), Encantada Island (34º 24′ 26″ S – 53º 45′ 56″ W), Torres Islet (34º 24′ 09″ S – 53º 44′ 59″ W); Marco Island (34º 20′ 59″ S – 53º 44′ 26″ W); 3) Verde Island and Coronilla's Islet (33º 56′ 21″ S – 53º 29′ 15″ W). Isobath data layer obtained from FREPLATA-Proyecto de Protección Ambiental del Río de la Plata y su Frente Marítimo (www.freplata.org).

1.4. Biology, population state and trends

Parturition and mating occur between November and January for South American fur seals and during January and February for South American sea lions (Franco-Trecu, 2005; Ponce de León, 2000, 2001; Ponce de León & Pin, 2006; Trimble, 2008). Gestation lasts around one year. In South American fur seals, lactation extends for several months and weaning begins between the 8th and the 12th month of age (Ponce de León, 1983, 1984, 2000; Ponce de León & Pin, 2006). From the 6th month of age, pups start eating fish and small mollusks as can be seen when analyzing stomach contents and gastrointestinal parasites of indirect cycle (Katz et al., 2012; Morgades et al., 2006). In some cases, South American fur seal lactation can be extended further, and the mother has to feed two pups from two consecutive breeding seasons: the yearling pup and the new one (Vaz-Ferreira & Ponce de León, 1987). In sea lions, there is a mother-pup relation for one year and in some cases for two (Vaz-Ferreira & Achaval, 1979) or possibly, up to three years (Soto, 1999). Little is known about the exact time of weaning, and whether pups are mixing milk with solid prey. It was suggested that weaning occurs when the mother actively rejects the older pup because a new one is born (Vaz-Ferreira, 1981; Vaz-Ferreira & Achaval, 1979).

South American fur seals have a lek reproductive system (Franco-Trecu, 2005). During the reproductive season, males fight each other to defend territories in very violent battles that can result in serious wounds and scars (Ponce de León, 2000; Ponce de León & Pin, 2006; Vaz-Ferreira, 1976, 1982; Vaz-Ferreira & Ponce de León, 1984, 1985, 1987). Females have no strong bonds with the areas defended by males. Fur seal colonies on islands are occupied by individuals from different age classes during the entire year. During the reproductive season there is a high density of animals in rocky areas as compared to sandy surfaces. As a consequence of the high environmental temperature, territorial males may abandon the reproductive areas in order to refresh themselves in the water (Vaz-Ferreira & Palerm, 1962; Vaz-Ferreira & Sierra de Soriano, 1962). After giving birth, South American fur seal females may remain with their pups for 6 to 11 days (Franco-Trecu, 2010) before starting short foraging trips that gradually become longer as the pups grow bigger and more independent (Ponce de León & Pin, 2006; Franco-Trecu, 2010).

The Uruguayan South American fur seal population is the biggest in South America (Vaz-Ferreira, 1982), with an annual growth rate of 3.3% (Páez, 2006) and an actual size estimated at 400,000 individuals. However, this species is included by the Convention on International Trade in Endangered Species of Wild Fauna and Flora (CITES) in the list of globally protected species because of population decline of other South American colonies (de Oliveira et al., 2006). For the International Union for Conservation of Nature (IUCN) the same species is listed as "Of low concern".

South American sea lions have a polygynous breeding system (Ponce de Leon & Pin, 2006; Trimble, 2008; Vaz-Ferreira, 1981; Vaz-Ferreira & Sierra de Soriano, 1962). The reproductive season extends from mid December to mid February. During this time males fight each other to establish territories and to defend females from other males (Campagna & Le Boeuf, 1988; Ponce de Leon & Pin, 2006). Pregnancy has an estimated duration of 363 days (Franco-

Trecu & Trimble, unpublished data). Territorial males display violent fights that may last at least one hour, and end with serious wounds. After parturition, sea lion females remain with their pups in order to suckle them for approximately one week and then start short foraging trips of three days, alternated with two-day suckling periods on land (Campagna & Le Boeuf, 1988). There is a decrease in the number of adult males and females at the end of the reproductive season because, after fasting during the breeding season, males begin their foraging period at sea. In addition, adult females alternate foraging trips at sea and suckling periods ashore, and move to other areas of the island (Franco-Trecu & Trimble, unpublished data; Ponce de León & Pin, 2006). Despite the fact that sea lions abandon the islands in order to move to feeding areas, there are generally some animals in the rookeries, even outside the reproductive season. In many other South American colonies this species maintains an increasing population growth (Grandi et al., 2008; Sepúlveda et al., 2006) and has been classified as low risk by the IUCN. However, sea lions in Uruguay are considered a highly endangered species due to their population decrease of 1.7 to 2% annually, with a total population estimated at only 12.000 individuals (Páez, 2006; Pedraza et al., 2009).

On Lobos Island and Lobos Islet, South American sea lion groups are found in small patches, surrounded by large groups of South American fur seals. However, on the Torres Islands (Rasa Island, Encantada Island and Islet), Marco Island and other islets, groups of sea lions are more numerous than on the bigger Lobos Island. According to Pedraza et al., (2009) and Ponce de León (unpublished data), sea lion populations are stable or increasing (2.4% annually) in Polonio´s Cape islands, while in Lobos Island their growth have a negative tendency. This is related to a positive trend in the *A. australis* population and may be an indicator of competition for territory (breeding areas), a process that occurs only on Lobos Island and Lobos Islet.

The breeding and reproductive areas of the elephant seal *Mirounga leonina* are located in sub Antarctic regions along the coast of South America (Campagna & Lewis, 1992; Lewis et al., 1998) and smaller colonies are formed in the Antarctic (Le Boeuf & Laws, 1994). The only large breeding colony of southern elephant seals on the South American continent is found in Península Valdés (42°04'S, 63°45'W) (Campagna & Lewis, 1992; Le Boeuf & Laws 1994; Lewis et al., 2004). Young, juvenile and adult animals migrate to northern regions (Lewis et al., 2006) and occur at different points all along the Uruguayan shoreline of the La Plata River and Atlantic Ocean beaches. During practically the whole year, elephant seals of both sexes are frequently seen in the flat areas of access to Lobos Island, Lobos Islet, Rasa Island and Coronilla Islet (Fig. 3 and 4). Mother and pup couples have been seen on Lobos Island in October, and in some odd cases, individuals have swum up the waters of the Uruguay River to the Departments of Rio Negro and Paysandú (Ponce de León & Pin, 2000).

1.5. Feeding, diet and diving behaviour

South American fur seals have mainly pelagic feeding habits (Naya et al., 2002; Ponce de León & Pin, 2000, 2006; Vaz-Ferreira, 1976) but also feed in shallower waters (Franco-Trecu, 2010). Their diet is basically composed of anchovies (*Engraulis anchoita, Anchoa marinii*), squid (*Illex argentinus, Loligo sanpaulensis*), hake (*Merluccius hubbsi*), striped weakfish (*Cynoscion guatucupa*), oceanic shrimp (*Pleoticus muelleri)* and cutlassfish (*Trichiurus lepturus*) (Frau & Franco-Trecu, 2010; Naya et al., 2002; Pin et al., 1996; Ponce de León et al., 1988; Ponce de León et al., 2000; Ponce de León & Pin, 2006; Vaz Ferreira 1976; Vaz-Ferreira & Ponce de León, 1984, 1987). Uruguayan fur seals usually do not interfere directly with artisanal and industrial fisheries, as they do not eat from nets nor destroy fishing gears (Ponce de León & Pin, 2006), though there are a few records of fur seal by-catch in artisanal (Franco-Trecu et al., 2009) and industrial fisheries (Szephegyi et al., 2010).

Diving records obtained by different researchers showed that during lactation, female fur seals perform dives of up to 186 m (media: 23.5 m ± 19.5 m) in depth with an average duration of 1.2 min ± 0.8 min (max. 5.3 min.) (Riet et al., 2010; York et al., 1998). These data suggest that females use both benthic and pelagic foraging strategies, and demonstrate their huge endurance for deep dives, apnea resistance and swimming ability. Diurnal dives were shallower and shorter than nocturnal ones (Riet et al., 2010). It was determined that lactating females consume different prey species, adapting their diving strategies to variations in food resources (Ponce de León & Páez, 1996; Ponce de León & Pin, 2006; Riet et al., 2010; York et al., 1998).

During early lactation, female sea lions perform dives of 21 m ± 8 m in depth with an average time of 1.9 min ± 0.7 min. Mean distance traveled per trip was 62.2 km ± 63.0 km. Foraging trips lasted 1.3 ± 0.8 days and did not exceed the continental shelf (>50 m of depth). Maximum distance from the colony was 98.60 km ± 31.3 km. These results indicate that during the breeding season females forage in coastal and shallow continental shelf areas (Riet et al., 2009, 2012). In autumn, foraging trips last 5 days (range: 1-14 days). Most animals seemed to complete round trips along the same tracks, meaning that each animal uses the same path on successive trips, with low overlap between individuals. Site fidelity to Lobos Island was highly remarkable for all animals, independently of their reproductive condition (Rodríguez et al., 2012).

Sea lions compete directly with small-scale coastal fishing and artisanal fisheries, feeding on species that are part of the fishermen's daily catch by stealing prey trapped in nets and longlines, and sometimes causing important damage or cracks in the gear (Franco-Trecu et. al., 2012; Lezama & Szteren, 2003; Ponce de León & Pin, 2006; Szteren & Páez, 2002). According to different authors, the sea lions' diet is mainly made up of coastal prey and some pelagic fishes: whitemouth croaker *(Micropogonias furnieri)*, striped weakfish (*Cynoscion guatucupa*), Brazilian codling (*Urophysis brasiliensis*), cutlassfish (*Trichiurus lepturus*), mackerel (*Trachurus lathami*), Argentinean conger (*Conger orbignyanus*), carangid (*Parona signata*), two species of anchovies (*Engraulis anchoita* and *Anchoa marinii*), and Argentinean croaker (*Umbrina canosai*) (Franco-Trecu, 2010; Naya et al., 2000; Pinedo & Barros, 1983; Ponce de León & Pin, 2006; Riet et al., 2011, 2012; Vaz Ferreira, 1981). As a consequence of interactions with sea lions, fishermen lost prey with high local commercial market value. Sometimes, fishermen find small shark specimens in their nets (*Mustelus schmitti, Galeorhinus galeus, Myliobatis* spp.) which have bite marks in their abdominal area

(Fig. 5) from sea lions that learned to exploit this energy reservoir (Ponce de León & Pin, 2006). Recent reports show that during the early lactation period, foraging home ranges of sea lion females overlapped with fishing effort areas of coastal bottom trawl fisheries (15%) and artisanal fisheries (>1%). For both fisheries the resource overlap per fisheries impact index identified the "hotspots" which are distributed along the coast, west of the breeding colony (56ºW - 55Wº) (Riet et al., 2011).

Figure 5. Artisan fishery capture of *Mylobatis* spp. The opened abdominal areas of the sharks are seen, from where sea lions have taken highly nutritive and energy rich livers and pancreas. Photograph: A. Ponce de León.

1.6. Future exploitation for tourism

South American fur seals and South American sea lions could represent an important tourism attraction. Since the seals are only a few meters away from visitors on Lobos Island and Polonio's Cape, seal watching in both of these popular natural areas could be exploited for tourism activities (Ponce de León & Pin, 2000, 2006; Ponce de León & Barreiro 2010). This type of exploitation should be regulated by serious and responsible Government rules in order to assure sustainable coastal management of environmental resources. New employment opportunities for local people in the Departments of Maldonado and Rocha would be created. This kind of offer would also contribute to public awareness-raising programs for conservation of these charismatic species and for conservation of aquatic ecosystems.

1.7. Management considerations

Nowadays, fur seals are an important nontraditional exportation item: between 60 and 80 living young fur seals are captured annually on Lobos Island and exported to aquaria and theme parks all over the world. The principal objective of these parks is to educate people about environmental issues and about the conservation of aquatic and marine resources and ecosystems. Although live sea lions had been sold by the Uruguayan Government since 1980, the exportation of living specimens of this Otariid was suspended in 2006 by DINARA-MGAP due to decreasing population numbers. Captures of animals are held in Lobos Island from mid March to mid November. This special period was defined in order to avoid disturbances and to be respectful of parturition, breeding and mating periods of the two Otariid species present in the island (Ponce de León, 2000; Ponce de León & Pin, 2006).

As far as research is concerned, DINARA maintains the old seal factory plant on Lobos Island, using parts of the buildings for providing accommodation to researchers who are developing studies in both species. Through the development of various research projects, the Government institution DINARA, students and graduates of the University of Uruguay (Universidad de la República) as well as from foreign countries, are gathering and collecting data to increase knowledge about the population dynamics of both species, which will also help in developing appropriate rules and guidelines for their management, and ensuring the conservation of Uruguayan natural resources.

2. Evidence of influenza virus infections in *Arctocephalus australis* individuals

Marine mammals are susceptible to a variety of pathogens including influenza viruses. In humans, influenza causes annual epidemics and occasional pandemic diseases, with a significant threat to human health. In wild animals, several outbreaks have been reported and especially marine mammals experienced several devastating episodes that highlight the importance of monitoring wild populations to perform conservation programs and to evaluate possible risks to human health.

Influenza viruses belong to the *Orthomyxoviridae* family and are enveloped viruses with a segmented, negative-sense RNA genome (Webster et al., 1992). Embedded in the lipid envelope, the hemagglutinin (HA) and neuraminidase (NA) proteins are responsible for virus attachment and release from host cells, respectively (Webster et al., 1992). This family of viruses is composed of four genera: influenza A, B and C viruses and Thogoviruses (Wright & Webster, 2001). While influenza B and C viruses are primarily "human" viruses, influenza A viruses infect a variety of avian and mammalian species including humans, horses, swine and marine mammals such as seals and cetaceans (Wright & Webster, 2001; Webster et al., 1992). Influenza B virus was isolated from a harbor seal (*Phoca vitulina*) for the first time in the year 2000 (Osterhaus et al., 2000) becoming a possible second reservoir of this virus.

Influenza viruses are unique among respiratory tract viruses as they undergo considerable antigenic variation. Both surface antigens of the influenza A viruses are subject to two types of variation: drift and shift. Antigenic drift involves minor changes in the hemagglutinin (HA) and neuraminidase (NA) and plays a role in influenza epidemics, which occur sporadically. Antigenic shifts involve major changes in these molecules resulting from replacement of the gene segment, producing new pandemic strains (Wright & Webster, 2001).

Phylogenetic evidence suggests that influenza epidemics in humans and other mammals, including seals, come from mutation and antigenic drift of viruses originating from aquatic birds (Webster et al., 1992). Several influenza events which have affected marine mammals have been described since the late seventies. The New England coast was the scene of an episode of influenza virus between December 1979 and November 1980. More than 400 harbor seals (*Phoca vitulina*) died of acute pneumonia associated with the influenza virus A/Seal/Massachusetts/1/80 (H7N7). This was the first evidence of an influenza virus antigenically and genetically related to avian viruses that could be associated with severe disease in wild animals (Geraci et al., 1982; Lang et al., 1981; Webster et al., 1981). This H7N7 strain was associated with an approximate 20% mortality of the seal population and also showed potential for causing conjunctivitis in humans. However, it was not transmitted among humans.

A new event was described along the New England coast from June 1982 through March 1983. This time the influenza virus isolation was an H4N5 subtype, which had previously been detected only in birds. It was recovered from harbor seals dying of viral pneumonia (Hinshaw et al., 1984). This strain, which caused an estimated mortality of 2 % to 4 %, was found to be genetically and serologically related to avian strains.

In January 1991 and January to February 1992, influenza A viruses were isolated from seals that died of pneumonia along the Cape Cod Peninsula in Massachusetts. Antigenic characterization identified two H4N6 and three H3N3 viruses. This was the first isolation of an H3 influenza virus from seals, although this subtype is frequently detected in birds, pigs, horses and humans (Callan et al., 1995). Genetic analysis indicated that the viruses were both of avian origin and that transmission from birds to seals was the most likely possibility.

Also, indirect evidence of influenza infection was reported from a variety of marine mammal species. In pinnipeds, antibodies against influenza A virus were detected in sera from harp seals (*Phoca groenlandica*) and hooded seals (*Cystophora cristata*) collected between 1991 and 1992 in the Barents Sea (Steuen et al., 1994), as well as from sea lions (Otariidae) and seals in the North and Bering seas (De Boer et al., 1990), and a ringed seal (*Pusa hispida*) in Alaska (Danner et al., 1998). A serological survey of influenza A antibodies from five species of marine mammals collected from Arctic Canada between 1984 and 1998, revealed that 2.5% of ringed seals (*Phoca hispida*) were serologically positive (Nielsen et al., 2001). A serological study of influenza virus infection in Caspian seals (*Phoca caspica*) detected antibodies to human-related (H3N2) virus in 36% of the seals (Ohishi et al., 2002). Two years later, another study suggested that human-related H3 viruses were prevalent in Baikal seals (*Phoca sibirica*) and ringed seals (*Pusa hispida*) inhabiting the central Russian Arctic (Ohishi et al., 2004). Serological evidence of influenza A virus infection was reported in Kuril harbor seals (*Phoca vitulina stejnegeri*) of Hokkaido, Japan, from samples collected between 1998 and 2005 (Fujii et al., 2007). In this study, antibodies to H3 and H6 subtypes of influenza A virus were detected. This was the first time that H6 antibodies were identified in seals (Fujii et al., 2007).

Indirect evidence of influenza A viruses has been reported in 27% of the South American fur seals sampled in Uruguay (Blanc et al., 2009). By Hemagglutination Inhibition Assay (HAI) it was found that all the positive samples reacted with A/New Caledonia/20/99(H1N1) antigen reaching HAI titer of 320 but none of the sampled serum reacted with A/Panamá/2007/99(H3N2) antigen. For the first time, the presence of influenza A in *A. australis* was confirmed (Blanc et al., 2009) (Fig. 6).

Influenza viruses have also been detected in whales. An H1N3 virus was isolated from a striped whale in the South Pacific (Lvov et al., 1978). In 1984 influenza A viruses of the H13N2 and H13N9 subtypes were isolated from a pilot whale (*Globicephala melas*) (Hinshaw et al., 1986). Serological, molecular, and biological analyses indicate that the whale isolates are closely related to the H13 influenza viruses from gulls (Hinshaw et al., 1986). In cetaceans, specific antibodies were observed in a low portion of sera from belugas (*Delphinapterus leucas*) in Arctic Canada (Nielsen et al., 2001).

Few studies have been reported regarding the detection of influenza B viruses in marine mammals. The first one reported the isolation of influenza B virus (B/seal/Netherlands/1/99) from a naturally infected harbor seal in the year 2000. Sequence analyses as well as serology indicated that this influenza B virus is closely related to strains that circulated in humans 4 to 5 years earlier. Retrospective analyses of sera collected from 971 seals showed a prevalence of antibodies of the influenza B virus in 2% of the animals after 1995, and in none before that year, suggesting that the virus was introduced in the seal population from a human source around 1995 (Osterhaus et al., 2000). Antibodies to influenza B viruses were detected by ELISA in 14% and 10% of serum samples collected from Caspian seals in 1997 and 2000, respectively (Ohishi et al., 2002).

Serologic evidence of influenza B virus has been reported from South American Uruguayan fur seals *A. australis* (Blanc et al., 2009). Thirty of the 37 serum samples assayed by HAI reacted against one of the three antigens used: 25/37 (68%) reacted against B/Beijing/184/93-like viruses, 20/37 (54%) reacted against B/Hong Kong/330/01, and 24/37 (65%) reacted against B/Sichuan/379/99. The results show that 17 sera reacted against all B antigens, only six reacted against two antigens and eight sera did not react against any of them. The highest titer reached was (640) against B/sichuan antigen. The results demonstrated influenza B virus circulation in South American fur seals for the first time in our country and in this species. The antigens assayed correspond to strains that circulated in humans between the years 1999 and 2001, 3 to 5 years after the study was carried out, confirming the hypothesis of other authors that marine mammals could be a reservoir of influenza strains that circulated in the past (Fig. 6).

It is important to consider that marine mammals share their habitat with several different wild shorebirds as well as with aquatic birds, the main influenza A virus hosts. The presence of bird feces in water, which can shed high concentrations of Avian Influenza viruses, and the close contact during feeding activities between birds and seals, increase the probability of fecal-oral transmission.

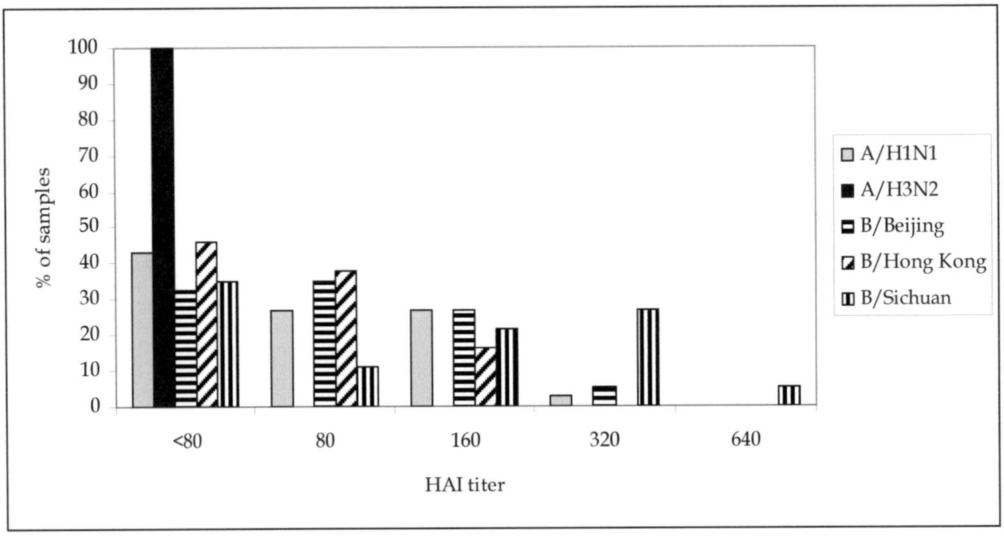

Figure 6. Antibodies to Influenza A and B virus by HAI in fur seal sera. Percent (number positive/number tested) of samples bearing antibodies vs. HAI titer for each Influenza antigen assayed. Titers ≥ 80 were considered positive. Antigens used: A/New Caledonia/20/99(H1N1), A/Panamá/2007/99(H3N2), B/Beijing/184/93-like viruses, B/Yamanashi/166/98, B/Hong Kong/330/01, and B/Sichuan/379/99.

Monitoring the distribution of the influenza virus in wild animal species including marine mammals is important for understanding the ecology and evolution of the virus, and also to understand how the virus can mutate and re-emerge more virulent, producing devastating epidemic diseases.

3. *Mycobacterium pinnipedii* infections in *Arctocephalus australis, Otaria flavescens* and *Mirounga leonina* individuals

Bacteria belonging to the Genera *Mycobacterium* are acid-fast bacilli (AFB) classified in different complexes and species according to biochemical, cultural and genetic features. The first communication of tuberculosis (TB) in captive seals dates from the early twentieth century (Blair, 1913). However, the diagnosis and study of tuberculosis and mycobacteriosis in different species of marine mammals is fairly recent. Ehlers (1965) reported a tuberculosis case in a Northern seal (*Cystophora cristata*). Subsequently, Kinne (1985) described tuberculosis cases in several marine mammal species. In 1986, the first isolates of *Mycobacterium* spp. were obtained in Australian fur seals and sea lions (*Arctocephalus pusillus doriferus* and *Neophoca cinerea* respectively) and New Zealand fur seals (*Arctocephalus forsteri*) in captive and wild conditions (Forshaw & Phelps, 1991; Woods et al., 1995). Successive isolations were made from wild pinniped species in the Southern Hemisphere (Bastida et al., 1999; Bernardelli et al., 1996; Cousins et al., 1993; Hunter et al., 1998; Romano et al., 1995; Woods et al., 1995; Zumárraga et al., 1999).

In Uruguay, the first isolation of *Mycobacterium* spp. in pinnipeds was conducted in 1987 from samples of South American sea lion *O. flavescens* specimens kept in "Villa Dolores" municipal zoo. Of the ten animals studied, one died and nine others were tuberculin-positive and were later euthanized. Seven animals showed typical histo-pathological lesions of tuberculosis, and a total of 6 strains were isolated. Initially, it was considered that the bacillus was *M. bovis* according to results from a smear, biochemical tests and culture features. The strains were inoculated to guinea pigs (0,1 mg) that developed characteristic lesions and subsequently *Mycobacterium* spp. were isolated, fulfilling Koch's postulates (Castro-Ramos et al., 1998). In 1997 *Mycobacterium* spp. was isolated from lung samples of an adult South American fur seal stranded on the coast of Montevideo. The animal was collected from the beach by a NGO and sent to quarantine in a zoo for recovery, but died four days after admission. The observed granulomatous lung lesions were typical of TB, and a *Mycobacterium* spp. strain was isolated (Castro-Ramos et al., 2001).

Between 2001 and 2006, pathological, microbiological and genetic studies were conducted on dead stranded animals of different species of pinnipeds found along ocean shores of Uruguay: South American fur seal (n = 129), South American sea lion (n = 24) and Southern elephant seal (n = 1). Necropsies were performed using standard methods (Dierauf, 1990). Samples from several organs with or without lesions (lung, mediastinal lymph nodes, spleen, liver) (n = 36) were stored at 4° C, frozen at -20° C or fixed in 10% formalin.

Formalin fixed samples were processed by standard histological methods: 4-5 cuts were made at 5-6 μm and stained with Hematoxylin-Eosin and Ziehl-Neelsen (ZN) (Luna, 1968). Mycobacteriological studies were performed according to the methodology described by the Pan American Zoonoses Center (Centro Panamericano de Zoonosis (OPS/OMS) (1979), Office International des Epizooties (OIE) (2000), Runyon et al., (1980) and Tacquet et al., (1967). Smears from single or pooled samples of each animal were performed and then cultured in Stonebrink and Lowenstein Jensen media. Cultures were kept for eight weeks at 37° C and periodically reviewed. Culture tests were based on microscopic features, morphology of the colony, growth temperature, time of development and cromogenicity of isolates. The identification was completed with the following biochemical tests: niacin, nitrate reduction, catalase at 22° C and 68° C, hydrolysis of Tween 80 at 5 and 10 days, reduction of potassium tellurite 0.2% at 3 days, urease and pyrazinamidase. A total of 14 strains were isolated (Table 1).

Strains isolated in 1987 (N° 01073, adult male *O. flavescens*), 1997 (N ° 01337, juvenile male *A. australis*) and 2002 (N ° 2493, juvenile female *O. flavescens*) were analyzed through amplification of 200 bp of the Internal Transcribed Spacer (ITS) region through Polymerase Chain Reaction (PCR) as in Roth et al., (2000). Sequences obtained were compared to those available at GenBank database through a maximum parsimony phylogenetic tree, and strains were grouped with sequences of *M. tuberculosis* / *M. pinnipedii*.

During the necropsies, granulomatous lesions were observed in only five animals: two South American sea lions (juvenile male and female) and three South American fur seals (two adult and one juvenile male) (Fig. 7), from which *M. tuberculosis* / *M. pinnipedii* complex strains were isolated. Isolates were also obtained from organs without gross lesions belonging to pups and juvenile fur seals (n = 9) and from a sub-adult male elephant seal (Castro-Ramos et al., 2005, 2006) (Fig. 9).

Code	Year of sampling	Species	Sex	Category	Origin	Baciloscopy	Culture
9/2001	2001	Aa	♂	Adult	w	---	---
2493	2002	Of	♀	Juvenile	w	+	+
0874	2002	Aa	♂	Adult	w	+	+
0873	2003	Aa	♀	Pup	w	+	+
0875	2003	Aa	♀	Pup	w	+	+
1405	2003	Of	♂	Juvenile	w	+	+
1332/3	2004	Aa	♀	Pup	w	---	+
2172/1	2005	Aa	♀	Pup	w	-	+
2172/2	2005	Aa	♀	Pup	w	-	+
2172/3	2005	Aa	♀	Pup	w	-	+
2172/4	2005	Aa	♂	Pup	w	-	+
2172/6	2005	Aa	♂	Pup	w	-	+
2172/7	2005	Aa	♂	Pup	w	-	+
2173	2005	Of	♀	Pup	w	-	+
2174	2005	Ml	♂	Juvenile	w	-	+
Aa= *Arctocephalus australis*; Of= *Otaria flavescens*; Ml= *Mirounga leonina*; c= captive, w= wild.							

Table 1. Isolation of *Mycobacterium spp / M. pinnipedii* in *Arctocephalus australis*, *Otaria flavescens* and *Mirounga leonina* in Uruguay.

Macro and microscopic lesions were recorded. Hydrothorax and hemothorax were found in two adult animals. Papillary and proliferative lesions in parietal and visceral pleura (Fig. 7) were associated with a chronic inflammatory process, mononuclear and lymphocytic infiltration, and in some cases congestion and hemorrhage. Lungs presented yellowish-white nodules on surface and deep pulmonary parenchyma, which corresponded to granulomas with AFB. Histology showed mononuclear infiltration throughout the parenchyma as well as congestion, emphysema and atelectasia near nodules. Granulomas showed the typical structure with a necrotic center surrounded by a mild fibroblastic reaction and mononuclear infiltration with AFB in single arrangements or small groups (Fig. 8). In bronchi and bronchioles mononuclear exudates at the lumen and lymphocyte/macrophage aggregates below the cartilage were present. Necrotic foci were also recorded in mediastinal lymph nodes with AFB. In one juvenile sea lion a mediastinal abscess and hematoma were found between the great vessels near the heart, which was associated with a chronic inflammatory process.

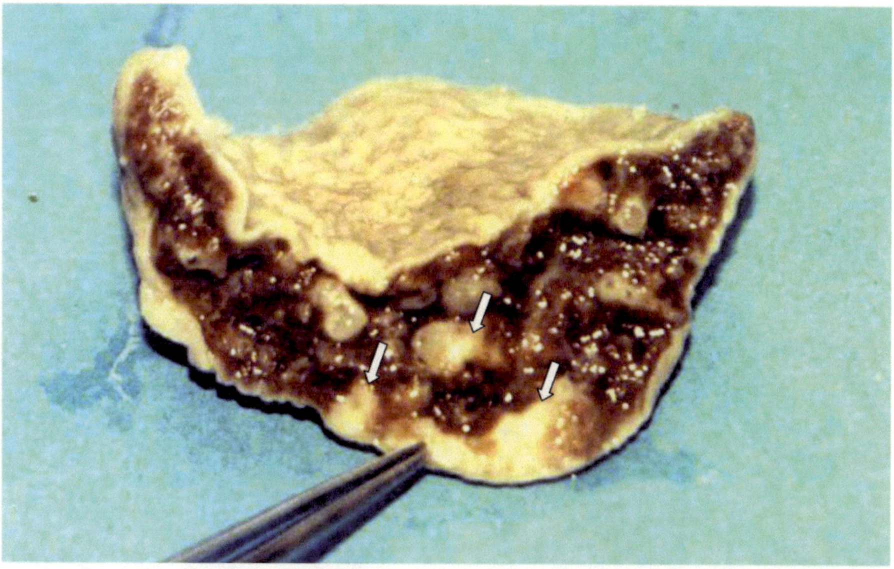

Figure 7. Macroscopic view of an *Arctocephalus australis* lung. Numerous granulomas were present on the surface and deep parenchyma (white arrows) and a significant thickening of the visceral pleura.

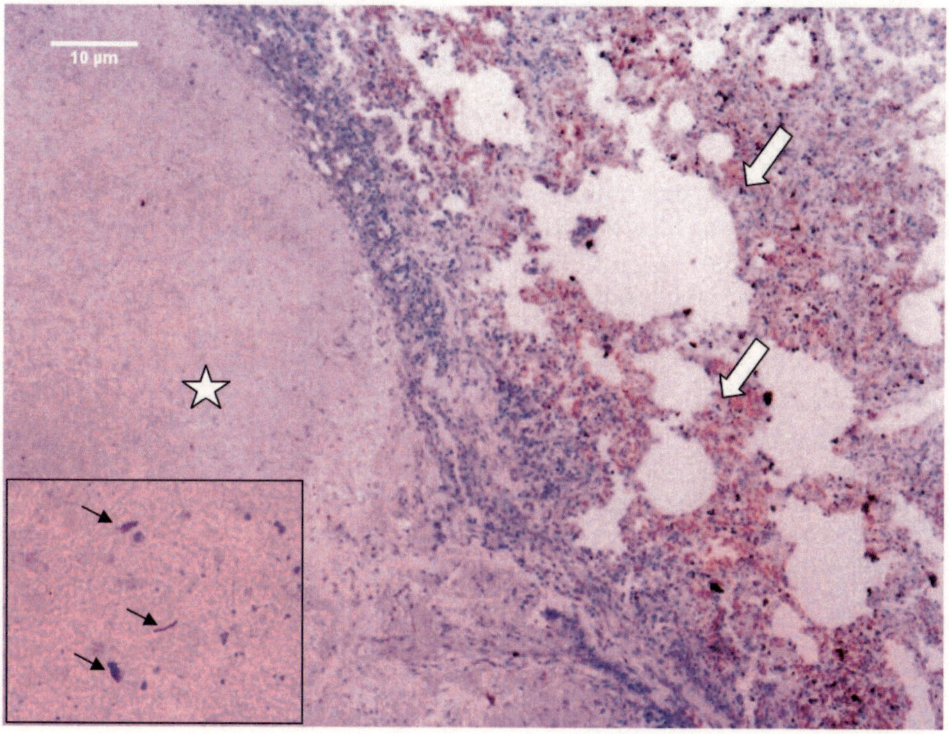

Figure 8. Histological section from an *Arctocephalus australis* lung. A granuloma in the lung parenchyma (star) surrounded by areas of congestion and atelectasis (arrows) can be observed. Inset: AFB groups within the granuloma (immersion) (black arrows). (Ziehl-Neelsen. 400x)

Figure 9. *Mycobacterium* spp. / *M. pinnipedii* colonies in Stonebrink medium culture. Sample obtained from a *Mirounga leonina* lung.

These results indicate a wide distribution of the disease in Uruguayan fur seal and sea lion colonies associated with a high prevalence of the disease in zoos and aquaria from South American native pinnipeds populations (Jurzinsky et al., 2011; Lacave, 2009; Lacave et al., 2009; Moser et al., 2008). The region used for genetic analysis can only discriminate between two large complexes within the Genera *Mycobacterium*, *Mycobacterium tuberculosis* complex (MTC) and *M. avium* (MAC). Based on these results we can conclude that the sequences studied are part of the MTC and correspond to *M. pinnipedii* as determined by more precise genetic studies (Cousins et al., 2003, Jurczinski et al., 2012; Kriz et al., 2011; Moser et al., 2008) which include a strain isolated in Uruguay (Cousins et al., 2003). Discrimination between different members of the MTC is essential for epidemiological investigations of wild populations, as well as the diagnosis of human cases associated with an adequate chemotherapy (de la Rua-Domenech, 2006).

The isolation of bacilli in pups and juvenile animals without apparent lesions indicates early transmission of the organism even though the animals do not show symptoms of respiratory disease. Similar findings were reported for wild carnivores for which positive cultures for *M. bovis* were not accompanied by gross or histological lung lesions (Bruning-Fann et al., 2001; Little et al., 1982). Furthermore, cattle with *M. bovis* developed a non-progressive disease, with small lesions in lymph nodes (retropharyngeal and mediastinal), which in turn are positive reactors to diagnostic tests for delayed cutaneous hypersensitivity (Tuberculin skin test). Most individuals of the population carrying the dormant bacilli may become ill with TB at some point in their life, if an immunodeficient situation develops. It is therefore necessary to consider all infected individuals as potentially diseased animals (Rider, 1999).

The fact that most gross and microscopic lesions are located in the respiratory system of pinnipeds (Bastida et al., 1999; Bernardelli et al., 1996; Castro-Ramos et al., 1998, 2001; Forshaw & Phelps, 1991; Katz et al., 2002; Moser et al., 2008; Woods et al., 1995) indicates that the main transmission is by aerosols. Less frequently, lesions were located in liver, spleen, kidney and abdominal lymph nodes (Bernardelli et. al., 1996; Forshaw & Phelps 1991; Kiers et al., 2008; Kriz et al., 2011), cases in which the bacilli probably spread through blood flow or swallowing of sputum containing mycobacteria (Forshaw & Phelps, 1991; Kriz et al., 2011). In wildlife, it is expected that transmission takes place mainly through spray during coughing and sneezing which is frequently observed in pups and adults (H. Katz, pers. obs.) or the typical naso-nasal contact between pinnipeds.

M. pinnipedii have been isolated from fur seals, sea lions and elephant seals from the Southern Hemisphere (*Arctocephalus australis, A. forsteri, A. pussilus doriferus, A. tropicalis, Otaria flavescens, Neophoca cinerea, Mirounga leonina, Phocarctos hookeri*). They have been found both in wild animals and specimens kept in European aquaria or theme parks taken from South American colonies (Bernardelli et al., 1996; Castro-Ramos et al., 2005; Cousins et al., 2003; Duignan et al., 2003; Kiers et al., 2008; Kriz et al., 2011; Lacave, 2009). Micobacteriosis was diagnosed in only one Otariid species of the Northern Hemisphere (California sea lion, *Zalophus californianus*, Ehlers, 1965; Gutter et al., 1987). Most diagnoses were made in captive *Otaria flavescens*, probably because it is the most common species in aquaria as it is easily trained. Uruguay exported this species destined for aquaria from 1980 to 2006, when captures were restricted due to population decline (Páez, 2006). Nowadays, only live juvenile specimens of *A. australis* are caught for exportation to different destinations (Asia, Latin America and Europe). As this species is also carrier of *M. pinnipedii* (Cousins et al., 2003; Castro-Ramos et al., 2006; Katz et al., 2002) it is important to establish accurate diagnostic methods. The isolation of *M. pinnipedii* from different mammalian species (cattle, Bactrian camels, Malayan tapirs, Guinea pigs and humans), indistinguishable from strains isolated from pinnipeds, suggests that the bacillus has significant potential to infect a wide range of hosts, particularly when animals are in captivity (Cousins et al., 2003; Kiers et al., 2008; Moser et al., 2008).

Presently, the diagnostic methods in living specimens have certain inaccuracies or deficiencies that make it difficult to establish a universal technique or golden standard. This is particularly important given that individuals carrying the microorganism can take years to show signs of the disease; only in terminal cases have nonspecific symptoms including anorexia, dysphagia, lethargy and weight loss been described. Coughing has not been described as a sign accompanying respiratory infection although significant lung lesions were present (Bernardelli, 1996; Castro-Ramos et al., 1998, 2006; Cousins et al., 1993; Kiers et al., 2008; Kriz et al., 2011; Lacave, 2009). In necropsied seals, tuberculosis diagnosis had been made based on mycobacterial isolation, histopathology and genetic characterization of strains. Imaging methods (radiography, computer tomography) have been used in captive animals, but in wild conditions these procedures are impractical. Chest radiographs were performed on pinnipeds of different sizes, but in cases of very small lesions in large animals with thick blubber, radiological images do not give appropriate information (Forshaw & Phelps, 1991; Jurczynski et al., 2012). In some zoos computer tomography has been used for detection of small calcified granulomas (Jurczynski et al., 2011, 2012), but it is very difficult to be used routinely. Different serological tests (rapid test, Elephant TB STAT-PAK, Chembio; multiantigen print immunoassay (MAPIA) Chembio; dual path platform assay (DPP Vet; Chembio) have been used in *O. flavescens* individuals in captivity, with DPP technique demonstrating greater sensitivity (87.5%) (Jurczynski et al., 2012). The tuberculin skin test (TST) with purified protein derivative (PPD bovine and avian) for screening has been done in *O. flavescens* and *A. australis* individuals (Bernardelli et al., 1990; Castro-Ramos et al., 1998; Kiers et al, 2008; Lacave, 2009) and reported by Forshaw & Phelps (1991) in *A. forsteri, A. pusillus doriferus* and *N. cinerea*. This technique is very sensitive, economical and easy to perform. The possible occurrence of false negatives must be taken into consideration in cases of advanced infection with anergy or very recent infections that have not yet generated an appropriate immune response (Jurczynski et al., 2011). False positives may also occur in nonspecific cases of exposure to non-tuberculous mycobacteria (*Mycobacterium avium, M. chelonae, M. fortuitum* and *M. smegmatis*) (Bernardelli et al., 1990; Forshaw & Phelps, 1991). Tissue and bronchial secretion smears have been used for diagnosis (Jurczynski et al., 2012), but the AFB may correspond to mycobacteria other than *M. pinnipedii* and, therefore, other confirmation methods must be used. Molecular techniques (PCR, spoligotyping and MIRU / VNTR) applied to samples from purified cultures, tissues and sputum, have produced quick results, allowing the identification of strains involved and their origin (Cousins et al., 2003; Jurczynski et al., 2011; Kiers et al., 2008; Kriz et al., 2011; Moser et al., 2008). In zoo collections, it is suggested that the final diagnosis should be based on the simultaneous use of three different methods, needing a minimum of two positive tests to increase the overall sensitivity when making the final decision for euthanasia. In case of wild animals, the possible diagnostic methods could include the TST, culture and molecular analysis of sputum (PCR) and serology (DPP). This would be extremely important in order to establish or confirm the endemic conditions of the disease in wild colonies and prevent the exportation or handling of carrier animals.

4. Importance of infectious diseases in Uruguayan pinniped colonies with zoonotic potential

It has been documented that the introduction of novel pathogens into a native animal population without previous exposure could result in epizootics. Human and wildlife populations share a wide range of diseases. While the most common disease transmissions reported are between wild and domestic animals (though many zoonoses do have wildlife origin), emerging diseases of animal origin represent one of the greatest potential threats to public health.

The emergence and re-emergence of over 30 agents have been reported in marine mammals (Miller et al., 2001). There are several reports about their susceptibility to virus, bacterial, fungal and parasitic agents, provoking diseases that result in mass mortality events (Gulland et al., 1996; Miller et al., 2004; Ostehaus, 2000). Once new or previous pathogens are established in the host, they can represent a health risk to other marine vertebrates, humans, or both. It is, however, very difficult to know which routes these zoonotic marine mammal infections take in the marine environment.

The dynamic governing the relationship between infectious diseases affecting humans and marine vertebrates, including sea mammals is very complex and generally poorly understood. It is probable that human activity is a greater threat to marine vertebrate health than vice versa (Mos et al., 2006). However, it is very difficult to establish the role of marine animals as vectors or carriers of zoonotic diseases. Although the role of sea mammals in transmission of potential zoonotic pathogens is not well established, several risk factors, including frequent and prolonged direct contact with live specimens, were clearly identified in workers exposed to these animals (Hunt et al. 2008). Thus, these animals may present a zoonotic potential and also the potential for epizootic events which could cause health problems in marine animals. Since the influenza virus is transmitted by direct contact with infected individuals, by aerosol, or contact with infected objects, several incidents of influenza A virus transmission have occurred from infected seals and whales maintained in captivity to humans (Webster et al., 1981).

Due to their life span, high trophic feeding and the continuous exposition to emerging pathogens, sea mammals might be considered as sentinel species of emerging/re-emerging diseases. Pinnipeds share these characteristics and may serve as effective sentinels providing information about public and aquatic ecosystem health, and indicating the current or potential negative impacts on animal health at the individual or population level.

Implementing a proper management and the accurate execution of conservation policies of wildlife requires the analysis of the vulnerability of the animals to infectious diseases. The majority of animals included in our studies belonged to colonies near cities or towns in proximity to shore (Punta del Este, Polonio's Cape), regions that are commonly used by humans, marine and terrestrial mammals and different avian species, establishing an optimal opportunity for zoonotic disease transmission and long-term disease maintenance. The dense aggregations of pinniped colonies make fur seals and sea lions vulnerable to epizootics. Also, movements between adjacent latitudinal domains are common in both pinniped species, which could transmit or acquire pathogens during feeding trips to other parts of the Atlantic Ocean and the La Plata River.

The diagnosis of influenza virus in high-density seal populations in Uruguay generated considerable concern about the potential impacts on South American fur seal and sea lion colonies, as well as potential health risks to humans and domestic animals. In the evolution and ecology of influenza viruses, interspecies transmission is an important factor; seabirds and marine mammals are conspicuous animals that integrate changes in the ecosystem and reflect the existing state of the environment (Aguirre & Tabor, 2004; Boersma, 2008; Moore, 2008; Thiele et al., 2004). Transmission of the influenza virus occurs between avian and several marine mammal species (Mandler et al., 1990) at least for influenza A virus, representing an important step in the evolution and emergence of new mammalian viral strains. Fur seals have the potential to serve as an influenza reservoir for other mammal species. However, more detailed studies are needed to elucidate the role of seals in the epidemiology of influenza along the Uruguayan coasts and in other South American countries.

Our findings confirm that fur seals can act as reservoirs of human influenza strains that circulated in the past, and also suggest that influenza A and B viruses may be transmitted from humans to seals as has been mentioned by other authors (Ohishi et al., 2002, 2004; Osterhaus, 2000). This transmission is due to the highly social lifestyle of pinnipeds, which congregate at sea and on land, and frequently associate with seals from other colonies. It is important to note that there is a strong interaction between seals and humans on Lobos Island and Polonio's Cape during live animal captures; nowadays, most interactions occur during capture and research activities, as well as in rehabilitation centers and sometimes with divers that swim near the seal islands. These events constitute opportunities for new influenza strains to jump between humans and seals, providing the potential for an epidemic event. Gaining information on the full spectrum of influenza viruses circulating in our seal colonies and detecting these viruses will remain an important task for its surveillance, outbreak control, and animal and public health.

As described previously, tuberculosis in pinnipeds had been recorded from the beginning of the twentieth century but it was not until mid 80's and later that several publications appeared with data from stranded animals and zoological collections. It is important to note that most records are from Otariids from the Southern Hemisphere, a single diagnosis is from a southern elephant seal and one from one native Northern Otariid species kept in captivity in the Northern Hemisphere.

Despite the fact that the seal harvest in Uruguay extended for 200 years, there are no records of macroscopic lesions observed in animals, nor of tuberculosis diagnosis from the staff working in the capture, slaughter or processing of the by-products, that could contribute retrospective data to this disease in Uruguayan pinniped colonies. The disease is considered endemic because of the numerous cases diagnosed in wild animals and the high prevalence of South American seals kept in aquaria and zoos. Unfortunately, to date non epidemiological studies have been conducted in any of the seal colonies from the Southern Hemisphere which could indicate the prevalence of the disease in its natural environment.

Information presented in this work, including the isolation of *M. pinnipedii* in pups and juvenile seals, indicates the early transmission of the organism. However, it is difficult to establish the course of the disease or immune mechanisms that may develop in each individual to control the infection or, in other cases, allow individuals to act as healthy carriers for several years until some trigger factor determines the development of the disease.

The most important mechanism of transmission of *M. tuberculosis*, and probably *M. pinnipedii*, is through the air by droplets produced when an individual with respiratory tract TB eliminates aerosolized microorganisms by coughing or sneezing. Large droplets fall quickly due to their weight and reach the ground without evaporation. Smaller drops evaporate and decrease in size, becoming infectious droplet nuclei, which remain suspended in ambient air for a long time and can be airborne for days. Successful transmission requires that these micro droplets charged with bacilli are of sufficiently small size (1 to 5 μm) to enable them to reach very deep into the lungs and alveoli (Rider, 1999). This phenomenon is more frequent in captive conditions in zoos and aquaria where spray released during enclosure hygiene, poor ventilation and exposure among animals in confined environments, constitute increased risk factors for infection as has been shown in different zoos (Cousins et al., 1993 and 2003; Kiers et al., 2008; Thompson et al., 1993). In wild pinniped colonies, natural ventilation and increased exposure to UV is a natural way of microorganism control that could be the reason for the less common occurrence of TB in these colonies.

In the early course of host-pathogen interaction, mycobacteria are phagocytized by alveolar macrophages. In case the macrophage cannot destroy the pathogen, it resides in a quiescent state with a relatively low multiplication until cell mediated immunity is compromised transiently or permanently. Among the risk factors for immune-compromise, are age, history of a spontaneous TB cure with residual fibrotic lesions, and the time elapsed since infection. Other medical conditions such as endocrine disorders, tumors, malnutrition and stress may influence the progression of infection to disease (Musser et al., 2000; Rider, 1999). This evolution of the disease is consistent with the findings by different authors who have isolated AFB in organs without lesions from pups and juveniles (Bruning-Fann et al., 2001), identified the presence of calcified granulomas in mediastinal lymph nodes in adult and some juvenile animals in captivity and documented tuberculosis lesions in older animals (Jurczynski et al., 2011, 2012).

Keepers and veterinarians at zoos, aquaria and rehabilitation centers are at increased risk of infection because of their extensive contact with the animals. The degree of risk depends on the type of accommodation and sanitizing procedures of the enclosures (Kiers et al., 2008; Thompson et al., 1993) in association with the presence of open tuberculosis cases. Given the zoonotic condition of the disease, it is important to take preventive measures in all personnel working near or with access to wild and captive pinnipeds (researchers, fishermen, seal hunters, rehabilitation centers, aquaria and zoos staff) (Kiers et al., 2008; Thompson et al., 1993) and to establish hygiene measures for reducing the chances of spreading the infection.

Since development of the disease is triggered under immunosuppressive conditions, it is important to assess which factors directly or indirectly affect the immune system and allow development of the disease in wild populations of the Southern Hemisphere. This also applies to other diseases that may be affecting the health status of wild colonies, particularly in *O. flavescens* from Uruguay, whose population is declining.

Further investigation is needed in order to establish the sources of zoonotic potential in the marine environment and to better understand the nature of health risks for sea mammals and human. Additional epidemiologic studies are also required, to investigate epizootic episodes assessing its impact and to elucidate how these diseases spread among and within marine mammal's populations.

Author details

Juan Arbiza and Andrea Blanc
Universidad de la República, Facultad de Ciencias, Sección Virología, Uruguay

Miguel Castro–Ramos
Ministerio de Ganadería Agricultura y Pesca, División de Laboratorios Veterinarios "Miguel C. Rubino", Departamento de Bacteriología, Laboratorio de Tuberculosis, Uruguay

Helena Katz
Universidad de la República, Facultad de Veterinaria, Área de Patología, Uruguay

Alberto Ponce de León
Ministerio de Ganadería, Agricultura y Pesca, Dirección Nacional de Recursos Acuáticos, Departamento Mamíferos Marinos, Uruguay

Mario Clara
Universidad de la República, Centro Universitario de Rivera and Instituto de Ecología y Ciencias Ambientales, Facultad de Ciencias, Uruguay

Acknowledgement

The present work has been conducted with research permission from DINARA-MGAP at Isla de Lobos and Cabo Polonio. We are extremely grateful to the fur seal keepers Leonardo Olivera, Nelson Veiga, Miguel Casella and César Barreiro who helped us to work with the animals at Isla de Lobos; Lic. Oscar Castro, Dr. Gustavo de Souza, Dra. Graciela Pedrana, Dr. Antonio Moraña, Dr. Francisco Gutiérrez, MSc. Valentina Franco and Dr. Federico Riet, who helped during field work and contributed with bibliography that enriched the present article. We thank Dr. Riet for the figure showing pinniped colonies distribution.

5. References

Acosta & Lara, F. (1984). La pesca de lobos. *Revista de la Sociedad Universitaria*. Montevideo. Vol. 1, (March, 1984), pp. 337-352

Aguirre, A. A. & Tabor, G. M. (2004). Introduction: marine vertebrates as sentinels of marine ecosystem health. *EcoHealth* Vol. 1, (September, 2004), pp. 236–238

Bartholomew, G. A. (1970). A model for the evolution of pinnipeds polygyny. *Evolution* (Lancaster, Pa), Vol. 24, (1970), pp. 546-559

Bastida, R.; Loureiro, J.; Quse, V.; Bernardelli, A.; Rodriguez, D. & Costa, E. (1999). Tuberculosis in a wild subantartic fur seal from Argentina. *Journal Wildlife Diseases*, Vol. 35, (October, 1999), pp. 796-798

Bernardelli, A.; Nader, A. J.; Loureiro, J.; Michelis, H. & Debenedetti, R. (1990). Micobacteriosis en mamíferos y aves marinas. *Revue Scientifique et Technique de l' Office International des Epizooties,* Vol. 9, No. 4, (1990), pp. 1121-1129

Bernardelli, A.; Bastida, R.; Loureiro, J.; Michelis, H.; Romano, M. I.; Cataldi, A. & Costa, E. (1996). Tuberculosis in sea lions and fur seals from the south-western Atlantic coast. *Revue Scientifique et Technique de l' Office International des Epizooties,* Vol. 15, (September, 1996), pp. 985-1005

Blair W. R. (1913). Report of the veterinarian on the mammals: In Seventeenth Annual report of the New York Zoological Society. *New York Zoological Society,* (1913), pp. 73-74

Blanc, A.; Ruchansky, D.; Clara, M.; Achaval, F.; Le Bas, A. & Arbiza, J. (2009). Serologic Evidence of Influenza A and B Viruses in South American Fur Seals (*Arctocephalus australis*). *Journal of Wildlife Diseases,* Vol. 45, N° 2, (April, 2009), pp. 519–521

Boersma, P. D. (2008). Penguins as Marine Sentinels. *BioScience.* Vol 58, N° 7, (July, 2008), pp. 597-607

Bruning-Fann, C. S.; Schmitt, S. M.; Fitzgerald, S. D.; Fierke, J. S.; Friedrich, P. D.; Kaneene, J. B.; Clarke, K. A.; Butler, K. L.; Payeur, J. B.; Whipple, D. L.; Cooley, T. M.; Miller, J. M. & Muzo, D. P. (2001). Bovine tuberculosis in free-ranging carnivores from Michigan. *Journal of Wildlife Diseases,* Vol. 37, No. 1, (January, 2001), pp. 58–64

Callan, R. J.; Early, G.; Kida, H. & Hinshaw, V. S. (1995). The appearance of H3 influenza viruses in seals. *Journal of General Virology,* Vol. 76, (January, 1995), pp. 199–203

Campagna, C. & Le Boeuf, B. (1988). Reproductive behaviour of Southern Sea Lions. *Behaviour,* Vol. 104, (January, 1998), pp. 233-261

Campagna, C. & Lewis, M. (1992). Growth and distribution of a southern elephant seal colony. *Marine Mammal Science,* Vol. 8, (October, 1992), pp. 387–396

Cappozzo, H. L.; Bernabeu, R. O. & Crespo, E. A. (1994). Discriminación de stocks de *Otaria flavescens* y *Arctocephalus australis* en Uruguay y Argentina. Sub Proyecto 1. Pp 1-9. In: *Problemas de conservación y manejo de los mamíferos marinos del Atlántico Sud Occidental en Uruguay y Argentina: un proyecto conjunto de investigación.* Programa de las Naciones Unidas para el Medio Ambiente (PNUMA). Informe final sobre los trabajos realizados durante 1992-1993

Castro-Ramos, M.; Ayala, M.; Errico, F. & Silvera, F. V. (1998). Aislamiento de *Mycobacterium bovis* en Pinnípedos *Otaria byronia* (Lobo marino común) en Uruguay. *Revista de Medicina Veterinaria.* Vol. 79, No. 3, (1998), pp. 197-200.

Castro-Ramos, M.; Leizagoyen, C.; Alves, F.; Cirillo, F. & Zipitría, R. (2001). Tuberculosis en *Arctocephalus australis* (Lobo marino de dos pelos) en la costa de Uruguay, América del Sur. *V Encuentro Nacional de Microbiólogos,* Noviembre de 2001. Montevideo, Uruguay.

Castro-Ramos, M.; Katz, H.; Castro, O. & Gutiérrez, F. (2005). Primera comunicación de tuberculosis en *Mirounga leonina* (Mammalia, Phocoidea) (elefante marino del sur) en Uruguay. *VII Encuentro Nacional de Microbiólogos.* Octubre de 2005. Montevideo, Uruguay.

Castro-Ramos, M.; Katz, H.; Moraña, A.; Tiscornia, M. I.; Morgades, D. & Castro, O. (2006). Tuberculosis en pinnípedos (*Arctocephalus australis* y *Otaria flavescens*) de Uruguay. In: *Bases para la conservación y el manejo de la costa atlántica uruguaya.* Menafra R., Rodríguez-

Gallego, L., Scarabino, F. & Conde D. (Editors). *Vida Silvestre Uruguay*. Montevideo, Uruguay. Pp.668

Centro Panamericano de Zoonosis (OPS/OMS). (1979). Serie de Monografías Científicas y Técnicas CPZ-11 Bacteriología de la tuberculosis humana y animal, 63 pp

Convention on International Trade in Endangered Species of Wild Fauna and Flora (CITES) (2012). http://www.cites.org

Cousins, D. V.; Williams, S. N.; Reuter, R.; Forshaw, D.; Chadwick, B.; Coughran, D.; Collins, P. & Gales, N. (1993). Tuberculosis in wild seals and characterization of the seal bacillus. *Australian Veterinary Journal,* Vol. 70, (January, 1993), pp. 92-97

Cousins, D. V.; Bastida, R.; Cataldi, A.; Quse, V.; Redrobe, S.; Dow, S.; Duignan, P.; Murray, A.; Dupont, C.; Ahmed, N.; Collins, D. M.; Ray Butler, W.; Dawson, D.; Rodríguez, D.; Loureiro, J.; Romano, M. I.; Alito, A.; Zumárraga, M. & Bernardelli, A. (2003). Tuberculosis in seals caused by a novel member of the *Mycobacterium tuberculosis* complex: *Mycobacterium pinnipedii* sp. nov. *International Journal of Systematic and Evolutionary Microbiology,* Vol. 53, (January, 2003), pp. 1305-1314

Danner, G. R. & Mcgregor, M. W. (1998). Serologic evidence of influenza virus infection in a ringed seal (*Phoca hispida*) from Alaska. *Marine Mammal Science* Vol. 14, (April, 1998), pp. 380–384

De Boer, G. F.; Back, W. & Osterhaus, A. (1990). An ELISA for the detection of antibodies against influenza A nucleoprotein in humans and various animal species. *Archives of Virology,* Vol. 115, N° 1-2 (March, 1990), pp. 47–61

de la Rua-Domenech, R. (2006). Human *Mycobacterium bovis* infection in the United Kingdom: Incidence, risks, control measures and review of the zoonotic aspects of bovine tuberculosis. *Tuberculosis,* Vol. 86, N° 2, (March, 2006), pp. 77-109

De Oliveira, L. R.; Arias-Schreiber, M.; Meyer, D. & Morgant, J. S. (2006). Effective population size in a bottlenecked fur seal population. *Biology of Conservation*. Vol. 131, (September, 2006), pp. 505-509

Dierauf, L. (1990). *Handbook of Marine Mammal Medicine: Health, Disease and Rehabilitation.* Vol. 2. C.R.C. Press. 735 pp

Dirección Nacional de Recursos Acuáticos, Ministerio de Ganadería, Agricultura y Pesca (DINARA-MGAP). (2006). Información disponible en: http://www.dinara.gub.uy/Mamíferos-Marinos/Explotación.htm

Duignan, P. J.; Dupont, C.; Cousins, D.; Collins, D. M. & Murray, A. (2003). Tuberculosis associated with *Mycobacterium tuberculosis subsp. pinnipedae* subsp. nov. in New Zealand fur seals and sea lions. *3rd International Wildlife Management Congress*. University of Canterbury, Christchurch. New Zealand. Accessible in www.landcareresearch.co.nz/.../PosterSession4.d

Ehlers, K. (1965). Records of the hooded seal *Cystophora cristata* Erxleben, and other animals at Bremerhaven Zoo. *International Zoo Yearbook,* Vol. 5, (January, 1965), pp. 148-149

Forshaw, D. & Phelps, G. R. (1991). Tuberculosis in a captive colony of pinnipeds. *Journal Wildlife Diseases,* Vol. 27, (April, 1991), pp. 288-295

Franco Trecu, V. (2005). Comportamiento maternal y aspectos reproductivos de *Arctocephalus australis* en Isla de Lobos, Uruguay. Tesis. Universidad de la República, Facultad de Ciencias. Montevideo, Uruguay

Franco-Trecu, V.; Costa, P.; Abud, C.; Dimitriadis, C.; Laporta, P.; Passadore, C. & Szephegy, M. (2009). By-catch of franciscana *Pontoporia blainvillei* in Uruguayan artisanal gillnet fisheries: an evaluation after a twelve-year gap in data collection. *Latin American Journal of Aquatic Mammals*. Vol. 7 (1-2) (December, 2009), pp. 11-22

Franco-Trecu, V. (2010). Éxito de crianza y hábitos alimenticios en hembras del lobo fino sudamericano (*Arctocephalus australis*) y su relación trófica con hembras del león marino sudamericano (*Otaria flavescens*). Tesis de Maestría. PEDECIBA. Universidad de la República. Facultad de Ciencias. Montevideo, Uruguay. 90pp

Franco-Trecu, V., Aurioles, D.; Lima, M. & Arím, M. (2012). Prepartum and postpartum trophic segregation between sympatrically breeding female *Arctocephalus australis* and *Otaria byronia*. *Journal of Mammalogy*, (2012) 93(2). In press.

Frau-Martínez, R. & Franco-Trecu, V. (2010). Especialización alimenticia en el lobo fino sudamericano (*Arctocephalus australis*) en Uruguay. *XVI Reunião de Trabalho de Especialistas em Mamíferos Aquáticos da America do Sul*. Florianopolis-SC-Brasil. 24 al 28 de octubre 2010.

Fujii, K. C.; Kakumoto, M.; Kobayashi, S.; Saito, T.; Kariya, Y.; Watanabe, Y.; Sakoda, H.; Kida, & Suzuki, M. (2007). Serological evidence of Influenza A Infection in Kuril Harbor Seals (Phoca vitulina stejneri) of Hokkaido, Japan. *Journal of Veterinary and Medical Sciences*, Vol. 69, (March, 2007), pp. 259-263.

Geraci, J. R.; ST Aubin, D. J.; Barker, I. K.; Webster, R. G.; Hinshaw, V. S.; Bean, W. J.; Ruhnke, H. L.; Prescott, J. H.; Early, G.; Baker, A. S.; Madoff, S. & Schooley, R. T. (1982) Mass mortality of harbor seals: pneumonia associated with influenza A virus. *Science*, Vol. 215, (February, 1982), pp. 1129–1131

Grandi, F.; Dans, S. & Crespo, E. (2008). Social composition and spatial distribution of colonies in an expanding population of South American sea lions. *Journal of Mammalogy*, Vol. 89, No.5, (October, 2008), pp. 1218-1228

Gulland, F. M. D.; Koski, M.; Lowenstine, L. J.; Colagross, A.; Morgan, L. & Spraker, T. (1996). Leptospirosis in California sea lions (*Zalophus californianus*) stranded along the central California coast, 1981-1994. *Journal of Wildlife Diseases*, Vol. 32, (October, 1996), pp. 572-580

Gutter, A. E.; Wells, S. K. & Spraker, T. R. (1987). Generalized mycobacteriosis in a California Sea Lion (*Zalophus californianus*). *Journal of Zoo Animal Medicine*, Vol. 18, N° 2-3 (September, 1987), pp. 118-120

Hinshaw, V. S.; Bean, W. J.; Geraci, J. R.; Fiorelli, P.; Early, G. & Webster, R. G. (1986). Characterization of two influenza A viruses from a pilot whale. *Journal of Virology* Vol. 58, No.2 (May, 1986), pp. 655–656

Hinshaw, V. S.; Bean, W. J.; Webster, R. G.; Rehg, J. E.; Fiorelli, P.; Early, G.; Geraci, J. R. & St Aubin, D. J. (1984). Are seals frequently infected with avian influenza viruses? *Journal of Virology*, Vol. 51, No. 3 (September, 1984), pp. 863–865

Hunt, T. D.; Ziccardi, M. H.; Gulland, F. M. D.; Yochem, P. K.; Hird, D. W.; Rowles, T. & Mazet, J. A. K. (2008). Health risks for marine mammal workers. *Diseases of Aquatic Organisms*, Vol. 81, (May, 2008), pp. 81–92

Hunter, J.E.; Duignan, P.J.; Dupont, C.; Fray, L.; Fenwick, S.G. & Murray A. (1998) First report of potentially zoonotic tuberculosis in fur seals in New Zeland. New Zeland Medical Journal 111:130-131.

Jurczynski, K.; Lyashchenko, K.P.; Scharpegge, J.; Fluegger, M.; Lacave, G.; Moser, I.; Tortschanoff, S. & Greenwald R. (2012). Use of Multiple Diagnostic Tests to Detect *Mycobacterium pinnipedii* Infections in a Large Group of South American Sea Lions (*Otaria flavescens*). *Aquatic Mammals,* Vol. 38, No.1, (2012), pp. 43-55 (in press) DOI 10.1578/AM.38.1.2012.43

Jurczynski, K.; Scharpegge, J.; Ley-Zaporozhan, J.; Ley, S.; Cracknell, J.; Lyashchenko, K.; Greenwald, R. & Schenk, J. P. (2011). Computed tomographic examination of South American sea lions (*Otaria flavescens*) with suspected *Mycobacterium pinnipedii* infection. *Veterinary Record,* Vol. 2, (December, 2011), pp. 169-608

Katz, H.; Moraña, A.; Morgades, D.; Castro, O.; Casas, L.; Le Bas, A. & Delbene, D. (2002). Hallazgos histopatológicos en lobo marino (*A. australis*) en Uruguay. Estudios iniciales de un caso de tuberculosis. *X RT & 4º Congreso SOLAMAC,* Valdivia, Chile. 14-19 de octubre, 2002

Katz, H.; Morgades, D. & Castro, O. (2012). Necropsies findings in South American Fur Seal pups (*Arctocephalus australis*) in Uruguay. *Florida Marine Mammal Health Conference IV.* April 24 to 27, 2012. Sarasota, USA

Kiers, A.; Klarenbeek, A.; Mendelts, B.; Van Soolingen, D. & Koeter, G. (2008). Transmission of *Mycobacterium pinnipedii* to humans in a zoo with marine mammals. *International Journal of Lung Diseases.* Vol. 12, (July, 2008), pp. 1469-1473

Kinne, O. (1985). Diseases of Mammalia: Pinnipedia, In: *Diseases of Marine Animals.* Volume IV Part 2. p. 698. Biologische Anstalt Helgoland, Hamburg

Kriz, P.; Kralik, P.; Slany, M.; Slana, I.; Svobodova, J.; Parmova, I.; Barnet, V.; Jurek, V. & Pavlik, I. (2011). *Mycobacterium pinnipedii* in a captive Southern sea lion (*Otaria flavescens*): a case report. *Veterinarni Medicina,* Vol. 56, No.6, (June, 2011), pp. 307–313

Lacave, G. (2009). Pinnipeds tuberculosis: the importance of medical training and access to animals for population survey during spreading of a potentially zoonotic disease. *37th IMATA.* Atlanta, United States of America. November, 2009

Lacave, G.; Maillot, A.; Alerte, V.; Boschiroli, M. A. & Lecu, A. (2009). Atypical case of *Mycobacterium pinnipedii* in a patagonian sea lion (*Otaria flavescens*) and Tuberculosis cases history review in pinnipeds. *40th IAAAM,* San Antonio, May 2009

Lang, G.; Gagnon, A. & Geraci, J. R. (1981). Isolation of an influenza A virus from seals. *Archives of Virology,* Vol. 68, No.3-4 (September,1981), pp. 189-195

Le Boeuf, B. J. & Laws, R. M. (1994). *Elephant seals: population ecology, behaviour and physiology.* Berkeley, CA: University of California Press, pp. 414

Lewis, M.; Campagna, C.; Marin M. R. & Fernandez, T. (2006). Southern elephant seals north of the Antarctic Polar Front. *Antarctic Science,* Vol. 18, No.2, (June, 2006), pp. 1–9

Lewis, M.; Campagna, C.; Quintana, F. & Falabella, V. (1998). Estado actual y distribución de la población del elefante marino del sur en la Península Valdés, Argentina. *Mastozoología Neotropical,* Vol. 5, (June, 1998), pp. 29–40

Lewis, M.; Campagna, C. & Zavatti, J. (2004). Annual cycle and interannual variation in the haulout pattern of a growing southern elephant seal colony. *Antarctic Science*, Vol. 16, (September, 2004), pp. 219–226

Lezama, C. & Szteren, D. (2003). Interacción entre el león marino sudamericano (*Otaria flavescens*) y la flota pesquera artesanal de Piriápolis, Uruguay. *II Jornadas de Conservación y Uso Sustentable de la Fauna Marina*. Escuela Naval. Montevideo, 1-3 de octubre de 2003.

Little, T. W. A.; Swan, C.; Thompson, H. V. & Wilesmith, J.W. (1982). Bovine tuberculosis in domestic and wild mammals in an area of Dorset. III. The prevalence of tuberculosis in mammals other than badgers and cattle. *Journal of Hygiene*. Vol. 89, (April, 1982), pp. 225–234

Luna, LG. (1968). *Manual of histologic staining methods of the Armed Forces Institute of Pathology* (Third Edition). American Registry of Pathology. Lee G. Luna Editor. McGraw-Hill Book Company. USA

Lvov, D. K.; Zhdanov, V. M.; Sazonov, A. A.; Braude, N. A.; Vladimirteeva, E. A.; Agafonova L. V.; Skljanskaja, E. I. & Kaverin, N. V. (1978). Comparison of influenza viruses isolated from man and from whales. *Bulletin WHO*, Vol. 56, No.6 (1978), pp. 923-930

Mandler, J.; Gorman, O. T.; Ludwig, S.; Schroeder, E.; Fitch, W. M.; Webster, R. G. & Scholtissek, C. (1990). Derivation of the nucleoproteins (NP) of influenza A viruses isolated from marine mammals. *Virology*, Vol 176, No.1 (May, 1990), pp. 255-261

Miller, D., Ewing, R. Y. & Bossart, G. D. (2001). Emerging and resurging diseases. In: *Marine Mammal Medicine*, L. Dierauf and F. Gulland, Eds, pp.15–25, CRC Press, Boca Raton, FL

Miller, M. A.; Grigg, M. E.; Kreuder, C.; James, E. R.; Melli, A. C.; Crosbie, P. R.; Jessup, D. A.; Boothroyd, J. C.; Brownstein, D. & Conrad, P. A. (2004). An unusual genotype of *Toxoplasma gondii* is common in California sea otters (*Enhydra lutris nereis*) and is a cause of mortality. *International Journal of Parasitology*. Vol. 34, (March, 2004), pp. 275–284

Moore, S. (2008). Marine mammals as ecosystem sentinels. *Journal of Mammalogy*, Vol. 89, No.3, (June, 2008), pp. 534–540

Morgades, D.; Katz, H.; Castro, O.; Capellino, D.; Casas, L.; Benítez, G.; Venzal J. M. & Moraña, A. (2006). Fauna parasitaria del lobo fino *Arctocephalus australis* y del león marino *Otaria flavescens* (Mammalia, Otariidae) en la costa uruguaya, pp. 89–96. In: *Bases para la conservación y el manejo de la costa uruguaya*. Menafra, R., Rodríguez-Gallego, L., Scarabino, F. & Conde, D. (Editors). Vida Silvestre Uruguay. Montevideo, Uruguay. Pp. 668

Mos, L.; Morsey, B.; Jeffries, S. J.; Yunker, M. B.; Raverty, S.; De Guise, S. & Ross, P. S. (2006). Chemical and biological pollution contribute to the immunological profiles of free-ranging Harbor Seals. *Environmental Toxicology And Chemistry*, Vol. 25, No.12, (June, 2006), pp. 3110–3117

Moser, I.; Prodinger, W. M.; Hotzel, H.; Greenwald, R.; Lyashchenko, K. P.; Bakker, D.; Gomis, D.; Seidler, T.; Ellenberger, C.; Hetzel, U.; Wuennemann, K. & Moisson, P. (2008). *Mycobacterium pinnipedii*: transmission from South American sea lion (*Otaria byronia*) to Bactrian camel (*Camelus bactrianus bactrianus*) and Malayan tapirs (*Tapirus indicus*). *Veterinary Microbiology*, Vol. 127, No. 3-4, (March, 2008), pp. 399-406

MTOP – PNUD - UNESCO. (1980). Conservación y mejora de playas. Ministerio de Transporte y Obras Públicas – Programa de las Naciones Unidas para el Desarrollo – UNESCO. Proyecto UNDP/URU/73/007. Informe Técnico. FMR/SC/OPS/80/214 (UNDP). Pp. 593

Musser, J. M.; Amin, A. & Ramaswamy, S. (2000). Negligible genetic diversity of *Mycobacterium tuberculosis* host immune system protein targets: evidence of limited selective pressure. *Genetics,* Vol. 155, (May, 2000), pp. 7–16

Naya D. E; Vargas R. & Arim, M. (2000). Análisis preliminar de la dieta del león marino del sur (*Otaria flavescens*) en Isla de Lobos, Uruguay. Boletín de la Sociedad Zoológica del Uruguay (2da época). Vol. 12, pp 14-21

Naya D. E.; Arim, M. & Vargas, R. (2002). Diet of South American fur seals (*Arctocephalus australis*) in Isla de Lobos, Uruguay. *Marine Mammal Science*. Vol. 18, No.3 (July, 2002), pp. 734-745

Nielsen, O.; Clavijo, A. & Boughen J. A. (2001). Serological evidence of Influenza A infection in marine mammals of Arctic Canada. *Journal of Wildlife Diseases* Vol. 37, (October, 2001), pp. 820-825

Office International des Epizooties (2000). Bovine tuberculosis. In: *Manual of Standards for Diagnostic Test and Vaccines.* Paris: World Organization for Animal Health, pp 267-275

Ohishi, K.; Kishida, N.; Ninomiya, A.; Kida, H.; Takada, Y.; Miyazaki, N.; Boltunov, A. & Maruyama, T. (2004). Antibodies to Human-related H3 Influenza A viruses in Baikal Seals (*Phoca sibirica*) and ringed seals (*Phoca hispida*) in Russia. *Microbiology and Immunology* Vol. 48, (August, 2004), pp. 905-909

Ohishi, K.; Ninomiya, A.; Kida, H.; Park, C.; Maruyama, T.; Arai, T.; Katsumata, E.; Tobayama, T.; Boltunov, A.; Khuraskin, L. & Miyazaki, N. (2002) Serological evidence of transmission of human Influenza A and B viruses to Caspian Seals (*Phoca caspica*). *Microbiology and Immunology* Vol. 46, No. 9 (July, 2002), pp. 639-644

Olsen, B.; Munster, V. J.; Wallensten, A.; Waldenström, J.; Osterhaus, A. & Fouchier, R. (2006). Global patterns of influenza A virus in wild birds. *Science,* Vol. 312, (2006), pp. 384–388

Osterhaus, A.; Rimmelzwaan, G. F.; Martina, B.; Bestebroer, T. M. & Fouchier, R. (2000). Influenza B virus in Seals. *Science* Vol. 288, (May, 2000), pp. 1051-1053.

Páez, E. (2006). Situación de la administración del recurso lobos y leones marinos en Uruguay. Pág. 577-583. *Bases para la conservación y el manejo de la costa uruguaya.* Menafra, R., Rodríguez-Gallego, L., Scarabino, F. & Conde, D. Eds. 668pp, Vida Silvestre. Montevideo, Uruguay

Pedraza, S. N., Franco-Trecu, V. & Ligrone, A. (2009). Tendencias Poblacionales de *Otaria flavescens* y *Arctocephalus australis* en Uruguay. Workshop "Estado de situación del lobo marino común en su área de distribución". Valparaíso, Chile, 15-17 Junio 2009

Pérez Fontana, H. V. (1943). Informe sobre la Industria Lobera (ciento diez años de de explotación de la industria lobera en nuestro país). Servicio Oceanográfico y de Pesca. (S.O.Y.P.). Monteverde. Montevideo. 69 pp

Pin, O.; Ponce de León, A. & Arím, M. (1996). Identificación de categorías alimentarias en contenido estomacal y fecas del lobo fino sudamericano *Arctocephalus australis* en Isla de Lobos, Uruguay. *VII Reunión de Especialistas en Mamíferos Acuáticos de América del Sur, 1er*

Congreso de la Sociedad Latinoamericana de Especialistas en Mamíferos Acuáticos. Viña del Mar, 22-25 de octubre de 1996

Pinedo, M. C. & Barros, N. (1983). Análises dos conteúdos estomacais do leão marinho *Otaria flavescens* e do lobo marinho *Arctocephalus australis* na costa do Rio Grande do Sul, Brasil. Pp 187-199. In: *Primera Reunión de Trabajo de Expertos en Mamíferos Acuáticos de América del Sur*. Buenos Aires, Argentina. 25 al 29 de junio de 1984

Ponce de León, A. (1983). Crecimiento intrauterino y postnatal del lobo de dos pelos sudamericano, *Arctocephalus australis* (Zimmermann, 1783), en las islas de Uruguay. *VIII Simposio Latinoamericano de Oceanografía Biológica*. Intendencia Municipal de Montevideo, Uruguay. 28 de noviembre al 2 de diciembre de 1983.

Ponce de León, A. (1984). Lactancia y composición cuantitativa de la leche del lobo fino sudamericano *Arctocephalus australis* (Zimmermann, 1783). ILPE: Industria Lobera y Pesquera del Estado. Montevideo, Uruguay. *Anales*, Vol. 1, No.3, pp. 43-58

Ponce de León, A. (2000). Taxonomía, sistemática y sinopsis de la biología y ecología de los pinipedios de Uruguay. Pp. 9-36. In: *Sinopsis de la biología y ecología de las poblaciones de lobos finos y leones marinos de Uruguay. Pautas para su manejo y Administración*. Parte I. Biología de las especies. Rey, M. & Amestoy, F. (Editors). Montevideo, Uruguay, pp. 117

Ponce de León, A. (2001). Explotación, manejo y situación actual de las especies de otáridos que crían en islas de Uruguay. *III Jornadas sobre animales silvestres, desarrollo sustentable y medio ambiente.* AONIKEN, Comisión Ambientalista. Asociación de Estudiantes de Veterinaria. Facultad de Veterinaria. Montevideo, Uruguay. Noviembre de 2001.

Ponce de León, A. & Páez, E. (1996). Análisis del buceo en hembras del lobo fino sudamericano *Arctocephalus australis* en aguas uruguayas. VII Reunión de Especialistas en Mamíferos Acuáticos de América del Sur, 1er Congreso de la Sociedad Latinoamericana de Especialistas en Mamíferos Acuáticos. Viña del mar, Chile. 22 al 25 de octubre de 1996

Ponce de León, A., Malek, A. & Pin, O. (1988). Resultados preliminares del estudio de la alimentación del lobo fino sudamericano, *Arctocephalus australis* (Zimmermann, 1783), Pinnipedia, Otariidae, para 1987-1988. *III Reunión de Trabajo de Especialistas en Mamíferos Acuáticos de América del Sur*. Montevideo, Uruguay. 26 al 30 de julio de 1988

Ponce de León, A.; Pin, O. & Arim, M. (2000). Identificación de presas en contenido estomacal y fecas de ejemplares de lobo fino *Arctocephalus australis* del rebaño de Isla de Lobos, Uruguay. Pp. 37–51. In: *Sinopsis de la biología y ecología de las poblaciones de lobos finos y leones marinos de Uruguay. Pautas para su manejo y Administración*. Parte I. Biología de las especies. Rey, M. & Amestoy, F. (Editors). Montevideo, Uruguay, pp. 117

Ponce de León, A. & Pin, O. D. (2000). Planificación y estrategias sugeridas para el desarrollo de visitas turísticas en la Isla de Lobos, Uruguay. Pp. 77–90. In: *Sinopsis de la biología y ecología de las poblaciones de lobos finos y leones marinos de Uruguay. Pautas para su manejo y Administración*. Parte I. Biología de las especies. Rey, M. & Amestoy, F., (Editors). Montevideo, Uruguay, pp 117

Ponce de León, A. & Pin, O. D. (2006). Distribución, reproducción y alimentación del lobo fino *Arctocephalus australis* y del león marino *Otaria flavescens* en Uruguay. Págs. 305–313. In: *Bases para la conservación y el manejo de la costa uruguaya*. Menafra, R., Rodríguez-

Gallego, L., Scarabino, F. & Conde, D. (Editors). Vida Silvestre Uruguay. Montevideo, pp 668

Ponce de León, A. & Barreiro, J. C. (2010). Estado de las poblaciones de los lobos marinos de Uruguay. Departamento de Mamíferos Marinos, Dirección Nacional de Recursos Acuáticos (DINARA). *10º Congreso Nacional y 8º Internacional de Profesores de Biología*, La Paloma, Rocha. 19 al 22 de septiembre de 2010

Reeves, R. R.; Stewart, B. S. & Leatherwood, S. (1992). *The Sierra Club Handbook of Seals and Sirenias*. Sierra Club Books, San Francisco. Pp. 359

Rider, H. L. (1999). *Bases epidemiológicas del control de la tuberculosis*. Primera edición. Ed: Unión Internacional contra la Tuberculosis y Enfermedades Respiratorias. 68, Boulevard Saint-Michel 75006 Paris

Riet-Sapriza, F.; Costa, D. P.; Franco-Trecu, V.; Hückstadt L. A. & Chilvers, L. (2009). Foraging Areas & Diving Behavior of lactating South American sea lions *Otaria flavescens* during the austral summer at Lobos Island, Uruguay. *18th Biennial Conference on the Biology of Marine Mammals*. Quebec, Canada. October 2009.

Riet-Sapriza, F. G.; Franco-Trecu, V.; Costa D. P.; Chilvers, L. & Hückstadt, L. A. (2010). ¿Existe plasticidad en el comportamiento de buceo en hembras lactantes del lobo fino sudamericano *Arctocephalus australis* de Isla de Lobos, Uruguay?. *XIII Congreso SOLAMAC*. Florianópolis, Brasil. Octubre 2010.

Riet-Sapriza, F.; Costa, P. D.; Franco-Trecu, V.; Marín, Y.; Chocca, J.; González, B.; Beathyate, G.; Chilvers, L. B. & Hückstadt, L. A. (2011). Spatial and resource overlap between the Uruguayan fisheries and lactating South American sea lions, *Otaria flavescens*. *IV International Science Symposium on Bio-logging Location*. Hobart, Tasmania, Australia-March 2011.

Riet Sapriza, F.G.; Costa, D.P.; Franco-Trecu, V.; Marín, Y.; Chocca, J.; González, B.; Beathyate, G.; Chilvers, L. B. & Hückstadt, L. A. (2012). Foraging behavior of lactating South American sea lions, *Otaria flavescens* and spatial-resource overlap with the Uruguayan fisheries. *Deep-Sea Research II*. In press

Rodríguez, D. H.; Dassis, M.; Ponce de León, A.; Barreiro, J. C.; Farenga, M.; Bastida, R. & Davis, R. (2012). Foraging strategies of Southern sea lion (*Otaria flavescens*) females in La Plata River Estuary (Argentina-Uruguay). Deep-Sea Research Part II: Topical Studies in Oceanography. (2012) In preparation. http://dx.doi.org/10.1016/j.dsr2.2012.07.012

Romano, M. I.; Alito, A.; Bigi, F.; Fisanotti, J. C. & Cataldi, A. (1995). Genetic characterization of mycobacteria from South American wild seals. *Veterinary Microbiology*, Vol. 47, (March, 1995), pp. 89-98

Roth, A.; Reischl, U.; Streubel, A., Naumann, L.; Kroppenstadt, R. M.; Habicht, M.; Fischer, M. & Mauch, H. (2000). Novel diagnostic for identification of Mycobacteria using genus-specific amplification of the 16S-23S rRNA gene spacer and restriction endonucleoses. *Journal of Clinical Microbiology*, Vol. 38, No. 3, (March, 2000), pp. 1094-1104

Runyon, E. H.; Karlson, A. G.; Kubica, G. P. & Wayne, L. G. (1980). *Mycobacterium*. In: *Manual of Clinical Microbiology* (Third Edition). Lennett Editor, pp. 150-179 American Society for Microbiology, Washington DC

Sepúlveda, M.; Hinostroza, P.; Pérez-Alvarez, M. J.; Oliva, D. & Moraga, R. (2006). Seasonal variation in the abundance of South American sea lions *Otaria flavescens* (Shaw, 1800) in Chañaral Island, Reserva Nacional Pingüino de Humboldt, Chile. *Revista de Biología Marina y Oceanografía*, Vol. 44, No.3, (December, 2006), pp. 685-689

Smith, H. M. (1934) The Uruguayan fur-seal islands. *Scientific Contributions of the New York Zoological Society. Zoologica*, Vol. 9, No.6, (1934), pp. 271-294

Soto, K. H. (1999). Efectos de El Niño 1997-1998 sobre el ciclo reproductivo del lobo marino chusco *Otaria byronia* en las islas Ballestas, Pisco, Perú. Tesis Universidad Nacional Agraria La Molina.

Szephegyi M.N.; Franco-TrecuV.; Doño F.; Reyes F.; Forselledo R.; Crespo E. (2010). Primer relevamiento sistemático de captura incidental de mamíferos marinos en la flota de arrastre de fondo costero de Uruguay. *XVI Reuniao de Trabalho de Especialistas em Mamíferos Aquáticos da America do Sul*. Florianopolis-SC-Brasil. 24 al 28 de octubre 2010

Szteren, D. & Páez, E. (2002). Predation by southern sea lions (*Otaria flavescens*) on artisanal fishing catches in Uruguay. *Marine and Freshwater Research*, Vol. 53, (2002), pp. 1161-1167

Tacquet, A.; Tison, F.; Devulder, B. & Ross, P. (1967). Tecniques for decontamination pathological specimens for culturing Mycobacteria. *Bulletin International. Union Against Tuberculosis*, Vol. 39, (June, 1967), pp. 21-24

Steuen, S.; Have, P.; Osterhaus, A.; Arnemo, J.M. & Mousgtaard, A. (1994). Serological investigation of virus infections in harp seals (*Phoca groenlandica*) and hooded seals (*Cystophora cristata*). *Veterinary Record*, Vol. 134, (May, 1994), pp. 502–503

Thiele, D.; Chester, E. T.; Moore, S. E.; Sirovic, A.; Hildebrandt, J. A. & Friedlaender, A. S. (2004). Seasonal variability in whale encounters in the western Antarctic Peninsula. *Deep-Sea Research* II, Vol. 51, Issues 17–19, (August–September, 2004) pp. 2311–2325

Thompson, P. J.; Cousins, D. V.; Gow, B. L.; Collins, D. M.; Williamson, B. H. & Dagnia, H. T. (1993). Seals, seals trainers, and mycobacterial infection. *American Review of Respiratory Diseases*, Vol. 147, (January, 1993), pp. 164-7

Trimble, M. (2008). Reconocimiento materno-filial en el león marino sudamericano *Otaria flavescens* en Isla De Lobos, Uruguay. Tesis de Maestría, PEDECIBA Biología, Subárea Zoología. Facultad de Ciencias, Universidad de la República. 95pp

Vaz-Ferreira, R. (1950). Observaciones sobre la Isla de Lobos. *Revista de la Facultad de Humanidades y Ciencias*. Montevideo. Vol. 5, (1950), pp. 145-176

Vaz-Ferreira, R. (1952). Observaciones sobre las Islas de Torres y de Castillo Grande. *Revista de la Facultad de Humanidades y Ciencias.* Montevideo. Vol. 9, (1952), pp. 237-258

Vaz-Ferreira, R. (1956). Características generales de las islas uruguayas habitadas por lobos marinos. Servicio Oceanográfico y de Pesca. Departamento Científico y Técnico. *Trabajos sobre Islas de Lobos y Lobos Marinos* (1): 23 pp

Vaz-Ferreira, R. (1976). *Arctocephalus australis* (Zimmermann) South American fur seal. *Advisory Committee on Marine Resources Research*, pp. 1-13

Vaz-Ferreira, R. (1981). South American sea lion *Otaria flavescens* (Shaw, 1800), In: *Handbook of Marine Mammals*. Volume I: The walrus, sea lions, fur seals and sea otters. Pp. 39-65. Ridgway, S. H. & Harrison, R. J. Eds. Academic Press Inc. London.

Vaz-Ferreira, R. (1982). *Arctocephalus australis* (Zimmermann), South American fur seal. Pp. 497-508. In: *Mammals in the Seas*. FAO Fisheries Series 5, Vol. 4. Small cetaceans, seals, sirenias and otters. Pp. 531

Vaz-Ferreira, R. & Achaval, F. (1979). Relación y reconocimiento materno-filial en *Otaria flavescens* (Shaw) "lobo de un pelo" y reacciones de los machos subadultos ante los cachorros. *Acta Zoologica Lilloana*, Vol. 35, (1979), pp. 295-302

Vaz-Ferreira, R. & Palerm, E. (1962). Efectos de los cambios meteorológicos sobre agrupaciones terrestres de Pinnipedios. Ministerio de Industrias y Trabajo. Servicio Oceanográfico y de Pesca. Departamento Científico y Técnico. *Trabajos sobre Islas de Lobos y Lobos Marinos* No. 4, pp. 12

Vaz-Ferreira, R. & Sierra de Soriano, B. (1962). Estructura de una agrupación social reproductora de *Otaria byronia* (de Blainville), representación gráfica. Ministerio de Industrias y Trabajo. Servicio Oceanográfico y de Pesca. Departamento Científico y Técnico. *Trabajos sobre Islas de Lobos y Lobos Marinos,* No. 3, pp. 12

Vaz-Ferreira, R. & Ponce de León, A. (1984). Estudios sobre *Arctocephalus australis* (Zimmermann, 1783), lobo de dos pelos sudamericano, en Uruguay. Facultad de Humanidades y Ciencias, Universidad de la República. Contribuciones del Departamento de Oceanografía. Montevideo, Uruguay. Vol. 1, No. 8, pp. 1-18

Vaz-Ferreira, R. & Ponce de León, A. (1985). Estructura de grupos de dos especies de Otariidae. *Actas de las Jornadas de Zoología de Uruguay*, pp. 75-77

Vaz-Ferreira, R. & Ponce de León, A. (1987). South American Fur Seal, *Arctocephalus australis*, in Uruguay. Pp 29-32 In: *Proceedings of an International Symposium and Workshop.* Status, biology and ecology of fur seals. Croxall & Gentry. (Editors). NOAA Technical Report NMFS 51. Cambridge, 23-27 April 1984.

Webster, R. G.; Hinshaw, V. S.; Bean, W. J.; Van Wyke, K. L.; Geraci, J. R.; St Aubin, D. J. & Petursson, G. (1981). Characterization of an influenza A virus from seals. *Virology*. Vol. 113, No. 2 (September, 1981), pp. 712–724

Webster, R. G.; Bean, W. J.; Gorman, O. T.; Chambers, T. M. & Kawaoka Y. (1992). Evolution and ecology of Influenza A viruses. *Microbiology Reviews*, Vol. 56, No.1 (March, 1992), pp. 152–179

Woods, R.; Cousins, D. V.; Kirkwood, R. & Obendorf D.L. (1995). Tuberculosis in a wild Australian fur seal (*Arctocephalus pusillus doriferus*) from Tasmania. *Journal of Wildlife Diseases*, Vol. 31, No.1, (January, 1995), pp. 83-86

Wright, P., & Webster, R .G. (2001). Orthomyxoviruses. In: *Fields virology*. B. N. Fields, D. M. Knipe and P. M. Howley, Eds. pp 1254–1292. Lippincott-Raven Publishers, Philadelphia, PA

York. A.; Lima, M.; Ponce de León, A.; Malek, A. & Páez, E. (1998). First description of diving females South American fur seals in Uruguay. Abstract Volume. *WMMSC*, Monaco, (January 1998), pp. 153

Zumárraga, M. J.; Bernardelli, A.; Bastida, R.; Quse, V.; Loureiro, J.; Cataldi, A.; Bigi, F.; Alito, A.; Castro Ramos, M.; Samper, S.; Otal, I.; Martin, C. & Romano, M. I. (1999). Molecular characterization of mycobacteria isolated from seals. *Microbiology*, Vol. 145, (May, 1999), pp. 2519-2526

Chapter 8

Host-Virus Specificity of the Morbillivirus Receptor, SLAM, in Marine Mammals: Risk Assessment of Infection Based on Three-Dimensional Models

Kazue Ohishi, Rintaro Suzuki and Tadashi Maruyama

Additional information is available at the end of the chapter

http://dx.doi.org/10.5772/53215

1. Introduction

Incidences of infectious diseases in marine mammals have been increasing [1]. Among them, morbillivirus infection is the greatest threat to marine mammals becuase it has caused mass die-offs in several pinniped and cetacean species in the past few decades [2,3]. The genus *Morbillivirus* belongs to the family Paramyxoviridae, and the viruses in this genus have a genome consisting of a single piece of negative-stranded RNA, which encodes eight viral proteins: a nucleocapsid protein (N); a phosphoprotein (P); two virulence factors (C and V); a matrix protein (M); a membrane fusion protein (F); a hemagglutinin binding protein (H); and an RNA polymerase (L) [4]. The two viral surface glycoproteins, H and F, play important roles during the viral infection of host cells. The H protein is required for viral attachment to the host cells, while the F protein mediates membrane fusion with the host plasma membrane and enables the entry of the virus.

Until the discovery of a new mobillivirus in marine mammals in 1988, only four morbillivirus species had been identified in land mammals: the measles virus (MV); rinderpest virus (RPV); peste des petits ruminants virus (PPRV); and canine distemper virus (CDV) [4]. The new morbillivirus was isolated from dead harbor seals (*Phoca vitulina*) in a mass die-off around the Baltic and North Sea coasts and was named phocine distemper virus (PDV) [5,6]. Two other new viruses originating in cetaceans were also isolated from dead harbor porpoises (*Phocoena phocoena*) and striped dolphins (*Stenella coeruleoalba*) and were named porpoise morbillivirus (PMV) and dolphin morbillivirus (DMV) [7,8]. Based on the similarities of the gene sequences, it was proposed that the cetacean-origin viruses be unified as a single species, cetacean morbillivirus (CMV) [9,10].

Morbilliviruses propagate primarily in lymphoid tissues and induce acute disease. They are usually accompanied by lymphopenia and immunosuppression, which often lead to secondary, opportunistic infections in the host. The distemper viruses, CDV and PDV, often invade the central nervous systems of their hosts, although acute encephalitis is not common in other morbillivirus infections [4]. A notable feature of morbilliviruses is their high host specificity. The natural host of MV is humans, but it can also infect monkeys. Ruminants are the targets of RPV and PPRV. RPV mainly infects cattle, while PPRV infects goats and sheep. Although these viruses have multiple host compatibilities, they induce more severe disease in the primary hosts than in others [11,12]. The natural host for CDV is dogs, but ferrets (*Mustela putorius furo*) have been used as an experimental model due to their high sensitivity to CDV. Recently, the host range of CDV has been shown to be wider than previously thought and expanded to include other wild carnivores, such as Baikal seals (*Phoca sibirica*) or lions (*Panthera leo*) [13-15]. PDV and CMV have been isolated only from seals and cetaceans, respectively. While no morbilliviruses have been isolated from sirenians, serologic evidence of exposure to morbillivirus was reported in manatees (*Trichechus manatus*) without showing clinical signs of disease [16,17].

The cellular receptor of a virus is one of the major determinants of host specificity and tissue tropism. The signaling lymphocyte activation molecule (SLAM) has recently been shown to be the principal cellular receptor for morbilliviruses in humans, cows, and dogs [18,19]. SLAM itself was first discovered in 1995 as a novel receptor molecule involved in T-cell activation [20]. It is expressed on various immune cells, such as thymocytes, activated T and B cells, mature dendritic cells, macrophages, and platelets [21,22]. SLAM is also a marker for the most primitive hematopoetic stem cells [23]. The distribution and function of SLAM are consistent with the cell tropism and immunosuppressive nature of morbilliviruses. This indicates that the host range of morbillivirus may be explained by key amino acid residues of SLAM on the interface with morbillivirus.

In this chapter, we review morbillivirus infection in marine mammals and its possible primary receptor in the host, SLAM. Further, we discuss host–virus specificities based on three-dimensional models of SLAM and risk assessment of morbillivirus infection in marine mammals.

2. Morbillivirus infection and its impact on marine mammals

2.1. Mass die-offs in marine mammals and discovery of new morbilliviruses

Since the late 1980s, many mass die-offs have been reported around the coasts of Europe and the USA (Table 1). Approximately 18,000 harbor seals and several hundred grey seals (*Halichoerus grypus*) were found dead on northern European coasts in 1988–1989. The distemper-like gross observations suggested that a morbillivirus could have been the causative agent [6]. Detailed serological, virological, and immunohistochemical examinations showed that the agent was a new member of the genus *Morbillivirus*, named phocine distemper virus (PDV) [5,24]. In 2002, PDV again killed at least 21,000 seals inhabiting the same locales [25,26]. About the same time as the first outbreak on northern European coasts, in 1987–1988 the deaths of approximately 18,000 Baikal seals were reported in Lake Baikal which showed clinical signs identical to those reported in European seals [13]. However, subsequent genomic characterization revealed that the cause of the mass die-off of Baikal seals was CDV [27-31]. CDV also induced another mass die-off among Caspian seals (*Phoca caspica*), in which many seals died in 1997 and 2000, near Azerbaijan on the western shores of the Caspian Sea [32,33].

Date	Site	Animal species	No. of dead	Virus
1987-1988	USA Atlantic coast	*Tursiops truncatus*	>2,500	CMV
1987-1988	Lake Baikal	*Phoca sibirica*	>18,000	CDV
1988	North & Baltic Sea	*Phoca vitulina*	>18,000	PDV
1990	Mediterrean Sea	*Stenella coeruleoalba*	>2,000	CMV
1993	Mexican Gulf	*Tursiops truncatus*	>1,000	CMV
1997	Caspian Sea	*Phoca caspica*	>2,000	CDV
2000	Caspian Sea	*Phoca caspica*	>10,000	CDV
2002	North & Baltic Sea	*Phoca vitulina*	>21,000	PDV

CMV, cetacean morbillivirus; CDV, canine distemper virus; PDV, phocine distemper virus.

Table 1. Mass die-offs of marine mammals caused by morbilliviruses.

The first evidence of morbillivirus infection in cetaceans was described in several stranded harbor porpoises with pathological changes on the Irish coastline in 1988 [7]. A new morbillivirus was isolated and termed PMV for "porpoise" [34]. Since 1990, a severe mass die-off began to affect the striped dolphin (*Stenella coeruleoalba*) population on the Mediterranean coast of Spain and rapidly spread throughout the western Mediterranean Sea, including the coasts of France, Italy, Greece, and Turkey [8,35] (Table 1). A new virus, named DMV for "dolphin" was isolated as the causative agent [9]. A retrospective serologic investigation indicated that PMV and DMV were the agents responsible for another epidemic in bottlenose dolphins (*Tursiops truncatus*) along the Atlantic coast of the USA, for which the causative agent had been initially thought to be brevetoxin produced by a marine dinoflagellate (*Ptychodiscus brevis*) [36-38]. PMV also induced a die-off of bottlenose dolphins in the Gulf of Mexico during 1993–1994 [37-39]. Molecular biological analyses of these cetacean morbilliviruses showed that their gene sequences were similar [9,10]. Based on the similarities, it was proposed that these cetacean morbilliviruses be classified as a single species called CMV.

Thus, morbillivirus infection has a strong impact on populations of marine mammals, as listed in Table 1. In addition to mass die-offs, many smaller-scale die-offs were reported in various oceans. Even if the scale is small, outbreaks of morbillivirus infection can have serious consequences for marine mammal populations, especially among endangered species at risk of extinction, such as Mediterranean monk seals (*Monachus monachus*). In 1997, approximately 50% of the population of Mediterranean monk seals residing along the coast of Mauritania in Africa died suddenly. Morbilliviruses were isolated from the dead seals, although distemper-like lesions were not detected in the animals [40]. Hence, it remains unclear whether morbillivirus was the agent responsible for the dramatic deaths. The involvement of an algal bloom was also suggested as the primary cause [41]. In any case, when populations of many marine mammal species are decreasing, morbillivirus infection may cause a fatal blow.

2.2. Transmission and maintenance reservoir of marine morbilliviruses

The origin, precise mode of transmission, and maintenance reservoir of morbilliviruses causing die-off epidemics in marine mammals remain to be elucidated. Morbilliviruses proliferate in the infected animal for a short time after infection but do not persist in the host [4]. After being shed from the infected host, they do not survive long in the environment [4]. For the transmission of morbilliviruses, therefore, close contact between acutely infected and susceptible animals is required. In addition, because a morbillivirus infection results in lifelong immunity in the infected animal, when the virus is maintained in an animal population, a constant supply of new susceptible animals is needed. It has been calculated that the minimal population size for MV maintenance is approximately 300,000 individuals [42]. In the mass die-off of European seals in 1988, the most likely viral source was an infected seal population in the Arctic region, which moved southward and made contact with the population on the European coast. This hypothesis was based on the results of serologic studies using archival seal sera. PDV-specific antibodies were not observed in European seal sera before 1988, indicating that the population was naive and had not been previously exposed to the virus [43]. However, specific antibodies were detected in sera obtained from arctic seals long before 1998 [44,45]. In addition, alterations in the migration patterns of Arctic harp seal (*Phoca groenlandica*) populations were recorded. They were seen much farther south than usual in northern European waters in the year prior to the epizootic of the harp seal population [46]. The harp seal population is extremely large, with four million individuals in Canadian waters alone, which is sufficient to maintain morbillivirus circulating within the population. Subclinically or subacutely affected animals might play an important role in the transmission of the virus.

It should be noted that morbillivirus transmission sometimes occurs between marine and land mammals. As described above, the mass die-offs of Baikal seals and Caspian seals were caused by infection with CDV [28,30,32]. The most likely source of infection was land animals infected with CDV, because outbreaks were common among the numerous feral and domestic dogs around the lake [47]. Accidental infection in the opposite direction was also reported. A farmed mink population fed infected seal meat was infected with PDV in Denmark during the 1988 epizootic of PDV [48].

3. SLAM, a receptor of morbillivirus

The characteristics of SLAM have been extensively studied in humans. SLAM (CD150) is a type I transmembrane protein, and there are many members of the SLAM family including the well-known 2B4 (CD244), Ly-9 (CD229), NTB-A, and CD84 [49]. All of the SLAM family members have an extracellular region composed of a membrane-distal immunoglobulin variable (V) domain and a membrane-proximal immunoglobulin constant-2 (C2) domain, along with a cytoplasmic region bearing multiple tyrosine-based switch motifs (ITSMs) that bind cytoplasmic Src homology-2 (SH2)-containing proteins such as SLAM-associated protein (SAP) [50]. Evidence is accumulating that the interaction between the cytoplasmic region of SLAM and SAP family molecules mediates a switch to positive or negative signaling in immune cells and plays a crucial role in multiple immune regulations [22]. Genes for the SLAM family receptors are located within a ~400-kb cluster on chromosome 1 in humans and mice [51]. This gene location, coupled with the conserved exon-intron structure of SLAM-related genes, implies that they were generated by the sequential duplication of a single ancestral gene. A SLAM family receptor forms a homophilic dimer by weak binding between the V domains and acts as a self-ligand, suggesting that the receptors can trigger homotypic or heterotypic cell–cell interactions [52]. The V domain of SLAM (CD150) also provides an interface for binding with morbilliviruses [53]. The viral H protein has a strong affinity for the V domain of SLAM, which is 400-fold higher than for self-ligand interaction [54]. The interaction between the viral H protein and SLAM V domain is the initial event in infection with morbilliviruses. The results of recent detailed structural studies have suggested that the interaction changes the microenvironment of the interaction zone for the fusion activity of the F protein, although the mechanism of membrane fusion mediated by the F protein is not fully understood [55-58].

3.1. History of the discovery of the morbillivirus receptor

The human CD46 molecule was first identified as a cellular receptor for Edmonston vaccine strains of MV [59,60]. The Edmonston strain was isolated from the blood and throat washings of a child with measles using primary human kidney cells in 1954 [61]. It was later adapted to chick embryo fibroblasts and is being used as an attenuated vaccine [62]. This strain grows well in many cell lines, such as Vero cells, and has become the most extensively studied MV strain in the laboratory. However, because CD46, a complement-regulatory molecule, is expressed on all human nucleated cells, its ubiquitous distribution cannot explain the lymphoid tropism of MV. At present, CD46 is thought to be a specific receptor of the Edmonston strain, which is presumed to acquire the ability to use CD46 by adapting to cultured human kidney cells.

On the other hand, many wild-type strains have been isolated from clinical samples using the marmoset B cell line (B95a) [63], but they do not grow on many CD46+ cell lines. In order to identify the receptor for wild-type MV, functional expression cloning of a cDNA library of B95a cells was carried out using the VSV pseudotype system. SLAM was shown to be a cellular receptor for wild-type MV [18]. CDV and RPV were also shown to use canine and bovine SLAMs for entry into host cells [19]. Thus, SLAM is thought to be the major receptor for wild-type morbillviruses.

Recently, Nectin 4, a cellular adhesion junction molecule, has been identified as the third receptor for MV in polarized epithelial cells [64,65]. Infection experiments in monkeys showed that MV initially targets SLAM-positive immune cells such as alveolar macrophages, dendritic cells, and lymphocytes, and later the viral infection spreads to the epithelial cells of the trachea, lungs, oral cavity, pharynx, or intestines, which are SLAM-negative cells [66,67]. Another infection experiment using epithelial cell receptor-blind MV, demonstrated that the mutant MV inoculated intranasally to monkeys shows virulence and infectivity toward lymphoid tissues, although the virus cannot cross the airway epithelium and cannot be shed in the air [68]. The molecule forms tight junctions on polarized epithelial cells and was shown to function as the receptor for effectively releasing MV to the apical side of epithelial cells [69,70]. This explains why MV is highly contagious. Thus, the wild-type MV posesses two types of receptor, SLAM for entry and propagation and Nectin 4 for viral release into the air.

3.2. Structure of SLAMs of marine mammals

Cetaceans and sirenians have achieved complete adaptation to the aquatic environment and spend all of their lives in water. Cetaceans belong to the order Cetartiodactyla, superorder Laurasiatheria, and are closely related to hippopotami or ruminants among land animals. Sirenians, including dugongs and manatees, are in the order Sirenia, superorder Afrotheria, and are evolutionarily related to elephants or hyraxes. Pinnipeds, belonging to the order Carnivora, superorder Laurasiatheria, are not completely adapted to the aquatic environment and they must deliver and nurse their young on land. This characteristic of pinnipeds makes it possible to transmit infectious diseases between aquatic and land mammals. To determine the structure of marine mammal SLAMs, we collected blood samples from taxonomically different animal groups, i.e., cetaceans, pinnipeds, and sirenians. White blood cells were obtained from: two species of cetacean, a Pacific white-sided dolphin (*Lagenorhynchus obliquidens*) and a killer whale (*Orcinus orca*); two species of pinniped, a spotted seal (*Phoca largha*), and a walrus (*Odobenus rosmarus*); and a sirenian, a West Indian manatee (*Trichechus manatus*). The blood of an Indian elephant (*Elephas maximus bengalensis*) was also collected. After immune stimulation with phytohemagglutinin, RNAs of the leukocytes were extracted. First, the complete nucleotide sequences of the SLAM genes were determined. Three-dimensional models were then generated based on the deduced amino acid sequences to compare the interface of their SLAM V domains [71].

3.2.1. *Primary structure of SLAM proteins*

The deduced amino acid sequences of marine mammal and elephant SLAMs indicated that they contain 336–339 amino acid residues, inducing six cysteine residues and six potential *N*-linked glycosylation sites (Figure 1). They have two immunoglobulin-like domains, V and C2, in the extracellular region, and two ITSM motifs (T-X-Y-X-X-V/I) and one ITSM-like sequence in the intracellular region. These molecular features are shared with all reported mammalian SLAMs (Figure 1). The cetacean and pinniped SLAMs showed the greatest homology with those of artiodactyla (cow and sheep, 84–85% identity at the amino acid level) and of dogs (84%), respectively. Manatee SLAM shared the greatest homology with that of the elephant (86%) [71].

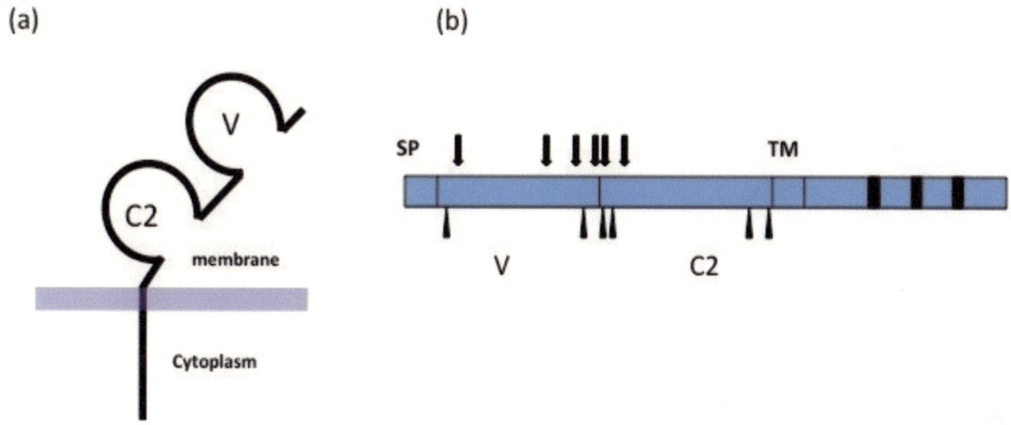

SP: signal peptide, TM: transmembrane region, V: immunoglobulin V-like domain, C2: immunoglobulin C2-like domain. Closed squares: ITSM and ITSM-like motifs, The *N*-linked glycosylation site and cysteine residues are indicated by arrows and triangles, respectively.

Figure 1. Schematic drawings of domain (a) and primary structure (b) of marine mammal SLAMs.

3.2.2. Phylogenetic analysis based on SLAM and morbillivirus H proteins

Phylogenetic trees based on SLAM and the morbillivirus H protein were constructed using the maximum-likelihood (ML) and Bayesian methods. In the phylogeny of SLAMs (Figure 2(a)), each taxonomic group, including primates (humans, chimpanzees, rhesus monkeys, and marmosets), cetaceans (Pacific white-sided dolphins and killer whales), artiodactyls (cows, buffalo, sheep, and goats), pinnipeds (spotted seals and walruses), and rodents (mice and rats), was monophyletic with a 100% ML bootstrap probability (BP) and a 1.00 Bayesian posterior probability (BPP). Manatee and elephant SLAMs, dog and pinipped SLAMs, and cetacean and artiodactyl SLAMs formed single clades, each with a 100% BP and a 1.00 BPP value, respectively.

Morbillivirus phylogeny based on MV H protein sequences reflected the host grouping, except for MV. CDV (dogs, Baikal seals) and PDV (seals), and PMV (porpoises) and DMV (dolphins), respectively, formed single clades each with 100% BP and 1.00 BPP support (Figure 2(b)). The monophyletic lineage of MV (human) (100% BP and 1.00 BPP) was within the grouping of ruminant viruses, RPV (cow) and PPRV (sheep and goat), with 100% BP and 1.00 BPP. These phylogenetic trees indicated that SLAMs and viral H proteins roughly co-evolved. However, the monophyletic lineage of MV and the ruminant viruses RPV and PPRV suggested that human MV may have originated from ancestral RPV in cattle by acquiring a binding affinity for human SLAM, as proposed in a previous report based on the morbillivirus P gene [9]

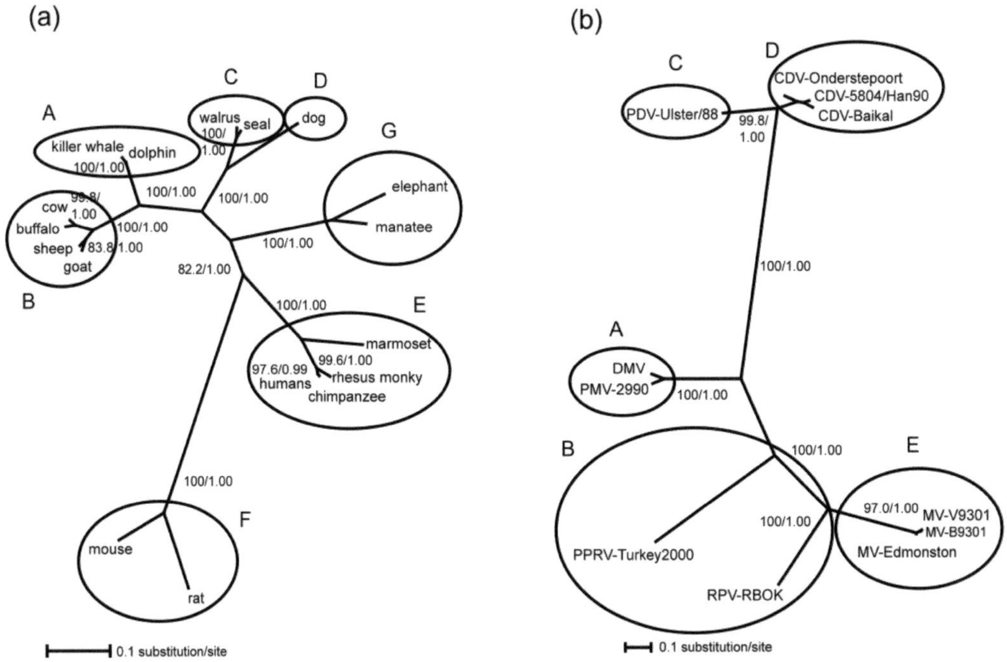

ML analyses were performed using PhyML [72], and an input tree was generated using BIONJ with the JTT model [73] along with amino acid substitution that incorporated invariable sites and used a discrete gamma distribution (eight categories) (JTT + I + G model). Bayesian phylogenetic analyses were conducted using MrBayes version 3.0 within the JTT + I + G model [74]. The ML bootstrap probabilities and BPPs are shown at the nodes. Host animals and the corresponding morbilliviruses are shown as circles with the same alphabetic notations: A, cetaceans; B, artiodactyla; C, pinnipeds; D, dogs; E, primates; F, rodents; and G, manatees and elephants. Morbillivirus has not been identified in rodents (F), or manatees and elephants (G). The following SLAM gene sequences were obtained from the Data Bank of Japan: Pacific white-sided dolphin (AB428366); killer whale (AB428367); spotted seal (AB428368); walrus (AB428369); Indian elephant (AB428370); American manatee (AB428371); human beings (*Homo sapiens*, NM_003037); chimpanzee (*Pan troglodytes*, XM_513924); marmoset (*Saguinus oedipus*, AF257239); Rhesus monkey (*Macaca mulatta*, XM_001117605); cow (*Bos taurus*, AF329970); buffalo (*Bubalus bubalis*, DQ228868); sheep (*Ovis aries*, NM_001040288); goat (*Capra hircus*, DQ228869); dog (*Canis familaris*, AF325357); mouse (*Mus musculus*, NM_013730); and rat (*Rattus norvegicus*, XM_001054873). Morbillivirus H protein sequences were obtained as follows: MV Edmonston AIK-C strain (AB046218); MV V9301 strain (AB012948); MV B9301 strain (AB012949); RPV RBOK strain (Z30697); PPRV Turkey 2000 strain (NC_006383); CDV Onderstepoort vaccine strain (AF305419); CDV Baikal seal strain (X84998); CDV 5804/Han90 strain (X85000); PDV Ulster/88 strain (D10371); PMV 2990 strain (AY586537); and DMV (NC_005283). The alignments of the deduced amino acid sequences from these genes were generated using ClustalW version 1.8, inspected visually, and edited manually. This figure was adapted from [71].

Figure 2. Phylogenetic trees of SLAM peptide sequences (a) and morbillivirus H proteins (b).

3.2.3. Three-dimensional models of marine mammal SLAM extracellular domain

In order to analyze the binding site for morbillivirus, three-dimensional (3D) models of marine mammal SLAM extracellular domains were generated by homology modeling. Previously, we constructed 3D homology models based on the crystallographic structure of the human NTB-A molecule, a member of the SLAM family, as a template [71,75]. In the present study, we generated a new version of the models by adding the recently determined crystal structure information of the bound complex of MV H and marmoset SLAM V [56]. Figure 3 shows the 3D model of the Pacific white-sided dolphin SLAM extracellular domain in a self-ligand form. In the models, the V and C2 domains are shown with rod-like structures, which are both constituted mainly of β-sheets containing several β-strands. The cysteine residues appear to be important in forming the basic 3D structures. On the basis of amino acid sequence similarity, this structure is shared with all of the SLAMs examined. The V domain possesses a two-layered β-sheet structure, and the front sheets provide an interface for binding with morbilliviruses.

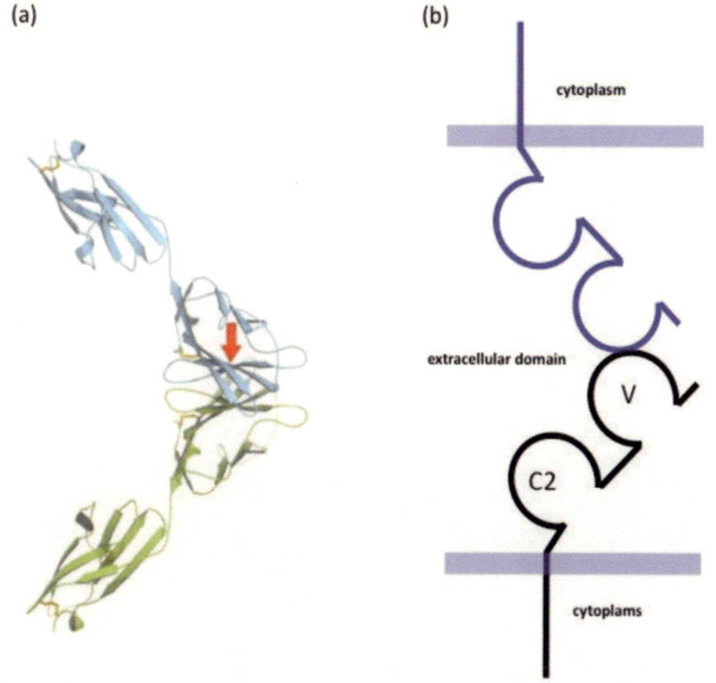

In (a), the blue and green models show respective SLAM extracellular domains forming a homophilic dimer. The β-strands are indicated by arrows and the disulfide bonds are shown as yellow bars. The thick red arrow indicates the direction of view of the front face of the morbillivirus binding site, as shown in Figure 4. In (b), the blue and black models indicate two SLAM molecules of two adjacent cells, respectively. The Protein Data Bank (PDB) entries for the marmoset SLAM in the complex (3ALW: A-D chains, 3ALX: A–D chains, 3ALZ: B chain; reference [56]) and for human NTB-A (21F7: A–D chains; reference [75]), were used as the template structure. The 3D model was constructed using the MODELLER 9.10 program [76] and visualized using PyMOL 1.4.1 (Schrodinger LLC) and PovRay (Persistence of Vision Pty. Ltd.).

Figure 3. Ribbon diagram of the 3D structure (a) and schematic drawing (b) of the SLAM extracellular domain from the Pacific white-sided dolphin.

Figure 4 shows top views of the front face of the modeled V domains of SLAMs of the spotted seal, Pacific white-sided dolphin, and West Indian manatee. The amino acid residues that have protruding side-chains on their front faces are likely a component of virus binding. We found such qualified 27 amino acid residues; 12 amino acid residues on the β-strands and 15 residues on loops (Figure 4). In addition, the amino acid residues positioned at 76, and the residues at 127-131 are thought to be important for the binding of the virus, because the side-chains of these residues are closely located to those of viral H protein in a crystal structure of the complex [56]. Particularly, the residues at 127-131 are thought to form an intramolecular β-sheet with the β-strand of MV H [56]. The overall 3D structures of the interfaces are similar among SLAMs, but several among the total 32 amino acid residues possibly contributing the binding affinity to the virus, differed among the three marine mammals.

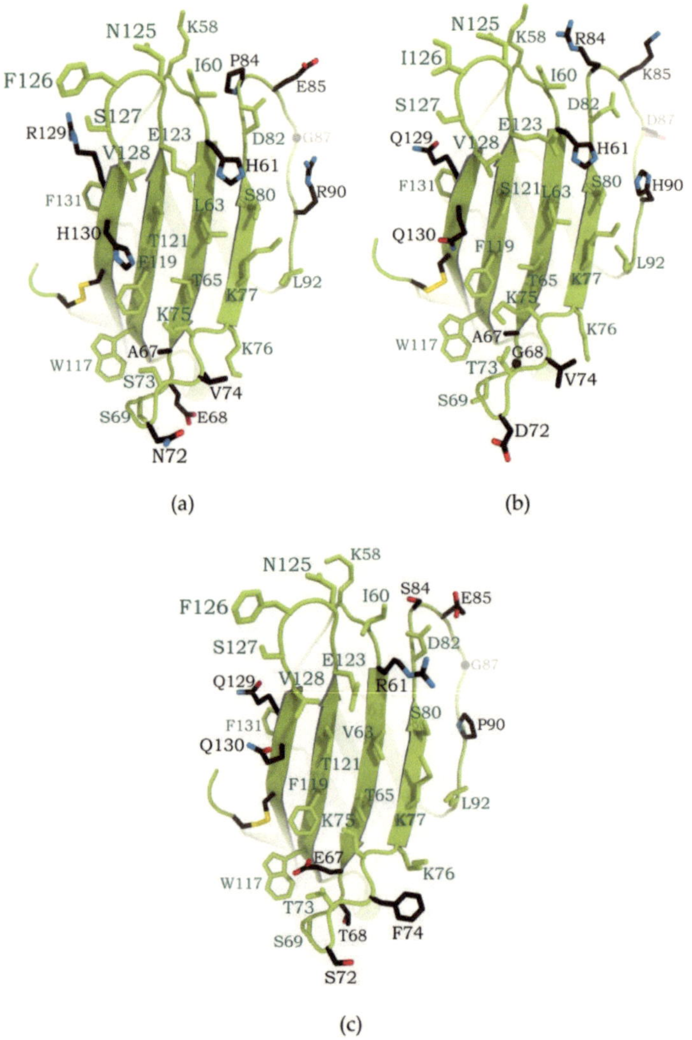

The interfaces are viewed from the direction shown by the thick red arrow in Figure 3. The amino acid residues that possibly interact with the viral H protein, are shown with their position numbers. The different amino acid residues among the three mammal SLAMs are indicated in black, and their side-chains are shown with the atoms colored (black for carbons, blue for nitrogens, and red for oxygens). The shared residues among the three are shown in blue. Disulfide bonds are shown in yellow.

Figure 4. Ribbon diagram of the 3D structure models of the SLAM interface for binding morbillivirus from the spotted seal (a), Pacific white-sided dolphin (b), and manatee (c).

3.2.4. Prediction of amino acid residues involved in virus binding and host–virus specificity

To identify amino acid residues that are important for host–virus specificity, we compared the 32 residues with those of land mammal SLAMs (Table 2). The difference in the SLAM interface was only two amino acid residues between seal and dog (Val and Ile at position 74, and Arg and Gln at position 129), between human and marmoset (Leu and Phe at position 119, Val and Ile at position 126, marmoset data not shown), and between cow and sheep (Asp and Gly at position 87, and Arg and His at position 90). This is consistent with the evidence that mass die-offs of Baikal seals and Caspian seals were caused by CDV; marmosets are highly sensitive to MV; and that RPV and PPRV can infect ruminants. It is noted that the identity of these 32 residues between dolphin and cow SLAMs is very high, although they are infected by different morbilliviruses, CMV and RPV, respectively. Four residues are different between the two animals, while eleven and fourteen residues are different between dolphin and seal, and between dolphin and humans, respectively.

a.a. No.	Seal	Dog	Dolphin	Cow	Sheep	Human	Manatee
58	K	K	K	K	K	K	K
60	I	I	I	I	I	I	I
61	H	H	H	H	H	H	R*
63	L	L	L	L	L	V	V
65	T	T	T	T	T	T	T
67	A	A	A	A	A	A	E*
68	E	E	G*	E	E	K*	T*
69	S	S	S	S	S	S	S
72	N	N	D	D	D	N	S
73	S	S	T	T	T	S	T
74	V	I	V	V	V	V	F
75	K	K	K	K	K	E*	K
76	K	K	K	K	K	N*	K
77	K	K	K	K	K	K	K
80	S	S	S	S	S	S	S
82	D	D	D	D	D	D	D
84	P	P	R	R	R	S	S
85	E	E	K	K	K	E	E
87	G	G	D	D	G	G	G
90	R	R	H	R	H	R	P*
92	L	L	L	L	L	L	L
117	W	W	W	W	W	W	W
119	F	F	F	F	F	L	F
121	T	T	S	S	S	T	T
123	E	E	E	E	E	E	E

a.a. No.	Seal	Dog	Dolphin	Cow	Sheep	Human	Manatee
125	N	N	N	N	N	N	N
126	F	F	I	V	V	V	F
127	S	S	S	S	S	S	S
128	V	V	V	V	V	V	V
129	R*	Q	Q	Q	Q	Q	Q
130	H	H	Q	H	H	R	Q
131	F	F	F	F	F	F	F
Viruses	PDV, CDV	CDV	CMV	RPV, PPRV	PPRV, RPV	MV	None

The residue position, which varies among animals, is shaded. The light- and dark-shaded boxes indicate a variation in chemically (charge, hydrophilicity, etc.) similar or different residues. The asterisk indicates the specific residue with a chemical change for the animal SLAM. See the legend of Figure 2 for the animal names and accession numbers of the amino acid sequences used in this table.

Table 2. Amino acid residues on the SLAM interface possibly involved in regulating the binding and specificity of morbilliviruses.

Among the 32 residue positions, variations in amino acids were found at 18 positions (Table 2, light and dark shading). At six positions (63,73,74,119,121, and 126), the changes are between chemically similar residues (light-shaded boxes in Table 2) and these do not seem to markedly affect binding with the viruses. On the other hand, the variations at the other twelve residue positions (61,67,68,72,75,76,84,85,87,90,129, and 130; dark-shaded boxes in Table 2) occur in amino acids with chemically different characteristics. In particular, the variations among amino acids with opposite charge may significantly alter the affinity for viruses (positions 68,75, and 85). The twelve amino acid residues are thought to be important in determining host–virus specificity. Almost of the twelve residues are located in the edge region of the interface. This may indicate that residues located in the central region of the interface play an important role in virus entry itself, rather than in host–virus specificity. Alternatively, they may be essential for a primary immunological function.

A detailed binding assay using surface plasmon resonance analysis was carried out between human SLAM mutants and the MV H protein [56,57]. The respective changes in the residues from H61, E123, and R130 of the human SLAM interface to serine residues completely abolished the binding ability to the MV H protein. In crystallographic analysis, R130 was suggested to form an intramolecular salt bridge with E75 [56]. Only human SLAM possesses these two residues on the interface. As shown in Table 2, K68, a strong positively charged residue, is also specific for human SLAM. These facts suggest that they are key residues for MV infection. On the other hand, H61 and E123 are conserved in all mammals, except for R61 of manatees, suggesting that these residues play a crucial role in viral infection, rather than in host–virus specificity.

Although a detailed analysis of SLAM–virus interaction has not been conducted in systems other than the human SLAM–MV complex, animal-specific residues can be seen in Table 2. For example, the markedly specific residue set R84-K85–D87-H90 is found in dolphin. These residues are located spatially near each other in the 3D homology model of the interface (Figure 4). To clarify the influence of the changes in the charge, electrostatic potentials on the surface of SLAM interfaces of the three marine mammals, are shown in Figure 5. It can be seen that the zone constituted by the residues at positions 84, 85, 87, and 90 are different among the SLAM interfaces of three marine mammals.

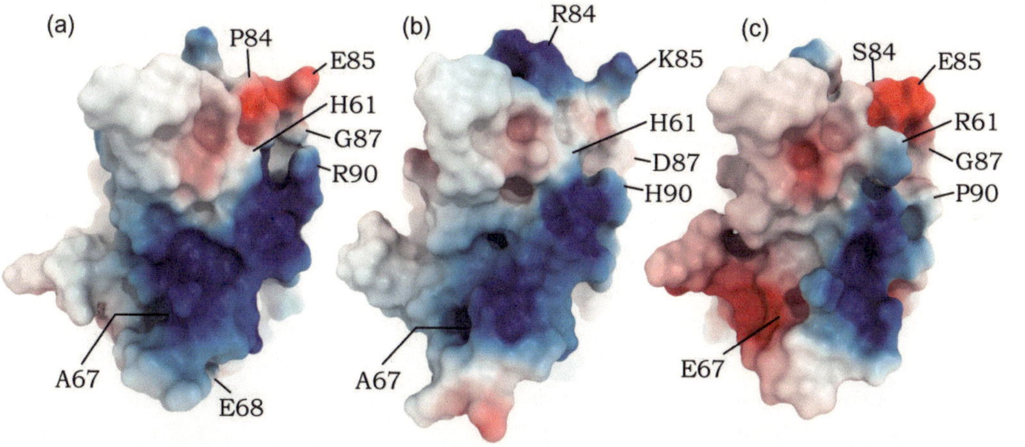

Electrostatic potential was calculated using DelPhi v.4 [77,78]. Positively and negatively charged surfaces are colored blue and red, respectively.

Figure 5. Electrostatic potential on the surface of the SLAM interfaces with morbilliviruses deduced from the 3D homology model structures of the SLAM interfaces from the spotted seal (a), Pacific white-sided dolphin (b), and manatee (c).

The amino acid residues at positions 67 and 68 are also highly variable among the three interfaces (Table 2). The manatee-specific residue E67 appears to contribute greatly to the formation of the negatively charged zone (Figure 5). The manatee SLAM has two another specific residues, R61 and P90, which induce an acquisition of stronger positive charge or a loss of positive charge on the interface (Table 2, Figure 4). These findings suggest that if there is a morbillivirus for the sirenians, it has the H protein with a very different SLAM binding interface.

Crystallographic analysis of the complex of MV H and marmoset SLAM V unexpectedly showed two different potentially tetramic configurations, form I and form II [56]. Residue N53 is located at the interface only in form II. Its replacement mutant changing to Q53 showed a reduction in molecular masses, meaning the loss of glycosylation, and an unexpected increase in MV entry into human cells. The reason for this is not fully understood. However, because only primate and manatee SLAMs possess residue N53, it may also be involved in host–virus specificity.

4. Risk assessment of morbillivirus infection based on SLAM interface structures

Human beings have a long history of diseases caused by morbilliviruses, which introduced devastating contagious diseases to humans and domestic animals. Since Jenner's seminal discovery of the concept of immunity and vaccines, vaccines against various pathogens have been developed. By virtue of that great endeavor, an effective vaccine against measles is now available [79,80]. It has reduced measles deaths worldwide by 74% between 2000 and 2010 (from 535,300 to 139,300), although measles is still a threat for children in developing countries [81]. Rinderpest induced by RPV, one of the oldest recorded livestock plagues, has been actually eradicated by the success of the Global Rinderpest Eradication Programme [82-84]. Thus, numerous efforts at eradication have achieved the control of human and domesticated animal diseases. Similar control of morbillivirus infections in wild animals will be one of the most important issues in the field of veterinary medicine in the 21st century. Marine mammals have large geographical ranges in the oceans. For example, baleen whales are known to migrate seasonally from the equator to the polar seas, indicating that they may be a dynamic vector for infectious diseases. In addition, recent global climate change may alter the ecology of marine mammals, such as their habitats, migration patterns, food, and behavior, and may increase opportunities for contact among previously geographically separate mammalian populations. These alterations may increase the possibility of viral transmission and the likelihood of outbreaks in susceptible mammalian populations.

In viral infection, disease incidence, and transmission, several factors in host cells play a key role. In addition, ecological factors such as animal distribution, population structure and size, and behavior are decisive factors in actual infection. In the present study, we used a new approach to assess the potential infectivity of morbilliviruses based on receptor structure predictions. The residues on the interface of the SLAM V domain probably contribute to virus binding, and some residues among them are key for host–virus specificities. The analysis of these residues on 3D models of the SLAM receptor is useful for estimating the risk of morbillivirus infection in wild animals. This approach is applicable to animals for which no information on infection and disease. For the animals, it is possible to predict potential infection with known morbilliviruses. This may reveal possible infection spectra of morbilliviruses and suggest which mammals are reservoirs that maintain the viruses and how these viruses spread among wild mammals in nature.

Control of infectious diseases in wild animals is very difficult. However, even though we cannot stop outbreaks from occurring in nature, information on the potential sensitivity of wild mammals against a virus may minimize the damage or prevent the spread of disease by such means as artificial transportation. Because marine mammals are positioned at the top of oceanic food chain, a decrease in marine mammal populations will affect marine ecosystems. We believe that the present study contributes to the conservation of marine mammals and their ecosystems.

5. Conclusion

Morbillivirus, a member of the family Paramyxoviridae, is a causative agent of mass mortalities of marine mammals. To date, four virus species, MV, RPV, PPRV, and CDV have been identified in land mammals, and two virus species, PDV and CMV, have been identified in seals and cetaceans, respectively [4,47]. The notable biological feature of morbillivirus is its high level of host specificity. The cellular receptor for a virus is a major determinant of its host specificity and tissue tropism. SLAM is the principal cellular receptor for morbilliviruses allowing entry and propagation [18,19]. SLAM contains two immunoglobulin-like domains, the V and C2 domains, in the extracellular region. The morbillivirus H protein binds to the V domain on the target cells, which triggers viral infection [53] To assess the host–virus specificity of morbillivirus in marine mammals, we determined the complete nucleotide sequences of SLAM from five species belonging to cetaceans, pinippeds, and sirenians, and generated 3D homology models. The results showed that the overall structures are similar in the mammals examined. We found 32 amino acid residues on the interface of SLAM V domain that are potentially involved in the interaction with viruses. Among them, a set of 18 amino acid residues is important for morbillivirus binding because some residues in the set differ among the mammal groups, which are susceptible to different morbillivirus species. A change in some residues in the set may cause an electrostatic change on the interface surface. These amino acid residues are thought to be important for host–virus specificity.

Analysis of these residues on the interfaces of SLAMs will be useful to assess the risk of morbillivirus infection in wild animals. Recent climate change may increase the opportunities for new contacts among wild mammals and for the transmission of viruses. In the present study, we propose a new approach to assess the viral sensitivities of wild mammals by analyzing the host receptors. This approach will contribute to the conservation of wildlife including marine mammals and ecosystems.

Author details

Kazue Ohishi* and Tadashi Maruyama
Japan Agency for Marine-Earth Science and Technology (JAMSTEC), Yokosuka, Japan

Rintaro Suzuki
Protein Research Unit, National Institute of Agrobiological Science, Tsukuba, Japan

Acknowledgement

The authors are grateful for the cooperation of the Yokohama Hakkeijima Sea-Paradise Aquamuseum (Yokohama), Kamogawa Sea World (Chiba), Churaumi Aquarium (Okinawa), and Kanazawa Zoo (Yokohama).

---

* Corresponding Author

6. References

[1] Dierauf AL, Gulland FMD. CRC Handbook of Marine Mammal Medicine: Health, Disease, and Rehabilitation. Florida: CRC Press;2001.

[2] Barrett T, Blixenkrone-Møller M, Di Guardo G, Doming M, Duignan P, Hall A, Mamaev L, Osterhaus ADME. Morbilliviruses in Aquatic Mammals: Report on Round Table Discussion. Veterinary Microbiology 1995;44(5) 261–265.

[3] Di Guardo G; Marruchella, G; Agrimi U, Kennedy S. Morbillivirus Infections in Aquatic Mammals: A Brief Overview. Journal of Veterinary Medicine Series A, Physiology, Pathology Clinical Medicine 2005;52(2) 88–93.

[4] Griffin DE. Measles Virus, In: Knipe DM, Howley PM. (eds.) Field's Virology. Philadelphia: Lippincott Williams & Wilkins;2001. p1401–1441.

[5] Mahy BWJ, Barrett T, Evans S, Anderson EC, Bostock CJ. Characterization of a Seal Morbillivirus. Nature 1988;336(6195) 115.

[6] Osterhaus AD, Vedder EJ. Identification of Virus Causing Recent Seal Deaths. Nature 1988;335(6185) 20.

[7] Kennedy S, Smyth JA, Cush PF, McCullough SJ, Allan GM, McQuaid S. Viral Distemper Now Found in Porpoises. Nature 1988a;336(6194) 21.

[8] Domingo M, Ferrer L, Pumaorola, M, Marco A, Plana J, Kennedy S, McAliskey M, Rima BK. Morbillivirus in Dolphins. Nature 1990;348(6296) 21.

[9] Barrett T, Visser IK, Mamaev L, Van Bressem MF, Osterhaus ADME. Dolphin and Porpoise Morbillivirus Are Genetically Distinct from Phocine Distemper Virus. Virology 1993;193(2) 1010–1012.

[10] Bolt G, Blixen Krone-Moller M, Gottschalck E, Wishaupt RGA, Welsh MJ, Earle JAP, Rima BK. Nucleotide and Deduced Amino Acid Sequences of the Matrix (M) and Fusion (F) Protein Genes of Cetacean Morbilliviruses Isolated from a Porpoise and a Dolphin. Virus Research 1994;34(3) 291–304.

[11] Diallo A., Barrett T, Barbron M, Subbarao SM, Taylor WPDifferentiation of Rinderpest and Peste des Petits Ruminants Viruses Using Specific cDNA Clones. Journal of Virological Methods 1989;23(2) 127–136.

[12] Anderson EC, Hassan A, Barrett T, Anderson J. Observation on the Pathogenicity for Sheep and Goats and the Transmissibility of the Strain of Virus Isolated during the Rinderpest Outbreak in Sri Lanka in 1987. Veterinary Microbiology 1990;21(4) 309–318.

[13] Grachev MA, Kumarev VP, Mamaev LV, Zorin VL, Baranova LV, Denikina NN, Belikov SI, Petrov EA, Kolesnik VS, Kolesnik RS, Dorofeev VM, Beim AM, Kudelin, VN, Nagieva FG, Sidorov, VN. Distemper Virus in Baikal Seals. Nature 1989;338(6212) 209.

[14] Appel MJG, Yates R.A, Foly Gl, Bernstein JJ, Santinelli S, Spelman LH, Miller LD, Arp LH, Anderson M, Barr M, Pearce-Kelling S, Summers BA. Canine Distemper Epizootic in Lions, Tigers, and Leopards in North America. Journal of Veterinary Diagonosis Investigation 1994;6(3) 277-288.

[15] Roelke-Parker ME, Munson L, Packer C, Kock R, Cleaveland, SM, Carpenter M, O'Brien SJ, Pospischil, A Hofman-Lehmann, R Lutz, H Mwamengele GL, Mgasa MN, Machange GA, Summers BA, Appel MJ. A Canine Distemper Virus Epidemic in Serengeti Lions (Panthera leo). Nature 1996;379(6564) 441–445.

[16] Duignan PJ, House C, Walsh MT, Campbell T, Bossart GD, Duffy N, Fernandes PJ, Rima BK, Wright S, Geraci JR. Morbillivirus Infection in Manatees. Marine Mammal Science 1995;11(4) 441–451.

[17] Grant RJ, Kelley KL, Maruniak JE, Garcia-Maruniak, A, Barrett T, Manire CA, Romero CH. Expression from Baculovirus and Serological Reactivity of the Nucelocapsid Protein of Dolphin Morbillivirus. Veterinary Microbiology 2010;143(2-4) 384–388.

[18] Tatsuo H, Ono N, Tanaka K, Yanagi Y. SLAM (CDw150) Is a Cellular Receptor for Measles Virus. Nature 2000;406(6789) 893–897.

[19] Tatsuo H, Ono N, Yanagi Y. Morbilliviruses Use Signaling Lymphocyte Activation Molecules (CD150) as Cellular Receptors. Journal of Virology 2001;75(13) 5842–5850.

[20] Cocks BG, Chang CJ, Carballido JM, Yssel H, de Vries JE, Aversa G. A Novel Receptor Involved in T-Cell Activation. Nature 1995;376(6537) 260–263.

[21] Schwartzberg PL, Mueller KL, Qi H, Cannons JL. SLAM Receptors and SAP Influence Lymphocyte Interactions, Development and Function. Nature Reviews Immunology 2009; 9(1) 39–46.

[22] Veillette A. SLAM-Family Receptors: Immune Regulators with or without SAP-Family Adaptors. Cold Spring Harbor Perspectives in Biology 2010;2(3) a002469.

[23] Kiel MJ, Yilmaz OH, Iwashita T, Tilmaz OH, Terhosrt C, Morrison SJ. SLAM Family Receptors Distinguish Hematopoietic Stem and Progenitor Cells and Reveal Endothelial Niches for Stem Cells. Cell 2005;121(7) 1109–1121.

[24] Kennedy S, Smyth JA, McCullough SJ, Allan GM, McNeilly F, McQuaid S. Confirmation of Cause of Recent Seal Deaths. Nature 1988b;335(6189) 404,

[25] Jensen T, van de Bildt M, Dietz HH, Andersen TH, Hammer AS, Kuiken T, Osterhaus A. Another Phocine Distemper Outbreak in Europe. Science 2002;297(5579) 209.

[26] Muller G, Wohlsein P, Beineke A, Haas L, Greiser-Wilke I, Siebert U, Fonfara S, Harder T, Stede M, Gruber AD, Baumgartner W. Phocine Distemper in German Seals, 2002. Emerging Infectious Diseases 2004;10(4) 723–725.

[27] Barrett T, Crowthe J, Osterhaus ADME, Subbarao SM, Groen J, Haas I, Mamaev IV, Titenko AM, Visser IKG, Bostock CJ. Molecular and Serological Studies on the Recent Seal Virus Epizootics in Europe and Siberia. Science of the Total Environment 1992;115(1–2) 117–132.

[28] Mamaev LV, Denikina NN, Belikov SI, Volchkov VE, Visser IKG, Fleming M, Kai C, Harder TC, Liess B, Osterhaus ADME, Barrett T. Characterisation of Morbillivirus Isolated from Lake Baikal Seals (Phoca sibirica). Veterinary Microbiology 1995;44(2–4) 251–259.

[29] Mamaev LV, Visser IK, Belikov SI, Denikina NN, Harder T, Goatley L, Rima B, Edginton B, Osterhaus AD, Barrett T. Canine Distemper Virus in Lake Baikal Seals (Phoca sibirica). Veterinary Record 1996;138(18) 437–439.

[30] Visser IKG, Kumarev VP, Orvell C, Vries PDE, Broeders HWJ, van de Bildt MWJ, Grosen J, Teppema JS, Burger MC, UytdeHaag FGCM, Osterhaus, ADME. Comparison of Two Morbilliviruses Isolated from Seals during Outbreaks of Distemper in Northwest Europe and Siberia. Archives of Virology 1990;111(3–4) 149–164.

[31] Visser IKG, van der Heijden RWJ, van der Bildt MWG, Kenter MJH, Orvell C, Osterhaus ADME. Fusion Protein Gene Nucleotide Similarities, Shared Antigenic Sites and Phylogenetic Analysis Suggest that Phocid Distemper Virus Type 2 and Canine Distemper Virus Belong to the Same Virus Entity. Journal of General Virology 1993b;74(9) 1989–1994.

[32] Forsyth MA, Kennedy S, Wilson S, Eybatov T, Barrett T. Canine Distemper Virus in a Caspian Seal. Veterinary Record 1998;143(24) 662–664,

[33] Kennedy S, Kuiken T, Jepson PD, Deaville R, Forsyth M, Barrett T, van de Bildt MW, Osterhaus AD, Eybatov T, Duck C, Kydyrmanov A, Mitrofanov I, Wilson S. Mass die-off of Caspian seals caused by canine distemper virus. Emerging Infectious Diseases 2000;6(6) 637-639.

[34] Visser KG, van Bressem M-E, de Swart RL, van de Bildt MWG, Vos HW, van der Heijden J, Saliki JT, Orvell C, Kitching P, Barrett T, Osterhaus ADME. Characterisation of Morbilliviruses Isolated from Dolphins and Harbour Porpoises in Europe. Journal of General Virology 1993a ;74(4) 631–641.

[35] Di Guardo G, Agrimi U, Morelli L, Cardeti G, Terracciano G, Kennedy S. Post Mortem Investigations on Cetaceans Found Stranded on the Coasts of Italy between 1990 and 1993. Veterinary Record 1995;136(17) 439–452.

[36] Lipscomb TP, Schulman FY, Moffett D, Kennedy S. Morbilliviral Disease in Atlantic Bottlenose Dolphins (Tursiops truncatus) from the 1987–1988 Epizootic. Journal of Wildlife Diseases 1994;30(4) 567–571.

[37] Krafft A, Lichy JH, Lipscomb TP, Klaunberg BA, Kennedy S, Taubernberger JK. Postmortem Diagonosis of Morbillivirus Infection in Bottlenose Dolphins (*Tursiops truncatus*) in the Atlantic and Gulf of Mexico Epizootics by Polymerase Chain Reaction-Based Assay. Journal of Wildlife Diseases1995;3(7) 410–415.

[38] Taubenberger JK, Tsai M, Krafft AE, Lichy JH, Reid AH, Schulman FY, Lipscomb TP. Two Morbilliviruses Implicated in Bottlenose Dolphin Epizootics. Emerging Infectious Diseases1996;2(3) 213–216.

[39] Lipscomb TP, Kennedy S, Moffett D, Krafft A, Klaunberg BA, Lichy JH, Regan GT, Worthy GA, Taubenberger JK. Morbilliviral Epizootic in Bottlenose Dolphins of the Gulf of Mexico. Journal of Veterinary Diagnosis Investigation 1996;8(3) 283–290, .

[40] Osterhaus A, Groen J, Niesters H, van de Bildt M, Martina B, Vedder L, Vos J, van Egmond H, Abou-Sidi B, Barham ME. Morbillivirus in Monk Seal Mass Mortality. Nature 1997;388(6645) 838–839.

[41] Hernández M, Robinson I, Aguilar A, González LM, López-Jurado LF, Reyero MI, Cacho E, Franco J, López-Rodas V. Costas E. Did Algal Toxins Cause Monk Seal Mortality? Nature 1998;393(6680) 28–29.

[42] Black EL. Epidemiology of Paramyxoviridae, In: Kingsbury D. (ed.) The Paramyxoviruses. New York; Plenum Press, New York 1991. p509–536.

[43] Osterhaus ADME, Groen J, Uytdehaag FGCM, Visser IKG, Vedder EJ, Crowther J, Bostock CJ. Morbillivirus Infections in European Seals before 1988. Veterinary Record 1989; 125(12) 326.

[44] Henderson GA, Trudgett A, Lyons C, Reonald K. Demonstration of Antibodies in Archival Sera from Canadian Seals Reactive with a European Isolate of Phocine Distemper Virus. Science of the Total Environment 1992;115(1–2) 93–98.

[45] Ross PR, Visser IKG, Broeders HW, van de Bildt MW, Bowen WD, Osterhaus AD. Antibodies to Phocine Distemper Virus in Canadian Seals. Veterinary Record 1992;130(23) 514–516.

[46] Dietz R, Ansen CT, Have P, Heide-Jorgensen M-P. Clue to Seal Epizootic? Nature 1989; 338(6217) 627.

[47] Barrett T, Rima BK. Molecular Biology of Morbillivirus Diseases of Marine Mammals. In: Pfeiffer CJ. (ed.) Molecular and Cell Biology of Marine Mammals. Florida: Krieger Publishing Company; 2006. P161-172.

[48] Blixenkrone-Moller MV, Svansson V, Have P, Botner A, Nielsen J. Infection Studies in Mink with Seal-Derived Morbillivirus. Archives of Virology 1989;106(1-2) 165–170.

[49] Veillette A, Cruz-Munoz ME, Zhong MC. SLAM Family Receptors and SAP-Related Adaptors: Matters Arising. Trends in Immunology, 2006;27(5) 228–234.

[50] Sayos J, Wu C, Morra M, Wang N, Zhang X, Allen D, van Schaik S, Notarangelo L, Geha R, Roncarolo MG, Oettgen H, De Vries JE, Aversa G, Terhorst C. The X-Linked Lymphoproliferative-Disease Gene Product SAP Regulates Signals Induced through the Co-Receptor SLAM. Nature, 1998;395(6701) 462–469.

[51] Morra M, Howie D, Grande MS, Sayos J, Wang N, Wu C, Engel P, Terhorst C. X-Linked Lymphoproliferative Disease: A Progressive Immunodeficiency. Annual Reviews Immunology, 2011;19 657–682.

[52] Mavaddat N, Mason DW, Aikinson PD, Evans EJ, Gilbert RJ, Stuart DI, Fennelly JA, Barclay AN, Davis SJ, Brown MH. Signaling Lymphocytic Activation Molecule (CDw150) Is Homophilic but Self-Associates with Very Low Affinity. Journal of Biological Chemistry 2000; 275(36), 28100–28109.

[53] Ono N, Tatsuo H, Tanaka K, Minagawa H, Yanagi Y. V Domain of Human SLAM (CDw150) Is Essential for Its Function as a Measles Virus Receptor. Journal of Virology 2001;75(4) 1594–1600.

[54] Hashiguchi T, Kajikawa M, Maita N, Takeda M, Kuroki K, Sasaki K, Kohda D, Yanagi Y, Maenaka K. Crystal Structure of Measles Virus Hemagglutinin Provides Insight into Effective Vaccines. Proceedings of the National Academy of Sciences of the USA 2007;104(49) 19535–19540.

[55] Navaratnarajah CK, Leonard VH, Cattaneo R. Measles Virus Glycoprotein Complex Assembly, Receptor Attachment, and Cell Entry. Current Topics of Microbiology and Immunology 2009;329 59–76.

[56] Hashiguchi T, Ose T, Kubota M, Maita N, Kamishikiryo J, Maenaka K, Yanagi Y. Structure of the Measles Virus Hemagglutinin Bound to Its Cellular Receptor SLAM. Nature Structural & Molecular Biology 2011a;18(2) 135–141.

[57] Hashiguchi T, Maenaka K, Yanagi Y. Measles Virus Hemagglutinin: Structural Insights into Cell Entry and Measles Vaccine. Frontiers in Microbiology 2011b; 2, Article 247.

[58] Plemper RK, Brindley MA, Iorio RM. Structural and Mechanistic Studies of Measles Virus Illuminate Paramyxovirus Entry. PLoS Pathogens 2011;7(6), e1002058

[59] Dorig RE, Marcil A, Chopra A, Richardson CD. The Human CD46 Molecule Is a Receptor for Measles Virus (Edmonston Strain). Cell 1993;75(2) 295–305.

[60] Naniche D, Varior-Krishnan G, Cervoni F, Wild TF, Rossi B, Rabourdin-Combe C, Gerlier D. Human Membrane Cofactor Protein (CD46) Acts as a Cellular Receptor for Measles Virus. Journal of Virology 1993;67(10) 6025–6032.

[61] Enders JE, Peebles TC. Propagation in Tissue Cultures of Cytopathogenic Agents from Patients with Measles. Proceeding of the Society of Experimental Biological Medicine 1954; 86(2) 277-286.

[62] Katz SL, Milovanovic MV, Enders JF. Propagation of Measles Virus in Cultures of Chick Embryo Cells. Proceeding of the Society of Experimental Biological Medicine 1958;97(1) 23–29.

[63] Kobune F, Sakata H, Sugimura A. Marmoset Lymphoblastoid Cells as a Sensitive Host for Isolation of Measles Virus. Journal of Virology 1990;64(2) 700–705.

[64] Muhlebach MD, Mateo M, Sinn PL, Prufer S, Uhlig KM, Leonard VH, Navaratnarajah CK, Frenzke M, Wong XX, Sawatsky B, Ramachandran S, McCray PB, Cichutek K, von Messling V, Lopez M, Cattaneo R. Adherens Junction Protein Nectin-4 Is the Epithelial Receptor for Measles Virus. Nature 2011;480(7378) 530–533.

[65] Noyce RS, Bondre DG, Ha MN, Lin L-T, Sisson G, Tsao M-S, Richardson CD. Tumor Cell Marker PVR4 (nectin 4) Is an Epithelial Cell Receptor for Measles Virus. PLoS Pathology 2011; 7(8):e1002240.

[66] Sakaguchi M, Yoshikawa Y, Yamanouchi K, Sata T, Nagashima K, Takeda K. Growth of Measles Virus in Epithelial and Lymphoid Tissues of Cynomolgus Monkeys. Microbiology and Immunology 1986;30(10) 1067–1073.

[67] de Swart RL, Ludlow M, de Witte L, Yanagi Y, van Amerongen G, McQuaid S, Yuksel S, Geijtenbeek TB, Duprex WP, Osterhaus AD. Predominant Infection of CD150+ Lymphocytes and Dendritic Cells during Measles Virus Infection of Macaques. PLoS Pathology 2007;11: e178.

[68] Leonard VH, Sinn PL, Hodge G, Miest T, Devaux P, Oezguen N, Braun W, McCray PB Jr, McChesney MB, Cattaneo R. Measles Virus Bound to Its Epithelial Cell Receptor Remains Virulent in Rhesus Monkeys but Cannot Cross the Airway Epithelium and Is Not Shed. Journal of Clinical Investigation 2008;118(7) 2448–2458.

[69] Shirogane Y, Takeda M, Tahara M, Ikegame S, Nakamura T, Yanagi Y. Epithelial-Mesenchymal Transition Abolishes the Susceptibility of Polarized Epithelial Cell Lines to Measles Virus. Journal of Biological Chemistry 2010;285(27) 20882–20890.

[70] Tahara M, Takeda M, Shirogane Y, Hashiguchi, T, Ohno S, Tanagi Y. Measles virus infects both polarized epithelial and immune cells by using distinctive receptor-binding sites on its hemagglutinin. Journal of Virology 2008;82(5) 4630-4637.

[71] Ohishi K, Ando A, Suzuki R, Takishita K, Kawato M, Katsumata E, Ohtsu D, Okutsu K, Tokutake K, Miyahara H, Nakamura H, Murayama T, Maruyama T. Host-Virus Specificity of Morbilliviruses Predicted by Structural Modeling of the Marine Mammal SLAM, a Receptor. Comparative Immunology, Microbiology and Infectious Diseases, 2010;33(3) 227–241.

[72] Guindon S, Gascuel O. A Simple, Fast, and Accurate Algorithm to Estimate Large Phylogenies by Maximum Likelihood. Systematic Biology, 2003; 52(5) 696–704.

[73] Jones DT, Taylor WR, Thornton JM. The Rapid Generation of Mutation Data Matrices from Protein Sequences. Computer Applications in the Biosciences 1992;8, 275–282.

[74] Ronquist F, Huelsenbeck JP. MrBayes 3: Bayesian Phylogenetic Inference under Mixed Models. Bioinformatics 2003;19(22) 1572–1574.

[75] Cao E, Ramagopal UA, Fedorov A, Fedorov E, Yan Q, Lary JW, Col, JL, Nathenson SG, Almo SC NTB-A Receptor Crystal Structure: Insights into Homophilic Interactions in the Signaling Lymphocytic Activation Molecule Receptor Family. Immunity, 2006;25(4) 559–570.

[76] Sali A, Blundell TL. Comparative Protein Modelling by Satisfaction of Spatial Restraints. Journal of Molecular Biology 1993;234(3) 779–815.

[77] Rocchia W, Alexov E, Honig B. Extending the Applicability of the Nonlinear Poisson-Boltzmann Equation: Multiple Dielectric Constants and Multivalent Ions. Journal of Physiology Chemistry B 2001;105 6507–6514.

[78] Rocchia W, Sridharan S, Nicholls A, Alexov E, Chiabrera A, Honig B. Rapid Grid-Based Construction of the Molecular Surface and the Use of Induced Surface Charge to Calculate Reaction Field Energies: Applications to the Molecular Systems and Geometric Objects. Journal of Computer Chemistry 2002;23(1) 128–137.

[79] Katz SL. John F. Enders and measles virus vaccine – a reminiscence, Current Topics of Microbiology and Immunology, 2009: 329; 3-11.

[80] Strebel PM, Cochi SL, Hoekstra E, Rota PA, Featherstone D, Bellini WJ, Katz SL. Journal of Infectious Diseases, 2011: 204(Suppl 1), 1-3.

[81] WHO Measles mortality reduction,
http://www.who.int/immunization/newsroom/measles_rubella/en/index.html (2011)

[82] FAO Global Rinderpest Eradication Programme (GREP).
http://www/fao.org/ag/againfo/programmers/en/grep/home.html/ (2011)

[83] OIE No More Death from Rinderpest. http://www.oie.int/for-the-media/press-release/detail/article/no-more-deaths-from-rinderpest/ (2011).

[84] Rweyemamu M, Roder PL, Taylor WP. Chapter 15, Towards the Global Eradication of Rinderpest, In: Taylor WP, Barrett T, Pastoret P-P. (Eds.) Rinderpest and Peste des Petits Ruminants, Virus Plagues of Large and Small Ruminants. New York: pp. 298-22, Academic Press;2005. p298-322.

Section 4

Exploitation

Chapter 9

Portuguese Sealing and Whaling Activities as Contributions to Understand Early Northeast Atlantic Environmental History of Marine Mammals

Cristina Brito

Additional information is available at the end of the chapter

http://dx.doi.org/10.5772/54213

1. Introduction

The 16th century maritime Atlantic journeys were one of the most ground breaking and prolific sources of scientific knowledge in nautical cartography, geography and ethnography, and also in the natural sciences [1]. The European pioneering exploration of Africa and Brazil resulted in writer and naturalist records of the new, exotic and useful, originating natural history studies in Europe based on observations in zoology, botany and tropical medicine [2]. Since then, explorers, travellers and traders have brought animals and natural objects to Europe. Previously, however, coastal inhabitants all over the world used near shore habitats and resources, altering marine ecosystems and their dynamics [3].

The marine environmental history refers to the mutual interactions between humans and the marine natural world [4], ambitioning to understand how humans have integrated the sea into their living style through the changes brought by time [5]. The environmental historical approach offers a multidisciplinary and long term research approach to anthropogenic interactions with marine life, albeit being largely incomplete when compared to its terrestrial counterparts [3]. Most of the available literature pertains to the study of invertebrates [6] and fish [7, 8], while historical research on the presence of cetaceans over time remains largely incomplete, particularly for the Northeast Atlantic [9]. Marine mammals are very useful tools to evaluate marine environmental changes as they are easily identified in historiography (particularly in letters, journey diaries, natural history treaties), making it possible to relate their presence records with their environmental conditions [10]. Marine mammals are of relatively large proportions that require surface visits to breathe, features that enhanced the awareness and interest in different human cultures around the world. Marine mammals were historically targets of hunting with economic purposes. During the 16th and 17th centuries, tortoiseshell, shark teeth, marine mammals' blubber and baleen, seal skin, ambergris [11], narwhal tusks, pearls and coral [12], and a host of other products were trade commodities of considerable value. Some exotic marine animals and products, especially if they were rare and difficult to catch and possess, were of particular interest and priceless to European royalty, scientists, and collectors. This was because of their perceived economical and spiritual values, but also due to their applications with mundane (medicine, food, condiments, aphrodisiacs), symbolic, talismanic or superstitious purposes. [13]. The transaction process, the local natural ecosystems on which they acted and perceptions towards the marine environment changed [3] [5]. Remains of seals, whales and dolphins are currently found at museums, universities, and even private collections [2].

The present work contemplates a large timeframe from early modern to modern times and is based on Portuguese sources. Aspects of marine environmental history will be addressed engaging with elements from the history of science and framing Portuguese maritime history within an Atlantic history. Atlantic history is an analytic construction and a specific category of historical analysis that helps to organize the study of the emergence of this ocean basin as a site for several and distinct forms of commercial exchange [14]. In order to approach the Atlantic history, a novel approach was chosen in this work, based on the exploitation patterns, trading features and scientific knowledge of Atlantic marine mammals collected in various sources over the years, all aspects still to be covered. Seals and whales are historically linked to local communities that exploited their body parts. This commerce was affected by changing markets, evolving technologies, scientific studies, regulations regarding access, and contradictory opinions regarding sustainability. Records on this commerce represent a valuable source of information about the characteristics and fluctuations of the exploitation market and indirectly about the condition of its inherent marine environment. Studies on whaling must include historical and an economics insight so as to provide a comprehensive interpretation of the natural resources shifts. By historically approaching marine mammals as a case study has a great potential to enhance our understanding of the interactions between the human culture and nature in the early modern and modern world.

2. Integrating natural and social sciences sources

Environmental history is an interdisciplinary field of History that receives inputs from different areas of human, social and natural sciences. Reliable data on historical occurrences provides relevant information on how communities, populations and species have shifted over long timelines [15, 16] and allows establishing baseline datasets [17]. Of paramount importance and more precisely, historical accounts on marine mammals may be extremely useful to add new data to their occurrence and distribution in poorly studied regions [17], and to compare past information with recent data [18]. The historical relationships between humans and marine mammals, such as sealing, whaling, use and trade of goods, strandings and naturalist sightings, have been regularly documented throughout time though various sources. However, for some regions such as for the Atlantic Ocean, data is still sparse and requires collection of the various sources and subsequent interpretation. There are several advantages in using historical data to extract biological information. Usually, in a historical source it is possible to find abundant and novel information on unspoiled populations, and on species (or groups) occurrence and distribution. This information can then be used as an indicator of past biodiversity, enabling the study of baseline levels and changes in cetaceans' presence and importance, as well as making the connection of biology with history, economics and culture.

In the present work, inter-cultural trade of exotic marine animals has a well-documented history that changed dramatically with the European Overseas Expansion. From the 16th century onwards many animals were brought to Europe mainly by the Portuguese, starting a new period in natural history [2]. The Portuguese reports correspond mainly to descriptions based on empirical knowledge accumulated in successive maritime routes, where the occurrence of species or animal groups was recorded. Navigators associated the presence of frequent animal to specific location so as to estimate their position.

Figure 1. Representation of whaling in Portuguese shores and the encounter of Portuguese navigators with sea wolves upon arrival at the island of Madeira. This image represents a clear manifestation of the 16th century Portuguese intention of domination over the seas and everything found in it, including all marine animals with economic value. It corresponds to a *fresco* from 2000 in the ceiling of the Church of Ponta Delgada, Madeira, representing all air, land and sea animals, with symbols related to the Portuguese Atlantic Discoveries. Photographed in 2009 by the author.

Portuguese sources for the 16th and 17th centuries Atlantic in the form of letters, chronicles, and scientific treaties, as well as illustrated broadsheets, leaflets, maps, images, paintings, objects and products were compiled and reviewed here (see Figure 1). The present research includes historical sources and accounts from the period between late 15th century and early 19th century, and will present two distinct case-studies for the Atlantic Ocean: (1) first encounter and posterior hunting of monk seals in the Atlantic; and (2) medieval and early modern whaling in the Iberian Peninsula and shifting of this activity to the Atlantic.

3. The Mediterranean monk seals in the Atlantic

The Mediterranean monk seal (*Monachus monachus*) is distributed across different countries but is also currently the most endangered pinniped, largely due to its small and isolated populations. The monk seal current distribution is severely contracted and fragmented due to Human pressures, however historic records indicate that it inhabited the entire Mediterranean Basin and the south-eastern North Atlantic, from the Azores islands to near the equator. In Portuguese territory, the Mediterranean monk seal persists to present days and was historically present both in the Azores and Madeira archipelagos, which was only very recently recognized [19]. The seal species displays a high site-fidelity, tends to occupy only a restricted part of the suitable habitat and, presently in the Atlantic, it can only be found in colonial aggregations [20, 21]. Similarly to the other two monk seal species, the living Hawaiian monk seal, *M. schauinslandi* and the extinct Caribbean monk seal *M. tropicalis*, the Mediterranean monk seal has been greatly impacted by human activities and has been exploited since ancient times. These seals have been hunted in the Mediterranean Sea for their oil and furs as early as since the classical Greek period [22, 23].

In the Atlantic, early captures began as soon as they were discovered by the Portuguese in the newly found Atlantic islands and in the then recently explored West African coast. Around 1420, new fauna and flora found during the maritime journeys were typically seen with scepticism and surprise, even though occasionally some animals appeared familiar to what was known in Europe [24] and their presence was normally recorded in explorer diaries.

The first new marine animal the Portuguese came across with was the Mediterranean monk seal, called sea wolf (*'lobo marinho'*), which was present in Madeira and in the West coast of Africa [26]. Early reports of monk seals show that, as for other sea animals, seals were interpreted as strange, monstrous and frightening beings [24]. In the case of sea wolves, however, shortly after acknowledging their docile characteristics, they started to be considered as an alternative and wealthy food source and resources. They were indeed the first product to be obtained and traded from the new Atlantic space to Europe [25].

When the first seals were recognized in Madeira in 1420, these animals were found in the form of a group of strange and calm animals with vocalizations very similar to those of wolves. Considering the medieval spirit, it is likely that men believed that all terrestrial animals had their equivalent on the ocean, and as such they named the seals sea wolves. Despite the initial surprise of finding such sea creatures, the subsequent encounters triggered a continuous period of intense captures [26]. Despite more than a century of continuous exploitation since the first capture, these animals are still present in the Madeira island by 1580 to 1590 [25]. Even though the number of sea wolves in the Azores was smaller than in Madeira in 1420, there are several reports and descriptions also testifying their historical presence in that archipelago during the 15th century [26, 27]. For example in the São Miguel (Azores) island: *'While fishing here with some fishermen from Ponta Delgada city, who ate every night at shore or better on the stone or over the sandbank, a big sea wolf appeared which looked like a calf. He leaned against the rocks and they gave him lots of fish bones of what the fishermen were eating. ... and sometimes the fishermen, who have seen this sea wolf, also known as white mesh because he brought a piece of mesh behind his ear'* [27].

Portuguese navigators found once again sea wolves on the West African coast, in the 15th century, where they were also very easily captured [25, 28]. These quiet animals were encountered in large groups and the following account from 1436 is most probably the first description for this species on the Atlantic coasts: *'... he saw at the brook entrance a big crowd of sea wolves (according to some they were up to five thousand), he killed as many as he could and took their entire hide to the ship. It was a great slaughter ... and this happened in the year of 1436'*. The abundance and economic value of sea wolves justified the risks and the delays in exploiting these animals in the *Rio do Ouro*'s coast [28].

The discovery of the multiple usages of their oil, fur and meat made the sea wolves very economically important. Their hides, after tanning, were used to manufacture clothes and shoes, and their fat was mostly used to manufacture soaps. In many soap-works, the quantities of fat were so high that it surpassed the quantity of olive oil used in this kind of product, being their oil frequently used [25, 28]. Machado [25] refers in his work that according to a list of imported products, sea wolves' hides were also quite solicited in European markets. The commercial value of sea wolves' products was extremely high [27]: *'As before, in Rio do Ouro, the Moors gave sea wolves' hides to Gomes Pires and promised him that they were going to search for gold and slaves if he came back there'*. Sea wolves were in fact a target of commercial marine exploiters, looking to obtain and trade their hides and oil.

To understand historical trends of presence and abundance, as well as the chronology of decline of sea wolves, we can estimate numbers for Madeira's population prior to the Portuguese exploitation of these resources judging by the capture levels and declining reports. This was achieved considering from published bibliography the historical known size of Africa's West coast population (population in *Rio do Ouro* was of five thousand individuals, according to the historical description by Zurara). In the present work, it was calculated that for Madeira in the early 15th century, i.e., prior to human arrival and exploitation, there was an estimated population size of almost 2000 monk seals. Sources and estimates show a historic baseline for the number of seals in both populations, prior to its exploitation, which allowed obtaining historical trends in a 580 years' time span (Figure 2 and 3). A chronology of decline was obtained reporting since the 15th century, and the trend of decline shows a reduction to half of its population during the first one hundred years and a smoother reduction over the last four centuries, until near extinction in present days. These events reflect heavy and intense exploitation of a marine resource preventing the natural recovery of the populations.

After the first encounter with sea wolves in *Rio do Ouro* in 1436, at least six trips were made to the African coast until 1447, with the exclusive purpose of capturing sea wolves. According to the chronicles, only one ship returned to the kingdom with no cargo [28]. The commercial value of sea wolves' products was extremely high and, at least in the Azores and Madeira, these resulted in a severe reduction in the population size. In the Azores, where probably only a small population lived, sea wolves were rapidly exterminated. In Madeira and for *Rio do Ouro* early takes resulted in a severe depletion during the subsequent couple of decades after the first encounter, which continued declining over the centuries, even though more smoothly [25,26, 28].

In certain cases, comparing abundance of animals from historical reports with present day animals could be considered by some as conjectural. For instance, historical descriptions of Madeira are very clear and are supported by their presence nowadays. However, for the Azores, there was no proof of sea wolves presence in the islands for a very long time. However, Azorean historical descriptions are quite similar to descriptions for Madeira, in terms of time and historical context, and there are clear resemblances in the morphological characteristics and behaviours described. We know very little about the sea wolves' life in the Azores in the past five hundred years and, as other authors have referred, this probably very small population (or vagrant individuals) did not survive due to the intense human exploitation in the 15th century [19, 29].

According to estimates for *Rio do Ouro* in 1550, the abundance of seals was likely to be considerable, allowing for a strong exploitation activity, however, there are no historical accounts reporting this activity. Although based on theoretical evidence, calculations of captures and boat capacity suggest that the reported abrupt decline in seals population between 1420 and 1450 was due to intense hunting. Since then, numbers are likely to have oscillated relating to hunting peaks from foreign ships or even from local people. Natural growth cannot overcome intense and continuous capture events and natural populations decrease severely under an anthropogenic predatory pressure. Typically, a species or population that endures a drastic event of overexploitation follows a path to extinction if no conservative measures are taken. Historical sources show that, following the discovery and commercial exploitation of the large West African sea wolf herds in the 15th century, monk seal colonies remained relatively undisturbed until the twentieth century. At the end of the 20th century, the largest population of the species lived in West Africa. The mean numbers, in the period 1993–97, were estimated to be 317 individuals and the population was thought to be stationary or changing at a very slow rate [29, 30].

All the documental sources comprised in the present work, led to the conclusion that sea wolves were abundant in Portugal and in the West African coast, from *Rio do Ouro* to Cape Blanco, and that the African coast was visited by the Portuguese navigators with the intentional purpose of exploiting its natural resources. It's also clear that sea wolves had an enormous commercial importance, and because of that were repeatedly and intensively captured [25, 28].

At the end of the 20th century, two major events regarding Portuguese and African monk seal colonies occurred. In West Africa, in the spring of 1997, a severe mass mortality reduced

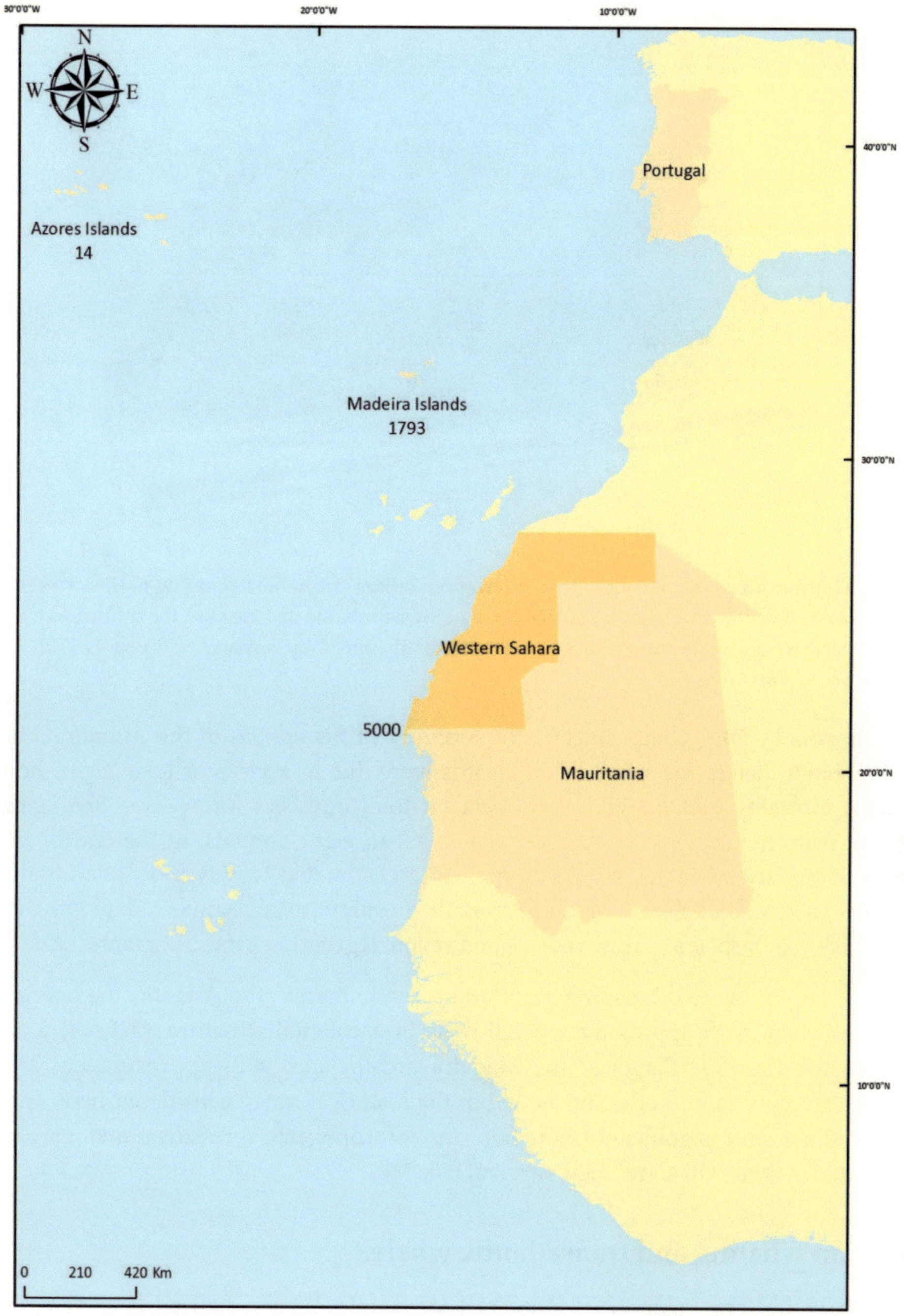

Figure 2. Map indicating the Mediterranean monk seal presence in some Atlantic locations in the 15th century and number of possibly pristine populations (known for the West African coast and the Azores, from historical sources, and estimated for Madeira).

Figure 3. Graphic representation of early modern populations, from the beginning of the 15th century, in the West coast of Africa, Madeira and Azores and their probable decline over the centuries. Image credits of Mediterranean monk seal illustration: authorship and © by Fernando Correia (www.efecorreia-artstudio.com).

their numbers by 70%, compromising the recovery of the species in the Atlantic. Scientists were unable to determine whether the deaths were due to a virus or toxic algae, however the mass mortality had a significant effect on total numbers for species' abundance. In Madeira, from a surviving population of just six to eight animals, at the end of the 20th century, monk seal numbers have increased to an estimated twenty-four individuals. This was due to a special conservation effort from the Portuguese government, beginning in May 1990, which established a marine reserve and severe legislation protecting monk seals [31].

Presently, only these two breeding populations are known in the Atlantic, the one in Cape Blanc peninsula, with approximately 120 seals in a colonial structure, [32] and a smaller group in the *Desertas* Islands at the Madeira archipelago, with approximately 25 seals. Sightings are now rare in other areas within the historical range considered here, and only an immediate and significant reduction in anthropogenic pressures and range-wide coordinated actions will allow their survival [30, 32].

4. Iberian whaling and transatlantic whales

Whales have for long being of great interest for people. Whales figured in ancient legends and visual representations as terrifying sea monsters, and inspired poets and artists with their strange grace and immense size. Over the years, an entire whale mythology thrived, inspired by the mystery surrounding these creatures' habits [33]. Whales have mainly been hunted throughout the centuries for their economically valuable oil, meat, bones and baleen.

Scientifically, the interest in the ecology of whales and their preservation are in fact relatively very recent. It was not until the 18th century that the whale was designated as a mammal rather than a fish, while cetology, as a scientific discipline and a branch of zoology, dates back only to the 1960s [33]. Historically, studies about whaling can be found since the early 20th century [e.g. 34] to the present day [e.g. 35].

Next is given an overview of two different perspectives for the starting of a similar maritime activity in two relative similar regions and by the same period. One based on transfer of technology from Europe to Portuguese South America, based on already known hunting techniques, and the other based mainly on local native knowledge in English North American colonies. It will be possible to understand that it was the implementation of the Basque whaling technique that allowed the early establishment of a continued and lucrative industrial activity in Brazil, since the beginning of the 17th century, in opposition to what happened in North America.

The Basques were the first western people to intentionally hunt large whales, establishing the characteristics of the industry for the following hundreds of years [36, 37]. The hunting procedure encompassed pursuits in small open boats and captures with hand harpoons and lances. This procedure began being used in 1050s and was used for many centuries [35, 38-40]. In mainland Portugal, the first references to whales date back to the 12th century, in the form of local records related to stranding, scavenging of whale remains or whaling related activities [10, 41]. The Portuguese history of cetaceans and humans has been documented in reports, descriptions, tales, legal documents, laws and regulations and tithes.

The Biscay right whale (*Eubalaena glacialis*) was the main targeted species in the Basque country [42], [37, 43]. Initially, The Basques expanded their whaling captures locally and regionally from the shores of the Bay of Biscay, to the south of France and north of Spain [36]. As shown by Aguilar [40], Basque historical sources indicate that whaling started in the Basque French country and continued south and west over the years, through a transfer process of information and technology. Basque navigators had shown the way on whaling enterprises and, for a long time, Basque whalers and pilots were employed on many ships, transmitting the experience gained during their activity [37]. Later, the Basques established permanent or semi-permanent shore stations for whale processing across the Atlantic Ocean, encompassing also the provinces of Santander, Asturias and Galicia [37, 40, 42, 44, 45]. For Basque whaling in European shores, many of the sources specify black whales as the target species. This is consistent with knowledge about the distribution and migration patterns of North Atlantic right whales, in medieval Basque times and early modern whaling. By the 16th century Basque whalers were regularly migrating and conducting expeditions to northern European seas and across the Atlantic to North America [46, 47]. They depleted right whale and bowhead populations in the Strait of Belle Isle, between Labrador and Newfoundland, by killing tens of thousands of whales from 1530 to 1620. And later, between 1660 and 1701, they were hunting whales in the western Arctic, reducing stocks considerably and affecting the whales' migratory patterns [3]. In later periods, especially after the 16th century, Dutch, British and other non-Basque entrepreneurs and whalers were also involved in the whaling trade from European ports and their overseas territories [for a review see 46 and 47].

The whaling activity was much more reduced in mainland Portugal. In the 20th century, there were two periods when short-lived enterprises operated whaling stations and used modern, Norwegian-type, whaling technology [48]. Also for the 20th century, besides the important presence of land based whaling in the archipelagos of Madeira and Azores (Figure 4) no other significant whaling episodes were historically recognized for Portugal. In previous works by the author [41, 48] a total of 38 historical sources recovered date from the 13th century, suggesting that Portuguese whaling began earlier than in the previously thought 20th century. There was a peak of whale-related events in the 13th and 14th centuries, contemporary to those found for the French and Spanish Basque countries, suggesting whaling started approximately at the same time in the both North and South Europe. Hence, it is now considered that whaling was not introduced in Portugal by the Basques, who instead spread westward from the French Labourd (11th century), via the Golf of Biscay, to Asturias, and southward to Galicia (14th century). The Portuguese whale captures is thus thought to have originated independently of Basque influence. The Portuguese sources do not clarify the species captured, numbers of whales taken, nor to the whaling technology used, but the activity was sufficiently well organized and developed to warrant the levying of tithes in the feudal system of 13th century Portugal [41].

In the 13th century, several whale products started to be utilized in Portuguese fishing villages [49]. The exploitation of large whales and small cetaceans in the near shore waters of mainland Portugal seems to have originated in the medieval times. However, a comprehensive study is still missing for Portugal shores. For instance, whaling and the presence of whales in the Basque shores has originated typical iconography and culture in that region over the centuries [36, 50]. Such historical pattern has not been observed in mainland Portugal and it is one important aspect to be considered in the future to understand on the species exploited and the intensity of this exploitation. During the 15th and 16th centuries, whaling expanded from the Portuguese shores, through the Atlantic Islands to overseas countries, particularly to Brazil. In Brazil and since the early 17th century, Portuguese settlers started a shore whaling business but a Basque crew was recruited for the first couple of seasons [51]. Since that time, and for two centuries, a structured shore based whaling enterprise developed in the coastal waters of Brazil, mainly devoted to the hunting of humpback whales during the calving season [51].

Hence, historical records allow to understand that Portugal together with the French and Spanish Basque Country were important whaling locations where a whale culture was developed and spread [41]. Portuguese whalers always conducted a land-based type of whaling while the Basques conducted a typical offshore whaling when moving into the Atlantic away from their Iberian shores (Figure 5).

For the West North Atlantic, a certain type of native whaling was referred for the North America by the 16th century. Following the descriptions by Acosta [52], the natives seemed to capture near shore whales with sticks and ropes. Additionally to the native resource exploitation, John Smith seems to have been the first European trying to systematically exploit this resource in North America [11]. He writes that the purpose of his 1614 New England journey was to hunt whales. This explorer, aware of the Basques expertise in whale hunting, asked in 1616 for permission to accompany their whaling expeditions to the North Sea, a goal he did not accomplish.

Since that time, settlers have tried to exploit this potentially profitable resource, although no records exist on the procedures and techniques used. Probably they have conducted a rather simple and primitive method of whaling, as shore-based whaling was not yet developed in West North Atlantic. In the Bermudas, for instance, the first whaling season only occurred in 1663, and only from that time onward did the operations intensified. Initially captures focused on the humpback whale (in 1700, 200 individuals captured), but as soon as these species became rare by the mid-18th century, captures were re-directed towards sperm whales [11]. Again, from the Bermudas example, it was only when the settlers from different European nationalities were permanently fixed in the overseas territories that it was possible to develop a whaling activity.

The beginning of whaling in the overseas regions was mostly supported on Basque expertise [36]. The Basque whaling model using small open boats and hand harpoons is particularly relevant in a context of globalization of techniques and ways of handling captured animals, their remains and products.

Hence, the historical information discussed here provides insight on the environmental history of whales and the importance of certain maritime extractive activities across the Atlantic. The interest on implementation of whaling overseas is indicative of the abundance and distribution of whales in those regions and of its value as local and global marine resources.

Figure 4. Oil painting (private collection) by an anonymous author, dated from 1876, showing a scene of sperm whale hunting in the Azores, with the subtitle: *"On the 30th of March of the year 1876, thanks to the Lord, this whale on my harpoon was struck"*. Photographed in 2005 by Cristina Picanço.

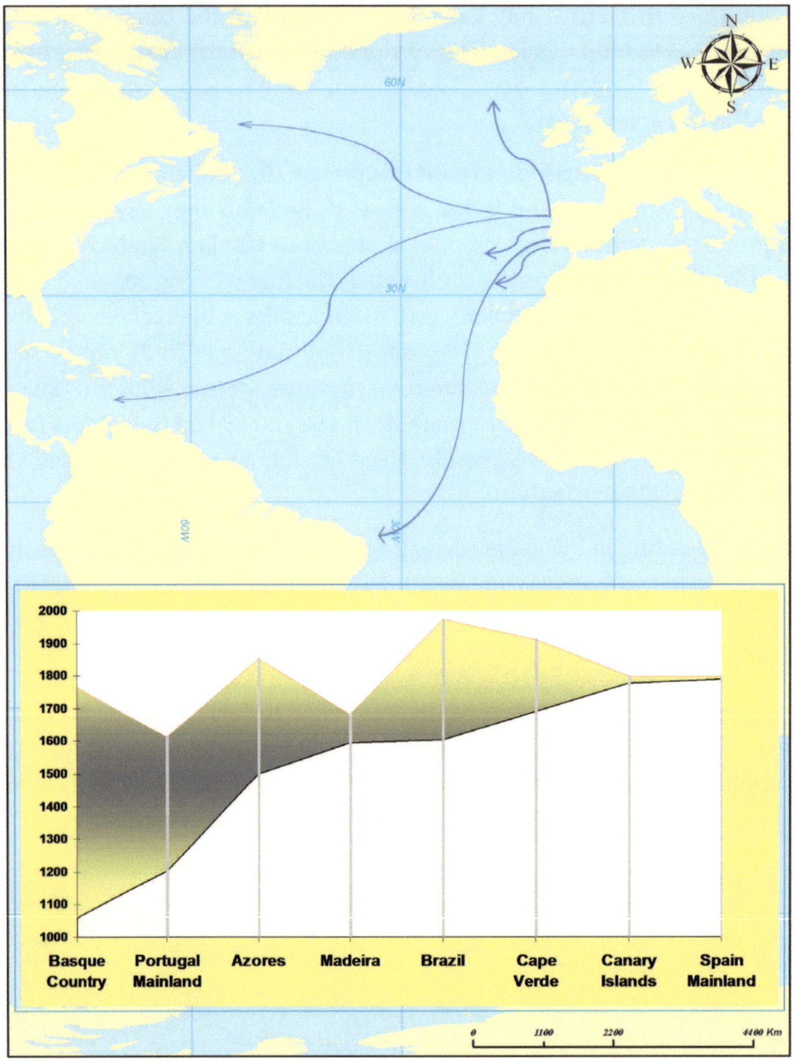

Figure 5. Geographical migration of Basque and Portuguese early modern whaling through the Atlantic (map) and temporal migration (graph) showing the years of the beginning and end of the activity in each area. Data compiled from Aguilar (1986), Reeves & Smith (2006) and Brito (2011).

5. A future glimpse into the environmental history of marine mammals

Many centuries ago, the Atlantic experienced a (near) pristine situation regarding ocean environmental equilibrium. Levels of the populations of predators and prey were relatively stable and fluctuated naturally. Natural disasters happened with much more localized effects and the impacts of climatic shifts were gradual and predictable. Marine resources were historically exploited, however, until the advent of industrialization, rapid depletion and ecological tipping effects were hindered by lack of technological advances. Industrial

Revolution provided the combination of sudden increases in manpower accompanied by new tools and technology which caused overfishing, produced changes in the environment, and cause shifts in ecosystem services and global cycles. The impacts of anthropogenic actions have accelerated in the last decades and are ubiquitous, fast and intense, and exceed the ability of the natural world´s adaptation potential. However, the history of human interactions with marine environments remains largely unstudied. The marine environmental history is, now, a useful tool to understand the past ecological and cultural Human driven transformations in oceans worldwide at small and large spatial and temporal scales.

Marine mammals (either whales, dolphins, and seals, or even manatees and dugongs) represent good case studies for the single-species approach to marine environmental history. The two case-studies presented in this chapter are an example of the valuable outcomes of an interdisciplinary analysis to recreate the environmental history of marine mammals. This history can then be used to frame present traditions as well as population levels of mammals.

The future of marine environmental historians, dedicated to the study of marine mammals, will greatly benefit from focusing on the research of the relation of people with species, with special interest in specific economic and/or cultural isolated situations, such as African and Brazilian manatees, or cetaceans' historical presence around oceanic islands. For instance, historically in some Atlantic cultures, cetaceans were considered "a different kind of fish", as their recognisable natural behaviour and some morphological characteristics are distinct from fish [53]. Studies related to local perceptions changing over time can also provide inputs to the environmental history of marine mammals and contribute for the implementation of long-term and continuous scientific research, interactive environmental education plans and conservation measures. This type of research may include different kinds of historical sources, such as written, iconographic or material sources, and all types of accounts from the period since the late 15th century. European reports of Atlantic (or other oceanic basins) journeys contain information about natural elements and marine mega fauna and represent invaluable sources of research. For a later period (from late 19th century onwards), scientific articles, newspapers, illustrations, maps, non-published scientific reports and some other grey literature, such as unpublished thesis, may also be used. Good history begins with good sources [3], but good marine environmental history needs also to be framed into interdisciplinary boundaries.

Author details

Cristina Brito
CHAM (Centre for Overseas History), Faculdade de Ciências Sociais e Humanas, Universidade Nova de Lisboa and Universidade dos Açores, Lisbon, Portugal
Escola de Mar, Edifício ICAT, Campus da FCUL, Campo Grande, Lisbon, Portugal

Acknowledgement

This research was supported by the Portuguese Foundation for Science and Technology (FCT), through a post-doctoral fellowship (SFRH/BPD/63433/2009) and a project

(HC/0075/2009), as well as the European Community's Programme "Structuring the European Research Area", under Synthesis at the Museo Nacional de Ciencias Naturales (CSIC) or Real Jardín Botánico. Thanks are due to Fernando Correia for the single authorization use of the Mediterranean monk seal illustration in the figure composition (authorship and © by Fernando Correia). The author would also like to thank her colleagues and friends Inês Carvalho, Cristina Picanço, Nuno Gaspar de Oliveira and Vitor Hugo, for helping in parts of this investigation, and her family members Nazaré Rocha, Armando Taborda, Susana Brito, Celso Pinto and Rafaela Maia, for embarking along in this scientific journey away from the shores of marine biology to the new world of maritime history.

6. References

[1] Canizares-Esguerra J (2004) Iberian Science in the renaissance: Ignored how much longer? *Perspectives on Science*, 12(1): 86-124.
[2] Ogilvie BW (2006) *The Science of Describing: Natural History in Renaissance Europe*. The University of Chicago Press, Chicago and London.
[3] Bolster WJ (2006) Opportunities in marine environmental history. *Environmental History*, 11: 567-597.
[4] Mcneill JR (2003) Observations on the nature and culture of environmental history. *History and Theory*, 42: 5-43.
[5] Hughes JD (2006) *What is Environmental History?* Polity Press, UK.
[6] Schwerdtner MK & Ferse SCA (2010) The history of makassan trepan fishing and trade. *PLoS ONE*, 5(6): e11346.
[7] Poulsen RT (2007) An environmental history of North Sea ling and cod fisheries, 1840-1914. *Fiskeri-og Sofartsmuseets Studieserie. Studieserien*: Esbjerg.99.
[8] Fortibuoni T, Libralato S, Raicevich S, Giovanardi O & Solidoro C (2010) Coding early naturalists' accounts into long-term fish community changes in the Adriatic Sea (1800–2000). *PLoS ONE*, 5(11): e15502.
[9] Sousa A & Brito C (2011) Historical strandings of cetaceans on the Portuguese coast: anecdotes, people and naturalists. *Marine Biodiversity Records*, 4: e102.
[10] Brito C & Sousa A (2011) The Environmental History of Cetaceans in Portugal: Ten Centuries of Whale and Dolphin Records. *PLoS ONE*, 6(9): e23951.
[11] Romero A (2006) More private gain that public good: whale and ambergris exploitation in 17th century Bermuda. *Bermuda Journal of Archaeology and Maritime History*, 17: 5-27.
[12] Jackson JBC (1997) Reefs since Columbus. *Coral Reefs*, 16, Suppl.: S23-S32.
[13] Findlen P (1996) *Possessing Nature: Museums, Collecting, and Scientific Culture in early Modern Italy*. The University of California Press, California.
[14] Morgan PD & Greene JP (2009) Introduction: The present state of Atlantic history. In Greene JP & Morgan PD (Eds) *Atlantic History: A critical appraisal*. Oxford University Press: 3-34.
[15] Rick TC & Erlandson JM (2008) Archaeology, historical ecology and the future of ocean ecosystems. In Rick TC and Erlandson JM (eds) *Human impacts on ancient marine ecosystems: a global perspective*. Berkeley. University of California Press: 297-308.

[16] Tingley MW & Beissinger SR (2009) Detecting range shifts from historical species occurrence: new perspectives on old data. *Trends in Ecology and Evolution*, 24: 625-633.

[17] Brito C & Vieira N (2010) Using historical accounts to assess the occurrence and distribution of small cetaceans in a poorly known area. *Journal of the Marine Biological Association of the United Kingdom*, 90(8): 1583-1588.

[18] Vieira N & Brito C (2009) Past and recent sperm whale sightings in the Azores based on catches and whale watching information. *Journal of the Marine Biological Association of the United Kingdom*, 89 (5): 1067-1070.

[19] Silva MA, Brito C, Santos SV & Barreiros JP (2009) Occurrence of pinnipeds in the Archipelago of the Azores: a checklist since Discovery until Present. *Mammalia*, 73: 60-62.

[20] Reeves R, Stewart BS, Clapham PJ & Powell JA (eds) (2002) *National Audubon Society Guide to Marine Mammals of the World*. New York: 150-153.

[21] Gilmartin WG & Forcada J (2009) Monk seals: *Monachus monachus, M. tropicalis, M. schauinslandi*. in W.F. Perrin and J.G.M. Thewissen (eds), *Encyclopedia of Marine Mammals*. Oxford: 741–744.

[22] Johnson W (2000) Sayings of 3000 Years: Book I, 900 B.C. to 1563 A.D. *The Monachus Guardian*, 3: 1-6.

[23] Johnson W (2001) Sayings of 3000 Years: Book II, 1601 A.D. to 2000 A.D. *The Monachus Guardian*, 4: 1-8.

[24] Almaça C (1998) *Baleias, Focas e Peixes-Bois na História Natural Portuguesa*. Lisboa: 11-12.

[25] Machado AJM (1979) *Os lobos-marinhos (Género Monachus, Fleming 1822), Contribuição para o seu conhecimento e protecção*. Madeira.

[26] Frutuoso G (2005) *Saudades da Terra*. Ponta Delgada 19.

[27] Barros J (1932) *Ásia - Dos feitos que os portugueses fizeram no descobrimento e conquista dos mares e terras do oriente. Primeira Década. Livro Primeiro*. Coimbra.

[28] Zurara GE (1989) *Crónica dos Feitos da Guiné*. Lisboa: Biblioteca da Expansão Portuguesa, 15.

[29] Larrinoa PF & Cedenilla MA (2003) Human disturbance at the Cabo Blanco monk seal colony. *The Monachus Guardian*, 6: 1-2.

[30] Forcada J, Hammond PS & Aguilar, A (1999) Status of the Mediterranean monk seal Monachus monachus in the Western Sahara and the implications of a mass mortality event. *Marine Ecology Progress Series*, 188: 249–61.

[31] Pires R (2001) Are monk seals recolonizing Madeira island? *The Monachus Guardian*, 4: 1-3.

[32] Forcada J (2000) Can population surveys show if the Mediterranean monk seal colony at Cape Blanco is declining in abundance? *Journal of Applied Ecology*, 37: 171-81.

[33] Cohat Y & Collet A (2001) *Whales: Giants of the seas and oceans*. New York: Harry N. Abrams, Inc.

[34] Jenkins JT (1921) A history of whale fisheries: From the Basque fisheries of the tenth century to the hunting of the finner whale at the present. London.

[35] Reeves RR & Smith TD (2006) A taxonomy of world whaling: Operations and eras. In *Whales, whaling, and ocean ecosystems*. London: 82-101.

[36] Ellis R (2002) Whales, Whaling, Early, Aboriginal. *In Encyclopaedia of Marine Mammals Perrin*, pp. 1310-1316. Ed. by William F., Würsig, Bernd and Thewissen, J.G.M., Academic Press, San Diego.
[37] Fontaine PH (2007) *Whales and Seals: Biology and Ecology*. Schiffer Publishing, Lda.
[38] Reguart AS (1791) Diccionario Histórico de las Artes de la Pesca Nacional. Madrid. Tomo I-IV.
[39] Ciriquiain M (1979) *Los Vascos en la Pesca de la Ballena*. Ed. Vascas Argitaletxea, San Sebastian.
[40] Aguilar A (1986) A review of old Basque whaling and its incidence on the right whales of the North Atlantic. *In Right Whales: Past and Present Status*, pp. 191-199. Ed. by R. L. Brownell Jr., P. B. Best & J. H. Prescott. Reports of the International Whaling Commission, Special Issue 10.
[41] Brito C (2011) Medieval and early modern whaling in Portugal. *Anthrozoos*, 24 (3): 287-300.
[42] Huxley MB (2000) La industria pesquera en el País Vasco peninsular al principio de la Edad Moderna: una edad de oro? *Itsas Memoria. Revista de estúdios marítmos del País Vasco*, 3: 29-75.
[43] Aguilar A (1981) The Black right whale, Eubalaena glacialis, in the Cantabrian Sea. *Reports of the International Whaling Commission*, 31: 457-459.
[44] Zubizarreta NS (1878) Introduccion, capítulo I y outras descripciones de la memoria acerca del oríen y curso de las pescas y pesquerías de ballenas y de bacalaos, así que sobre el descubrimiento de los bancos é isla de terranova. Imprenta de los Hijos de Manteli, Vitoria.
[45] Azpiazu JA (2000) Los balleneros vascos en Cantabria, Asturias y Galicia. *Itsas Memoria. Revista de estúdios marítmos del País Vasco*, 3: 77-97.
[46] du Pasquier T (2000) *Les baleiniers basques*. Collection Kronos, 31. Editions S.P.M.,. Paris.
[47] Romero A & Kannada S (2006) Comment on "Genetic analysis of 16th-century whale bones prompts a revision of the impact of Basque whaling on right and bowhead whales in the western North Atlantic". *Canadian Journal of Zoology*, 84: 1059-1065.
[48] Brito C (2008) Assessment of catch statistics during the land-based whaling in Portugal. *Marine Biodiversity Records*, 1: e92.
[49] Calado M (1994) *Da ilha de Peniche*. Edição de Autor, Portugal.
[50] Barthelmess K (2009) Basque whaling in pictures, 16th-18th century. *Itsas Memoria. Revista de estúdios marítmos del País Vasco*, 6: 643-667.
[51] Ellis M (1969) *A baleia no Brasil colonial: feitorias, baleeiros, técnicas, monopólio, comércio, iluminação*. Edições Melhoramento, São Paulo.
[52] Acosta J (1590) *Historia natural y moral de las Índias*. Casa de Juan de Léon, Sevilla.
[53] Sousa SP & Begossi A (2007) Whales, dolphins or fishes? The ethnotaxonomy of cetaceans in São Sebastião, Brazil. *Journal of Ethnobiology and Ethnomedicine*, 3: 9.

Chapter 10

Yankee Whaling in the Caribbean Basin: Its Impact in a Historical Context

Aldemaro Romero

Additional information is available at the end of the chapter

http://dx.doi.org/10.5772/50792

1. Introduction

Utilization of marine mammals in general and of whales in particular has varied greatly across historical periods and geographical locations. From prehistoric times (e.g., Stringer *et al.* 2008), the use of these animals was opportunistic by taking advantage of animals either beached (animals arrived dead on the coast) or stranded (animals arrived live on the coast). Later shore whaling (active whaling using small boats launched from the coast for a few hours) took place, and later came the development of industrial whaling, which engaged larger vessels embarked in whaling expeditions that might have lasted up to several years at a time. The heyday of industrial whaling took place during the mid-nineteenth century and was epitomized by Yankee whaling (Sanderson 1993).

For the purposes of this chapter, I define Yankee whaling as an offshore fishery carried out by American whaling vessels between 1712 and 1925 (Starbuck 1876, p. 20; Hegarty 1959, p. 47). The geographical area considered as Caribbean Basin for the purpose of this chapter is defined as all the coasts (clockwise) of Venezuela, northern Colombia, eastern Central America, the Yucatán Peninsula, and all of the coasts of the Antilles from Cuba down to Trinidad including the Bahamas and Bermuda. The Bahamas are commonly included as part of the Caribbean Basin from a geological and cultural viewpoint. Bermuda, although being geographically an oceanic island in the western North Atlantic, has cultural ties to many of the Caribbean Basin countries including being the springboard for numerous whaling operations in the region (Romero 2006).

The aim of this chapter is to describe Yankee whaling in the Caribbean Basin in a historical context to understand its development, cultural, and ecological impact.

2. Methods

As a general source of data on Yankee whaling expeditions I used Lund (2001). Of the known 14,864 voyages of Yankee whalers there were at least 1101 voyages to the North Atlantic of which 454 were voyages to the Caribbean Basin (including Bermuda). Many of those voyages included whaling operations in more than one location in the Caribbean. All information about those voyages was tabulated by vessel name, year of departure, and locality visited. Many of the logbooks of the vessels involved in this activity were examined at the New Bedford Whaling Museum, the Free Public Library, New Bedford, and the Providence Public Library. Other information of this activity for this area was compiled from numerous sources cited throughout this chapter.

3. Quantitative results

The places visited by Yankee whalers and the numbers of visits per locality are shown in Figure 1. Barbados and Bermuda are the places most visited by Yankee whaling vessels.

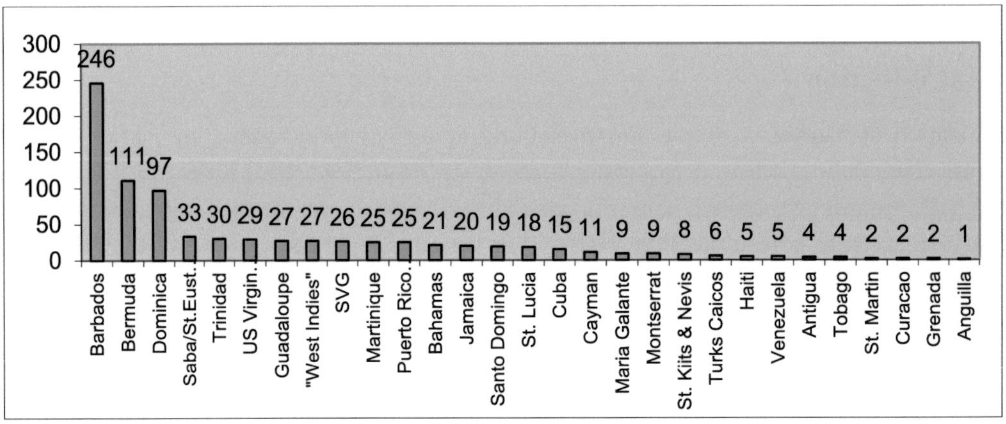

Figure 1. Places visited by Yankee whalers and their frequency.

The frequency with which that activity took place is shown in Figure 2. To that figure other historical information was added in order to put the activity in historical context. That context is interpreted in the Discussion section.

I combined this information with other historical records of the political, economic and social circumstances that might have influenced whaling in that geographic area. This chapter shows how a holistic description of a whaling activity requires understanding of the interplay among numerous factors.

4. Historical narrative of marine mammal exploitation in the Caribbean Basin

Native Americans exploited cetaceans and other marine mammals in the Caribbean Basin since before the arrival of the Europeans (Acosta 1590, Romero et al. 1997, Romero and

Hayford 2000, Romero et al. 2002). The first whaling operations in this area, as defined earlier by Europeans or people of European descent, were in Bermuda. As soon the first English colonists arrived in those islands in 1609 they tried to hunt humpback whales (*Megaptera novaeangliae*) but it was not until 1663 when the first successful attempt took place in those waters (Romero 2006).

The earliest record of an attempt to whale in the Caribbean Basin area by New Englanders was in 1688 when there was a petition to the Governor of New York asking for permission to carry on "a fishing Design about the Bohames Islands and Cap florida for sperma Coeti whales and Racks: And so to returned for this Port" (Starbuck 1876, p. 15). The term "sperma Coeti" refers to the sperm whale (*Physeter macrocephalus*) and "Racks" was a spelling used in the seventeenth century for wrecks. Although there is no record that this expedition ever took place, this is an interesting record because it is dated 24 years before the first actual successful hunt of a sperm whale took place by New Englanders in 1712 near Nantucket (Hawes 1924, p. 57).

The earliest known logbook that refers to a successful Yankee whaling expedition to the Caribbean Basin is that of *Two Brothers*, from Nantucket, MA. This brig visited Barbados in 1775 (logbook at the New Bedford Whaling Museum under the former collection of the Kendall Whaling Museum). The last one was of the schooner *Athlete* out of New Bedford that visited St. Thomas in 1921. Yet, there are indications that some Yankee whalers had been visiting the area for many years before that (Clark 1887, p. 64-65) (see Table 1).

Year	Locality	Source(s)
1730's	Bahamas	Sanderson 1993, pp. 212-213
1750-1784	Bahamas	Tower 1907, p. 33
1762	"French West Indies"	Starbuck 1876, p. 41
1762	"Bermuda Ground"	Stackpole 1953, p. 50, Lund 2001, p. 651
1762	Barbados	Stackpole 1953, p. 51
1763	Barbados	Stackpole 1953, p. 23
1768	"West Indies"	Stackpole 1953, p. 48-49
1775	St. Eustatius, Barbados	Stackpole 1953, p. 73

Table 1. Sources of earlier visits of the Caribbean Basin by American whaleships.

Although occasional expeditions also took place primarily between 1830 and 1860, the heyday of Yankee whaling in the Caribbean Basin occurred roughly between 1860 and 1880. Below is the narrative of how Yankee whaling interplayed with the majority of the countries visited.

4.1. Barbados

American whaling vessels frequently visited Barbados. This island has the largest number of visits registered in this study. Yankee whalers engaged in whaling and the trans-shipment of whale oil and utilized Barbadian ports for the re-stocking of provisions. Additionally, some Yankee whaling vessels were taken there after being captured by British vessels during the British-American War (1812-1815); others were simply abandoned there. This provided ample opportunity for Barbadians to have direct contact with whalers and to acquire whaling skills. For example, an unspecified number of locals joined Yankee whaling vessel crews in order to fulfill the need for hands on board because of death and desertion. Since Yankee whalers recruited many Barbadians, they gained the necessary skills to hunt whales they later applied to shore whaling after returning to Barbados. The combination of Yankee and shore whaling led to the local extinction of humpback whales in those waters (Romero and Creswell 2010 and references therein).

4.2. Bermuda

There are very few traces of intense contact between Bermudans and Yankee whalers. That was due to a combination of factors: (1) Bermuda was a stronghold of the British during the Revolutionary War, which made its waters off-limits to New Englanders; (2) during the British-American war of 1812, the English utilized Bermuda as a major base for their naval operations and any American vessel in those waters (whaler or otherwise) was captured and taken there; (3) during the American Civil War Bermudans who had historical ties with the South, particularly Virginia and the Carolinas, sided with the confederates, making of Bermudan waters hostile territory to Yankee whalers and (4) by the time of the heyday of Yankee whaling the local populations of humpbacks were already severely depleted since shore whaling began around 1663 in those waters.

Thus, despite the overall large number of Yankee whalers visiting Bermuda, it seems that those visits were more a matter of convenience for obtaining provisions for ships either heading to the Eastern Atlantic grounds or heading south to the Caribbean. My survey of archival material in Bermuda yielded no information about relationships between Yankee whalers and the locals (Romero 2009).

4.3. Trinidad and Tobago

Activities by Yankee whaling ships for Trinidad and Tobago have been summarized elsewhere (Romero et al. 2002). All indications are that there was never much interaction between Yankee and land-based whalers. Yankee whaling in the area did not start until the 1830s, when their Trinidadian counterparts were already fully engaged in whale hunting. If anything, Yankee whaling may have furthered the whale population decline since humpbacks are virtually extinct in that area at the present time.

4.4. St. Vincent and the Grenadines (SVG)

Yankee whalers began whaling in the waters of SVG in the early 1800's and their activity peaked in 1864. They hunted humpback, sperm, and short-finned pilot whales (*Globicephala macrorhynchus*). Both pilot and humpback whales were chosen as target species because of the seemingly abundant populations and the products that can be extracted from them. Humpback whales produce a high volume of oil (approximately 25 barrels per adult animal), whereas pilot whales produce two types of oils: one from the blubber and one from the melon (a bulbous area located on the head of the animal). Oil from the latter is of high quality as characterized by its ability to retain stable physical/chemical properties under conditions of extreme temperature and pressure. This oil was used to lubricate precision instruments and was exported to the United States. In addition, local residents utilized the oil from both species as well as other body parts for either human consumption of the meat or for the manufacturing of some goods. Sperm whale hunting did not persist because the demand for its oil declined and the meat was considered inedible. In addition, local fishermen found that sperm whales were difficult and dangerous to catch.

From archival records there were 25 were voyages to SVG for whaling, that took place between 1864 and 1886. In the 10 logbooks I examined 196 entries (daily records kept by the captain or designated crew member) regarding whale hunting. These entries documented that Yankee whalers sighted whales and lowered their boats 117 times. Of those attempts, 34.2% were successful at harpooning one or more humpback whales, but only 40 whales were landed. At least six of those landings were mother/calf pairs. Nine whales were hit but lost. With the addition kill/loss correction factor, Yankee whalers killed an estimated 75 humpback whales between 1864 and 1871.

The local residents adopted boat designs and equipment from Yankee whalers for shore whaling. In addition, terminology of the Yankee whalers such as 'sea-guaps' for sperm whales and 'blackfish' for short-finned pilot whales are names still used locally today. Cultural influence on SVG whalers by Yankee whalers can also be seen in the transfer of New England whaling shanties, or songs that helped the whalers keep rowing rhythm (Kannada 2006 and references therein).

4.5. St. Lucia

This island was occasionally visited by Yankee whalers and sometimes was used as a base for their whaling. These vessels mostly pursued humpbacks, but occasionally took some short-finned pilot whales (Reeves, 1988). The last report of Yankee whaling for St. Lucia is dated 1883 (Reeves and Smith, 2002).

4.6. Grenada

The presence of Yankee whalers was not uncommon in Grenadian waters during the second half of the nineteenth century. In the early months of 1857 as many as eight American whalers might have been seen anchored off St. George's, Grenada's capital, with their boats fully employed. Whaling ships primarily hunted humpbacks, but occasionally landed sperm whales, and a high percentage of their catch was cow-calf pairs. The ships provided whale meat to the local market of Grenada and the neighboring southern Grenadines (Romero and Hayford, 2000, and references therein).

4.7. Venezuela

Yankee whalers visited the area of the Gulf of Paria, between Venezuela and Trinidad, between 1837 and 1871 but may also have visited other localities and at other times. They predominantly hunted humpbacks, but occasionally they would strike a sperm whale or a 'blackfish,' (*G. macrorhynchus*). Yankee whalers also visited other coastal areas in eastern Venezuela. Since data are incomplete, the only quantitative statement I can make, based on the summary provided by Reeves *et al.* (2001), is that Yankee whalers, captured at least 25 whales, during at least nine whaling voyages. There was very little, if any, interaction between the whaling crews and Venezuelans. Therefore, there is no evidence that they ever influenced any marine mammal exploitation practice in Venezuela. Further, the presence of Yankee whaling ships created some stir in the local press, because the locals saw this operation as a breach of their national sovereignty (Romero et al 1997 and references therein).

5. Discussion

Fig. 2 shows Yankee whaling activity based on tonnage (blue line), number of trips to the Caribbean Basin (red line), and the historical factors that contributed to the fluctuation in the intensity of Yankee whaling overall. The first noticeable aspect between Yankee whaling activities in general with that in the Caribbean Basin is the asynchrony between the two. While Yankee whaling intensity reached a peak between the U.S. industrial boom of the 1830s and the industrial exploitation of mineral oil in the early 1860s, the expansion of the Yankee whaling activity in the Caribbean Basin took place between the mid 1860's and the mid 1870s. Therefore we need to examine economic factors to understand this phenomenon.

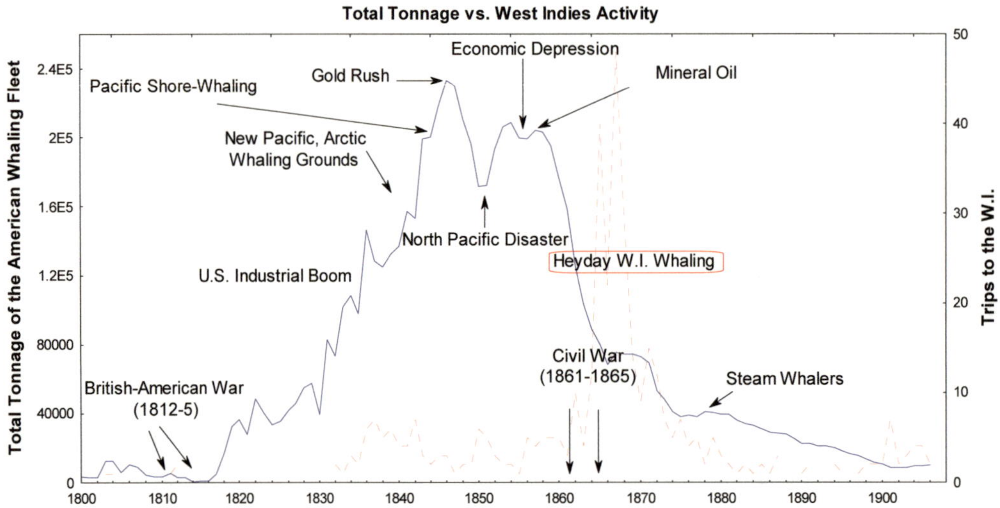

Figure 2. Total activity of Yankee whaling by tonnage (blue line) and by number of trips to the Caribbean basin (red line). There is an asynchrony between the two suggesting that the Yankee whaling activity in the West Indies was a marginal operation that took place after the traditional whaling grounds had been depleted. Arrows represent historical events to give a context to better understand how political and economic factors influenced these activities in general.

First we need to recognize that Yankee whalers shifted both whaling grounds and species targets as resourses became scarcer in different geographic areas. Romero and Kannada (2006), using historical catch records, report that populations of bowhead (*Balaena mysticetus*) and right whales (*Eubalaena glacialis*) in the North Atlantic became severely depleted by the 19th century. This depletion caused Yankee whalers to seek new hunting grounds in the Caribbean. Yet that happened shortly after substitute products such as kerosene and mineral oil became available in the market making. As a consequence the demand for whale oil declined as soon as subsurface mineral oil was discovered at Titusville, Pennsylvania, in 1859 (Coleman 1995). Although there was again an increase in the price of whale oil during the American Civil War (1861-1865) due to increased demand, whale oil prices declined severely after that.

The Yankee whale oil industry responded by trying to lower their production costs that were becoming higher as their vessels had to navigate to more difficult grounds. To that end they tried to make their operations more efficient by improving whaling technology with the introduction of the exploding harpoon head in 1864 and by reducing labor costs by hiring more and more crews from countries in the Caribbean Basin, particularly English-speaking ones. By this time, whalers were earning one-third to one-half of what merchant seamen earned and one-fifth of a shore laborer (Coleman 1995).

Thus a combination of the depletion of whale stocks in the historical whaling grounds of the North Atlantic together with lower labor costs by hiring natives from the West Indies shifted Yankee whaling activities to the Caribbean basin (and later to Artic and Antarctic waters). As Brandt (1940, p. 54) put it "Slowly the crews had to be composed more and more of halfcastes from all parts of the West Indies and of Central and South America."

These circumstances increased interaction between Yankee whalers and West Indies locals, which led to both technology transfer to the countries they were recruiting crews and depletion of local populations of whales, particularly humpbacks. Yet, the cultural influence of Yankee whalers on that part of the world was uneven. The large number of voyages to Barbados and Bermuda may be due not only to the presence of whales in those waters, but also because (1) Barbados is the first island a ship traveling from the east Atlantic encounters when sailing with the aid of the trade winds and (2) Bermuda is the only island between the North American continent and other whaling grounds in the eastern Caribbean such and Cape Verde.

Yet, by the time Yankee whalers initiated a significant activity in the Caribbean Basin, two localities –both under British sovereignty at that time: Bermuda and Trinidad and Tobago has already developed a local shore whaling industry: Bermuda in 1663 (Romero 2006) and Trinidad in the 1820's (Romero et al., 2002). This contradicts the generalization made by Caldwell & Caldwell (1971) that Yankee whalers directly influenced shore whaling in the Caribbean.

Yankee whaling activity in Trinidad and Tobago have been summarized elsewhere (Reeves *et al.*, 2001, Reeves and Smith, 2002). All available data indicate that there was never much interaction between Yankee and the already established shore whaling industry of Trinidad. Yankee whaling in the area did not start until the 1830s, when Trinidadian shore whaling was already in full swing. In fact, the owners of one of the whaling stations in Trinidad asked the Governor of the Island to refuse authorization for the American Schooner

Harmony, of Nantucket, to whale in the Gulf of Paria, for fear of competition (de Verteuil, 1994). It is not known how the Governor decided in this matter. Still some Yankee whaling vessels continued occasionally visiting Trinidad's waters until at least 1867. There are records of Yankee whaling ships visiting Tobago waters during 1877, although it is unclear whether or not they actually captured any whales (Reeves and Smith, 2002).

Yet, in other places such as SVG, the story was different. Despite the fact that the number of voyages by Yankee whalers these and other surrounding islands was relatively low in comparison to those to Bermuda and Barbados, their cultural influence in undeniable. For example, a SVG resident named William Wallace took interest in whaling and participated as a crewmember on several Yankee whaling expeditions. He later left Bequia and moved to New Bedford, Massachusetts, the center of the Yankee whaling fleet. While in New Bedford, he learned whaleboat design, tool production and maintenance, and hunting methods. Upon returning to SVG, he applied this new knowledge and began whaling. Whaling represented an opportunity for economic development for men of European ancestry returning to an impoverished island. Thus, the development of the whaling industry in SVG was the result of cultural contacts rather than a direct correlation of intensity of Yankee whaling operations in those waters (Kannada 2006).

The reason why Yankee whalers did not intensively exploit whales in SVG had to do with yield per unit of effort. The efficiency with which Yankee whalers caught whales in SVG was less than ideal: 38.5 % success rate (killing and hauling in). This was probably due to the limitations of the boats, which were rowboats, and the inaccuracy of hand-thrown harpoons. It seems that landing a humpback whale was a difficult task. These same limitations may have played a role in Yankee whaler's choice not to hunt pilot whales unless times were desperate. In contrast to the large, slow moving, and obvious humpback whales; the short-finned pilot whales are small and quick. Chasing after the smaller and faster pilot whales would have been extremely costly in terms of time and profit potential (Kannada 2006).

Prior to the 1986 International Whaling Commission (IWC) moratorium on commercial whaling, SVG and the rest of the world for that matter, whaled without regard to depletion issues. Despite a negative correlation between global oil value and the number of humpback whales caught in SVG, it is unlikely that the global market drove the industry. SVG exported oil and meat of humpback whales to neighboring countries in only small quantities. It is more likely that the persistence of the humpback whale fishery in SVG was due to local tradition and local demand for whale products as there are no longer exports of these products outside the country. The SVG market followed the typical supply and demand curve in that as the amount of humpback whale oil became available, the less it cost which lead to a higher demand for the product. Since the late 1930's, there has been little fluctuation in the number of humpback whales killed and that number has remained low (no more than 4 killed in any one year) (Kannada 2006).

6. Conclusions

The intensification of Yankee whaling in the Caribbean was due to a combination of factors such as (1) depletion of stock whales in traditional whaling grounds of the North Atlantic and (2) higher costs of whaling which led to the search of lower labor costs by hiring crews in the West Indies, particularly given that wages for crews of merchant vessels and shore industries were higher.

The discovery and development of a replacement commodity (mineral oil) together with the increase risks of investment on the whaling activity due to longer and longer voyages would ultimately signify the end of whale oil as a major commercial commodity.

The bulk of Yankee whaling in the Caribbean Basin was short-lived and out of sync with the heyday of Yankee whaling. It concentrated in the southeastern Caribbean most likely because of the abundance of humpback whales in that area. The cultural influence of Yankee whalers varied by location mostly determined on whether or not their arrival took place before the development of local shore whaling. Therefore, a combination of factors, including whale stocks, political events, and labor and other economic and social issues influenced Yankee whaling activities in the Caribbean.

Both Yankee whaling and local shore whaling led to the depletion of humpbacks in the Caribbean Basin as it has been quantified elsewhere (Swartz et al. 2003, Smith and Reeves 2003).

Author details

Aldemaro Romero
College of Arts and Sciences, Southern Illinois University Edwardsville, Peck Hall, Edwardsville, IL, USA

7. References

Acosta, J. de. 1590 (1940). Historia Natural y Moral de las Indias. Fondo de Cultura Económica. México.

Brandt, K. 1940. Whale oil. An economic analysis. Stanford University.

Caldwell, D. K. and Caldwell, M. C. 1971. Porpoise fisheries in the southern Caribbean – present utilizations and future potentials. In *Proceedings of the 23rd Annual Session of the Gulf and Caribbean Fisheries Institute,* Higman, J. B. (ed.), Rosenstiel School of Marine and Atmospheric Science: Coral Gables, FL; 195-206.

Clark, A. H. 1887. The whale-fishery. 1. History and present condition of the fishery. In *The Fisheries and Fishery Industries of the United States, Section V, Volume II, History and Methods of the Fisheries*, Goode, G. B (ed.), U.S. Government Printing Office: Washington, D.C.; 3-218.

Coleman, J.L. 1995. The American whale oil industry: a look back to the future of the American petroleum industry? Natural Resources Research 4(3):273-288.

de Verteuil, A. 1994. The Germans in Trinidad. Port-of-Spain: The Litho Press.

Hawes, C.B. 1924. Whaling. New York: Doubleday, Page and Company.

Hegarty, R.B. 1959. Returns of whaling vessels sailing from American ports. New Bedford: The Old Dartmouth Historical Society.

Kannada, S.D. 2006. Environmental history and current practices of marine mammal exploitation in St. Vincent and the Grenadines, W.I. MS Thesis. Jonesboro, AR: Arkansas State University.

Lund, J.N. 2001. Whaling masters and whaling voyages sailing from American ports. A compilation of sources. New Bedford, MA: New Bedford Whaling Museum.

Reeves, R.R. 1988. Exploitation of cetaceans in St. Lucia, Lesser Antilles, January 1987. Report of the International Whaling Commission 38:445-447.

Reeves, R.R. and T.D. Smith. 2002. Historical catches of humpback whales in the North Atlantic Ocean: an overview of sources. Journal of Cetacean Research and Management 4(3):219-234.

Reeves, R.R., Swartz, S.L., Wetmore, S. and Clapham, P.J. 2001. Historical occurrence and distribution of humpback whales in the eastern and southern Caribbean Sea, based on data from American whaling logbooks. *Journal of Cetacean Research and Management,* 3: 117-129.

Romero, A. 2006. "More private gain than public good": whale and ambergris exploitation in seventeenth-century Bermuda. *Bermuda Journal of Archaeology and Maritime History* 17 (in press).

Romero, A. 2009. Chasing fools' gold: whaling in 19th and 20th-century Bermuda. *Bermuda Journal of Archaeology and Maritime History* 19: 141-163.

Romero, A.; I. Agudo & S. Green. 1997. Cetacean exploitation in Venezuela. *Reports of the International Whaling Commission.* 47:735-746.

Romero, A., R. Baker, J. E. Creswell, A. Singh, A. McKie & M. Manna. 2002. Environmental history of marine mammal exploitation in Trinidad and Tobago, W.I. and its ecological impact. *Environment and History* 8(3):255-274.

Romero, A. & J.E. Creswell. 2010. Deplete locally, impact globally: environmental history of shore-whaling in Barbados, W.I. *The Open Conservation Biology Journal* 4:19-27.

Romero, A. & K. Hayford. 2000. Past and Present Utilization of Marine Mammals in Grenada. *Journal of Cetacean Research and Management* 2(3):223-226.

Romero, A. & S.D. Kannada. 2006. Comment on "Genetic analysis of 16th-century whale bones prompts a revision of the impact of Basque whaling on right and bowhead whales in the western North Atlantic". *Canadian Journal of Zoology* 84:1059-1065.

Sanderson, I.T. 1993. A history of whaling. New York: Barnes & Noble.

Smith, T.D. and R.R. Reeves. 2003. Estimating American 19(th) century catches of humpback whales in the West Indies and Cape Verde Islands. Caribbean Journal of Science 39:286-297.

Stackpole, E.A. 1953. The Sea-Hunters. The New England Whalemen during two centuries 1635-1835. Philadelphia: J.B. Lippincott Co.

Starbuck, A. 1876 (1989). History of the American Whale Fishery from its earliest inception to the year 1876. Reprinted by Castle Books, Secaucus, NJ.

Stringer, C.B.; J. C. Finlayson, R. N. E. Barton, Y. Fernández-Jalvo, I. Cáceres, R. C. Sabin, E. J. Rhodes, A. P. Currant, J. Rodríguez-Vidal, F. Giles-Pacheco & J. A. Riquelme-Cantal. 2008. Neanderthal exploitation of marine mammals in Gibraltar. *Proceedings of the National Academy of Sciences* (USA) 105:14319-14324.

Swartz, S.L., T. Cole, M.A. McDonald, J.A. Hildebrand, E.M. Oleson, A. Martinez, P.J. Clapham, J. Barlow and M.L. Jones. 2003. Acoustic and visual survey of humpback whale (*Megaptera novaeangliae*) distribution in the eastern and southeastern Caribbean Sea. Caribbean Journal of Science 39:195-208.

Tower, W.S. 1907. A history of the American whale fishery. Philadelphia University of Pennsylvania.

Necrologue